Variable Air Volume Systems for Environmental Quality

Variable Air Volume Systems for Environmental Quality

Steve Y. S. Chen

Stanley J. Demster

McGraw-Hill

New York   San Francisco   Washington, D.C.   Auckland   Bogotá
Caracas   Lisbon   London   Madrid   Mexico City   Milan
Montreal   New Delhi   San Juan   Singapore
Sydney   Tokyo   Toronto

Library of Congress Cataloging-in-Publication Data

Demster, Stanley J.
   Variable air volume systems for environmental quality / Stanley J.
Demster, Steve Chen.
      p.   cm.
   ISBN 0-07-011085-9
   1. Variable air volume systems (Air conditioning)   I. Chen,
Steve. II. Title.
   TH7687.95.D46   1995
   697.9'3—dc20   95-22732
                                                                CIP

1 2 3 4 5 6 7 8 9 0   DOC/DOC   9 0 0 9 8 7 6 5

ISBN 0-07-011085-9

*The sponsoring editor for this book was Robert W. Hauserman, and
the production supervisor was Pamela A. Pelton. This book was set in
Century Schoolbook by North Market Street Graphics.*

Printed and bound by R. R. Donnelley & Sons Company.

McGraw-Hill books are available at special quantity discounts to use as
premiums and sales promotions, or for use in corporate training programs. For more information, please write to the Director of Special
Sales, McGraw-Hill, Inc., 11 West 19th Street, New York, NY 10011. Or
contact your local bookstore.

To my wife Sue for her understanding,
encouragement, and patience, and to my daughter
Helen and my son Michael.

—Steve Y. S. Chen, P.E.

To Dr. David A. Pistenmaa, Ph.D., M.D., whose skill
and talent helped save my life and give me the
opportunity to live and work after my encounter
with cancer.

—Stanley J. Demster, P.E.

Contents

Part 3   Construction of VAV Systems

Chapter 12. Conformance to Design   299

Chapter 13. Coordination of Trades   307

Chapter 14. Protection of Equipment   311

Chapter 15. VAV Controls   315

Preface

The energy crisis of the 1970s stimulated the development of new technologies to reduce energy usage in buildings. Computer-aided energy management techniques as well as rigorous applications of the variable air volume (VAV) concept were all initiated during this period. Since then, many new VAV components, such as terminals, fans, air-regulating devices, and controls have been developed. These components have all contributed to the enormous advancement of VAV technology. The increasing application of direct digital control (DDC) since the early 1980s has also intensified this trend.

Today VAV technology is universally accepted as a means of achieving an energy-efficient and comfortable building environment. In fact, VAV systems are one of the most popular air-conditioning systems for commercial and institutional buildings in the United States. Yet not all VAV systems are successful. There are many nonworking VAV systems. Multiple factors contribute to this unfortunate situation. The most significant one is insufficient understanding of VAV concepts and the interaction of system components with control loops under the continuously changing flow conditions. More specifically, there is a lack of appreciation by building professionals of the dynamic interplay among system components and their controls in a given VAV environment. This interplay must be carefully analyzed, measured, and controlled throughout the entire life cycle of the VAV system. The design, construction, and operation phases must all be taken into account.

Traditionally, the design, construction, and operation of a VAV system are separate and independent processes handled by different groups of building professionals: typically, system designers, contractors, and operators. These processes are further complicated by the delegation of responsibilities among different engineering disciplines, including HVAC, electrical, control, acoustics, and facility management. This fragmentation of the building design and construction processes makes the total integration of the VAV system function very difficult.

This book addresses the issue of building and VAV system integration in four major sections: Introduction, VAV System Design, VAV System Construction, and VAV System Operation.

Part 1: Introduction. This section discusses the history, classifications, components, advantages, and disadvantages of VAV systems. The objective is to familiarize the reader with the VAV concept and the intricacy of VAV components and their interactions with controls.

Part 2: VAV System Design. This section emphasizes the importance of designing VAV systems for total environmental quality. Chapter 6 explores the impact of VAV design on five major environmental factors: namely, thermal comfort, indoor air quality (IAQ), variable volume air distribution, acoustics, and building pressurization. Total environmental quality is achieved through integrated design considerations, careful system and component selection, and a continuously interactive design process. These subjects are covered in detail in Chapters 7 through 10. Although seldom emphasized in conventional design, it is the interactive process of design and analysis that often determines the quality of VAV systems. Chapter 11 discusses this subject and shows 10 types of analyses with specific examples to illustrate their effectiveness in VAV design.

Part 3: Construction of VAV Systems. This section discusses realization of the original design intent through the proper construction and installation of system components.

Part 4: Operation of VAV Systems. This section is the final link to ensure the success of a VAV system in a given building environment. It covers a wide range of subjects, from the understanding of design intent to VAV routine maintenance. In particular, it emphasizes practical considerations for the proper operation of VAV systems, such as zoning, actual versus design, making modifications and corrections, and troubleshooting.

The glossary following Part 4 includes terms and concepts which are most frequently used in this book which require further explanation and clarification of their meanings.

In summary, this book provides a comprehensive overview of VAV technology based on the best currently available information, and on the authors' own experience in the design, construction, and operation of modern VAV systems. It is the sincere hope of the authors that this book will aid the reader in utilizing VAV technology to achieve a quality building environment.

Acknowledgments

The authors wish to thank McGraw-Hill Senior Editor Robert W. Hauserman and Christine H. Furry at North Market Street Graphics.

This book could not have been completed without their invaluable sug-
gestions and constructive criticisms. The authors also would like to
extend their sincere thanks to ASHRAE, EPRI, Carrier Corporation,
The Trane Company, Titus, Honeywell, Baltimore Aircoil, and many
others for permission to use their published materials. Special thanks
also to Karen Demster for an invaluable contribution that went far
beyond the typing of manuscript.

Steve Y. S. Chen, P.E.
Stanley J. Demster, P.E.

Variable Air Volume
Systems for
Environmental Quality

Introduction to VAV Systems

Overview

1.1 Why VAV?

The energy crisis of the 1970s and the need to improve interior environmental control in a more cost-effective, high-quality manner has increased the popularity of variable air volume (VAV) heating, ventilating, and air-conditioning (HVAC) systems in the United States. VAV systems have steadily evolved over the last 20 years, and they now represent the accepted state of the art in providing environmental control for larger office buildings, public facilities, and laboratories. However, VAV acceptance and success has been limited by:

1. Lack of an in-depth appreciation of VAV technology by design professionals

2. Operational difficulties resulting from misapplication of VAV technology

3. Equipment limitations and construction deficiencies

This book offers the engineer, designer, architect, contractor, technician, and operator/owner a better understanding of VAV technology. Existing VAV systems can be improved and used more effectively to deliver a higher-quality interior environment with less associated cost of installation and operation. However, to achieve this goal the overall process of HVAC design, installation, and operation must be better appreciated, and potential problems properly addressed. This book focuses on developing a better understanding of VAV systems from a process perspective, utilizing analytic techniques to resolve potential limitations and problems.

1.2   Definition of a VAV System

HVAC systems maintain indoor space air quality within specific design limits. Minimum and maximum design conditions and operational requirements are established for:

1. Space dry bulb temperature (thermal comfort)

2. Relative humidity

3. Air changes (IAQ)

4. Outdoor ventilation air (IAQ)

5. Space relative pressure

6. Acoustic levels generated by the HVAC system (noise)

Due to the complexity of HVAC applications and the ingenuity of engineers and system manufacturers, a wide variety of systems, devices, and equipment configurations are used to meet these six operational requirements of an HVAC system. The VAV system is one of the most popular, and sometimes controversial, of these systems. Throughout this book, VAV systems will be defined as air-handling systems that use an intentionally variable flow of air to satisfy several, or ideally all, of the six HVAC operational requirements.

Most VAV systems perform the dry bulb thermal control function using a room thermostat to adjust the heat energy added to or removed from the conditioned space. This simplistic approach may be adequate in many applications, but it also produces limitations that have caused VAV systems to sometimes fail the overall HVAC mission. More emphasis on the other five HVAC system requirements can result in a better approach to environmental quality produced by the VAV systems.

VAV systems have so many configurations that it may be more effective to define VAV by exception. A system that by design moves a constant volume of air to or from the conditioned space is not a VAV system. Systems that provide environmental control by radiation or convection methods are not VAV systems. All HVAC systems—large or small, one terminal or hundreds of terminals—that vary the air volume are VAV systems. Thus, a small 250-cfm fan coil unit with a variable-speed fan is a type of VAV system, and so is a 250,000-cfm built-up air-handling system that uses multiple fans. Also, specialized variable-flow exhaust systems and other ventilation-only systems should be included in this definition. Figure 1.1 illustrates several common VAV systems that fit this definition.

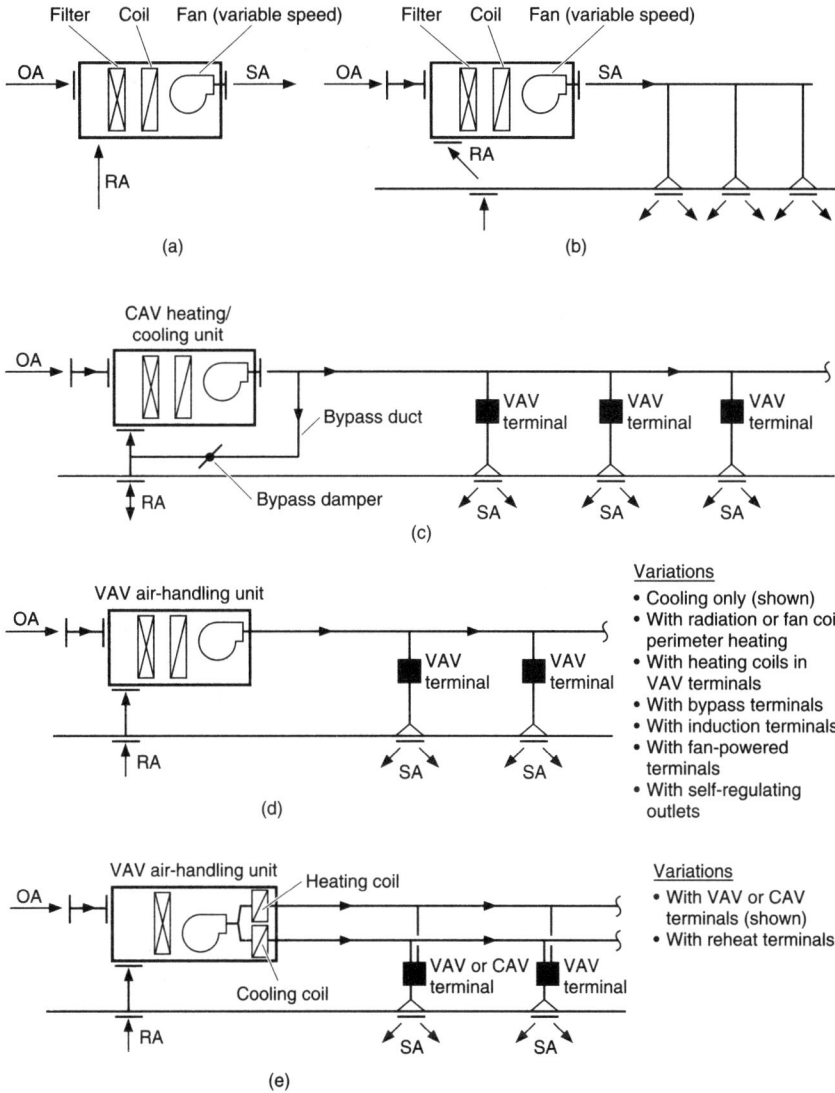

Figure 1.1 Various VAV systems. (*a*) VAV fan coil unit; (*b*) Single-zone VAV system; (*c*) Changeover/bypass VAV system; (*d*) Single-duct VAV system; (*e*) Dual-duct VAV system.

1.3 History of VAV Systems

Most of the original all-air HVAC systems were conceived as constant volume systems that varied the temperature of the delivered air to maintain space conditions. A constant volume of air was and remains easy to control, and problems like diffuser dumping, poor ventilation, and overcooling were less likely to occur. Because energy cost was not considered significant, the all-air constant volume system often operated continuously, regardless of occupancy, and used cooling and reheating of the air to control temperature and humidity. However, the consumption of fan energy combined with the reheat energy caused these systems to become less desirable in the 1970s, when energy costs rose dramatically. It is intuitive that as the load is reduced, it is possible to reduce the HVAC system capacity by merely reducing the air quantity. Building owners and operators decided to shut down the HVAC equipment (as they do space lighting) to save energy when there were periods of limited or no occupancy. Reducing the air quantity saves both the energy of air delivery and the energy involved in heating/cooling the air. Though simple and intuitive, this new HVAC technology was not as easy to implement as it would seem. Complex, interdependent technical issues have to be considered in using a VAV system—issues that were not considered with the previous constant volume technology.

Though it is difficult to document the first "inventor" of large-scale multizone VAV technology, one of the earliest large systems was created using a device known as the *HAM box*—"HAM" being an acronym for the initials of the design engineers who encouraged a vendor to build one of the first VAV terminal devices. Many others also like to take credit for this device, but a search of patents reveals that no clear-cut inventor of VAV can be found. VAV technology was very intuitive, and the first systems were rather unsophisticated adaptations of existing HVAC technology.

VAV technology has been refined for over 20 years. In the United States, it is today one of the dominant HVAC systems installed in most modern office buildings. Even laboratories, hospitals, and many industrial applications are using VAV systems due to the energy-savings benefits and affordable comfort these systems can provide.

1.4   Effects of Load Profile

The VAV concept envisions an air-handling system that adapts the air delivery volume in a manner that follows or tracks the load profile. When the VAV system is serving only interior zones, this profile will closely follow the occupancy of the space or process loads rather than weather effects. If the VAV system includes a large percentage of perimeter zones that are influenced by solar and transmission loads, the VAV system must respond to the change in the weather. If the HVAC system is shut down during periods of little or no occupancy, the VAV system must accommodate a warm-up or cooldown cycle which may be unrelated to the actual real-time loads.

Understanding the load profile is key to the proper selection and design of the VAV system. If the use of the space indicates a relatively constant load profile, a VAV system may not offer any energy savings, and if the ventilation requirements are very high, a constant volume system may be more effective and less costly. Likewise, if the warm-up and cool-down times are critical, the sizing of the HVAC system might have to be enlarged to make the system more responsive. Zone-to-zone variances in load profile encourage the designer to accommodate diversity by downsizing the main system components to save first cost, while similar load profiles in each zone diminish the effects that diversity will have on the system sizing. The required rate of change in the load profile in a particular zone may require the designer to increase the capacity of the terminal devices in that zone to raise the number of air changes so the system can respond faster to changing load. This condition is a common need in seldom-used assembly buildings, classrooms, and conference facilities that must go from zero to full occupancy in a short period.

Too often, HVAC design engineers design VAV systems using a computer program that provides only a cfm requirement based on the peak design-day thermal requirements. The actual zone or system load profile may not be modeled accurately. This design-day view of the system is often a steady-state rather than a load-profile-driven dynamic estimate of the air requirements. Adjustments must be made to accommodate how the system will be operated, and how the load profile may create requirements for airflow that are different from peak thermal design-day requirements. Programs also are restricted generally to the dry bulb requirements, and the other five HVAC requirements must be addressed outside the load program. Air changes, ventilation, humidity, and acoustic effects caused by the load profile must be included in the system design.

1.5 Control and Operation of VAV Systems

The size, construction, and controls of a well-designed VAV system are transparent for the maintenance of the proper space conditions. However, though the space conditions should not vary between systems, the energy efficiency and operating economy may vary widely based on the system selected. The use of specific cycles such as a dry bulb or enthalpy economizer can substantially change operating cost. The ability of the dampers to regulate flows and pressures and the use of other energy-saving measures can yield many benefits. For example, if a system uses electric reheat, the use of a control program to reset the discharge air temperature of the unit with the average or peak cooling load can greatly affect the electrical consumption. With so many climate, load, and operating factors affecting performance, the control and operation of a simple VAV system can become very complex if environmental quality and operating economy are to be optimized in addition to maintaining space dry bulb conditions.

Many VAV applications have encountered problems not because VAV technology was wrong for the application or because the design was flawed, but due to poor selection and application of VAV components. VAV systems are more sensitive to leaking dampers, miscalibrated controls, and short cycling of air from the exhaust to the fresh air intake. These packaging failures can create the impression that the VAV system is not adequate to meet the HVAC functions.

1.6   Advantages of VAV Systems

A properly designed VAV HVAC system can be one of the most energy-efficient and comfortable systems for the space occupants. Interior air quality, noise, and overall comfort are generally excellent with a properly designed, installed, and operating system. The flexibility and adaptability of the system to changing load conditions is a key feature. Because the systems are generally capable of moving large volumes of air, many problems such as interior painting, smoke, and other sources of air contamination can be removed more effectively than with any other system. The ability to add new zones and easily retrofit or modify the existing zoning provides the system with the ability to change as the HVAC needs within the building change. VAV technology can be successfully applied to a wide range of building types in all climates. With the advance of control technology, it is now possible to build intelligent VAV systems that balance and rebalance themselves on a dynamic basis to fine-tune their performance. Control of all six HVAC functions can be automated in a manner that provides an optimized system with maximized indoor environmental quality. VAV technology has the primary advantage of flexibility and adaptability that no other system can offer.

VAV technology allows a single system to provide simultaneous heating and cooling without a seasonal changeover. The noise produced by the VAV system is reduced at off-peak load periods, and drafts are also less of a problem at off-peak airflows.

1.7   Disadvantages of VAV Systems

There are several historical characteristics of VAV systems that often are viewed negatively. VAV systems can cause poor ventilation and indoor air quality if the terminal devices do not maintain adequate air circulation and the exchange of outdoor air is not maintained independent of thermal load. Supply diffusers at lower-than-design flow rates can mix the supply air with the space air less efficiently, and dumping of cool air from overhead diffusers is often a source of occupant complaints. The required accuracy and complexity of the control system is often a problem when the control system is not properly designed or installed. Large all-air systems require ductwork that may be larger than the space available above ceilings, and compromises in duct sizing lead to noise under design conditions and even inadequate supply duct pressure to operate the terminal devices correctly. An undersized system cannot meet peak loads, and/or control limitations produce uncomfortable space conditions. Fan-powered terminals and diffusers that are subject to wide variances in airflow often produce a change in the room background sound level which is noticed by the occupants. Use of ceiling-mounted VAV diffusers in perimeter zones for both heating and cooling often results in an inadequate heating cycle that causes drafts and loss of uniform effective heating in colder climates. Large systems with return fans seldom have adequate fan-tracking control and control of outside air. Drafts and indoor wind between zones can cause occupant discomfort. Some systems operate so poorly that building doors stand open from overpressure, or the outside air is sucked into every crack of the building from excessive negative pressure. Occupant complaints of being too cold or too warm are often the result of inadequate or limited zoning that places the control thermostats in locations that cannot allow the system to respond to occupant comfort. Zealous efforts to conserve energy may cause loss of proper dehumidification/humidification of ventilation air, and lack of proper maintenance can result in systems overburdened by dirty filters.

VAV systems that use fan-powered terminals introduce additional disadvantages. Each terminal includes a fan and electric motor that must be periodically serviced at additional cost. If a reheat coil is included, additional maintenance is required for filter changing and coil cleaning. The installation of the terminal requires an electrical branch circuit to power the fan, which raises installation cost. If ceiling plenums are used for return air, the fans can pull contaminated air from adjacent zones, causing an IAQ problem. The installation of the fan-powered terminals in a ceiling plenum requires the terminals to meet local code requirements for not just mechanical but also electrical codes that require independent laboratory certification for construction materials and wiring.

In general, these complaints and shortcomings of the applied VAV technology are not intrinsic or generic, and they should have limited impact if the application, design, construction, and operation of the VAV system are properly addressed. In almost all cases, problems and complaints result from errors or omissions in the design, construction, and operation of the system that can and should be corrected.

2

Types of VAV Systems

VAV technology has been applied to a wide range of air-handling unit, duct, and terminal device configurations. The following descriptions are of the most common configurations, and hybrids of these configurations are also possible, though less popular.

2.1 Single Zone

The smallest VAV system (Fig. 2.1) is the single-zone chilled/hot water fan coil or small air-handling unit that is mounted nearby or within the conditioned space. These units are popular because they are generally low in cost to install, can support perimeter heating requirements, and are accessible to the user. These units typically have a variable-speed fan that controls the quantity of air delivered, and a thermostat device may be used to control the air delivery temperature and/or the fan speed. Some U.S. engineers have not considered small fan coil units as true VAV systems because they are installed in the conditioned space and are served by piping (not all-air delivery means), but in many installations such as those found in other countries (e.g., Japan) the fan coil unit is fitted with limited ductwork and is a true single-zone air-handling system. These systems should be classified as VAV because the principles and potential problems associated with other, larger VAV systems apply to these small variable-flow units. With small VAV systems, the VAV unit is devoted to a single control zone, and the variable volume is produced by regulating the supply air fan volume that is proportionately delivered by all connected supply air diffusers and grilles. Typically, this single-zone system ranges in size from 500 to 5000 cfm.

Similar to the fan coil unit are the unitary unit and the heat pump, which are identical to the fan coil unit except that the water coils are

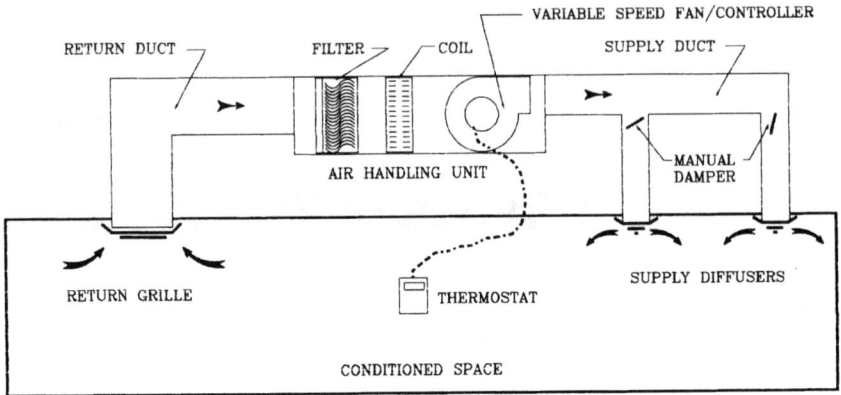

Figure 2.1   Typical single-zone VAV unit.

replaced by a refrigerant coil which provides heating and cooling of the air. The airside portion of these three systems is identical and the refrigeration/heating means is the only difference.

2.2 Bypass

The next type of VAV system is the bypass VAV that is used frequently for light commercial buildings. Three equipment configurations of bypass VAV are in general use:

1. *Duct bypass,* with supply or primary air diverted from the duct

2. *Terminal bypass,* with the supply air diverted at the terminal

3. *Outlet / diffuser bypass,* with the outlet device diverting the air back to the return

These systems are sometimes referred to as *variable volume and temperature changeover / bypass VAV* because they change over from heating to cooling and consequently change the supply air temperature from heating to cooling. (See Fig. 2.2a and b.) With these systems the overall air supply is provided at a relatively constant pressure and temperature from a single constant volume air-handling unit to a control device serving each of the connected zones. These multiple-zone systems are distinguished by the type of zone control device employed to regulate the space conditions. The bypass VAV system uses a single supply duct from the fan that terminates in an air valve assembly that either directs the air to the conditioned space or diverts it back to the air-handling return system via a return plenum or duct. The bypass VAV system maintains a constant fan delivery volume and the only airflow that varies is the flow within the conditioned zone. The fan deliv-

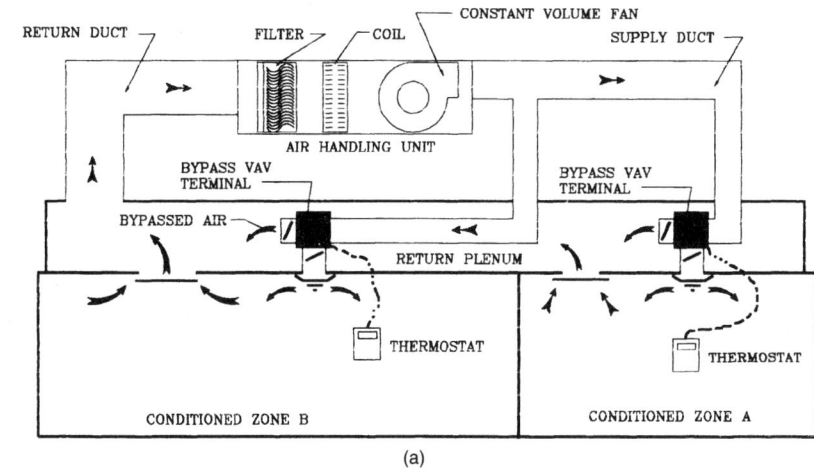

(a)

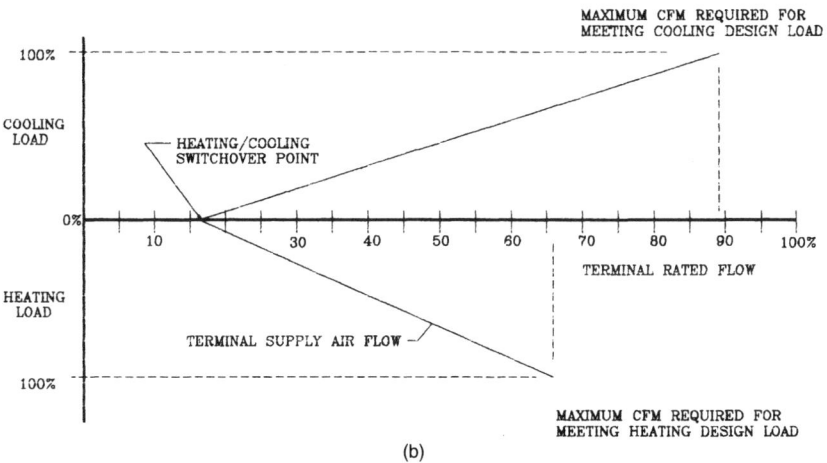

(b)

Figure 2.2   (a) Bypass VAV—Variable volume and temperature changeover. (b) Diagram of supply terminal airflow to the conditioned space for a bypass VAV terminal with switchover.

ery pressure and energy consumption remain constant because the flow is essentially constant. Typical size ranges for these systems are from 1000 to 10,000 cfm, 5 to 20 tons, and from two to sixty-four zones. Their popularity originates in new, light commercial construction using packaged units and in the retrofit market that permits existing constant volume air-handling systems to be retrofit to VAV service without major changes to the main fan unit, ductwork, and controls. Sometimes a hybrid system is created by adding VAV zones to a constant volume system. Bypass VAV systems have become less popular in recent years due to the limited fan energy savings offered by this VAV concept, as well as noise generation and control difficulties encountered in using this type of VAV system.

2.3 Dual Duct

A dual-duct VAV system (Fig. 2.3*a*) provides control of the conditioned zone temperature through the use of a zone control device that modu-

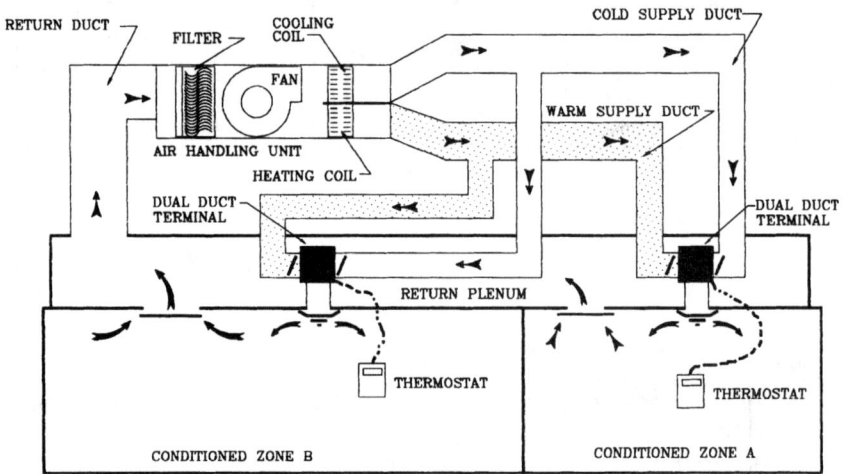

(a)

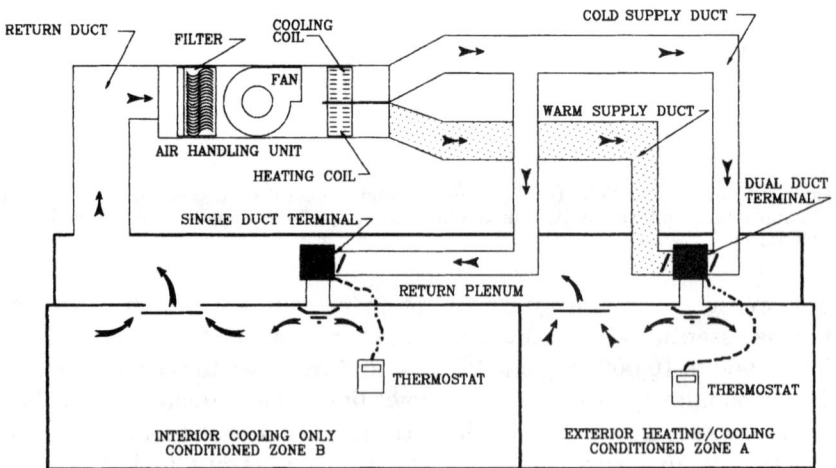

(b)

Figure 2.3 (*a*) Dual-duct VAV system. (*b*) Hybrid single- and dual-duct VAV system.

MAXIMUM CFM REQUIRED FOR
MEETING COOLING DESIGN LOAD

100% ┄┄┄┄┄┄┄┄┄┄┄┄┄┄┄┄┄┄┄┄┄┄┄┄┄┄┄┄

COOLING   VENTILATION
LOAD   MINIMUM
               FLOW

0%

        10   30   40   50   60   70   80   90   100%

                                          TERMINAL RATED FLOW

HEATING
LOAD

        TERMINAL MIXED SUPPLY AIR FLOW

100% ┄┄┄┄┄┄┄┄┄┄┄┄┄┄┄┄┄┄┄┄┄┄┄┄┄┄┄

                                    MAXIMUM CFM REQUIRED FOR
                                    MEETING HEATING DESIGN LOAD

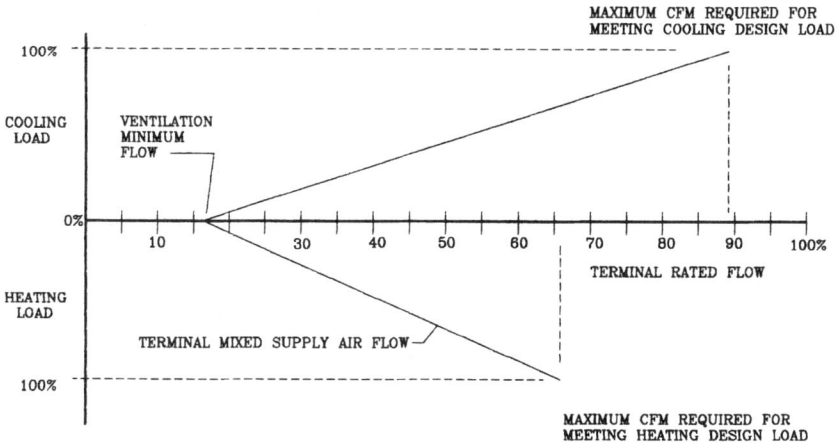

Figure 2.4   Diagram of airflow compared to load for dual-duct VAV terminal.

lates the amount of air delivered from each of the ducts. One duct provides warm supply air and the other duct provides cold air. By mixing the warm and cold air or by lowering the total amount of air delivered, or both, this system can regulate the conditioned space temperature. The supply fan delivery may or not vary depending on the control scheme used. Dual-duct systems are not popular because of the amount of ductwork required, which can be up to twice the size and cost of a single-duct system. Because the dual-duct delivery method can provide effective heating and cooling on a simultaneous basis, it is possible to use the dual-duct concept on a limited basis to provide perimeter heating capacity. (See Fig. 2.3b.) With this configuration, a limited number of exterior zones use a heating duct and the interior zones use only the cooling duct in a hybrid configuration. This hybrid configuration is more cost-effective because heating of the main duct can be limited to the exterior zones only. The changeover from heating to cooling typically involves the reduction in supply airflow to some heating minimum, at which time the cooling volume begins to increase up to the maximum allowed (Fig. 2.4). The dual-duct VAV systems can be implemented with one or two supply fans, and the choice is generally a matter of economics and available space, with the single-fan systems requiring less space and cost.

2.4   Single-Duct Pressure Independent

The next classification of VAV system is the more popular single-duct pressure-independent terminal system that modulates the flow of air

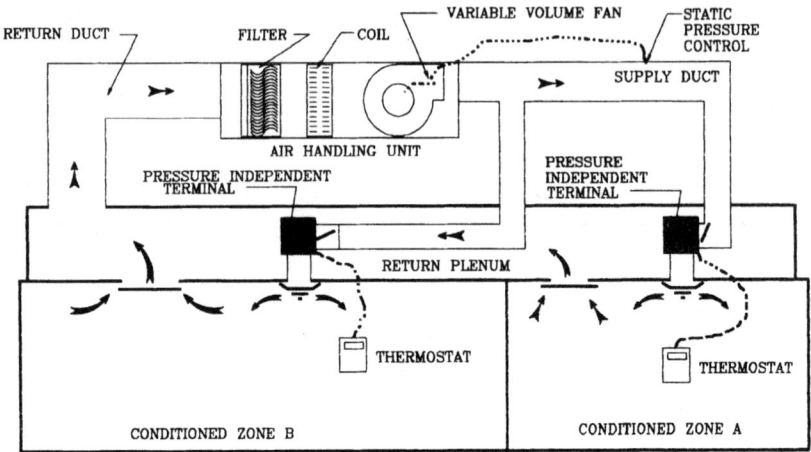

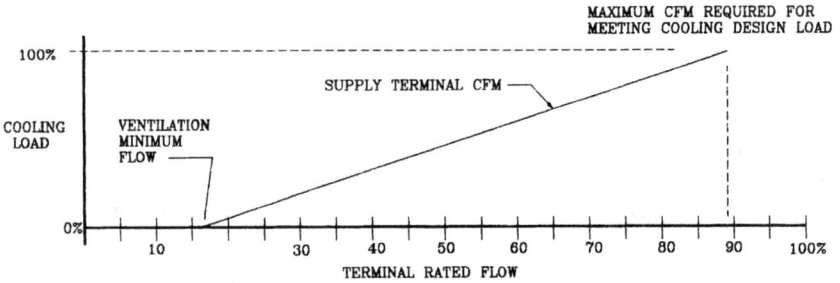

Figure 2.5 (a) Single-duct pressure-independent VAV system. (b) Diagram of airflow compared to load for single-duct pressure-independent VAV system.

to each control zone through a terminal device that regulates the flow of supply air in response to the zone temperature. (See Fig. 2.5a and b.) Simultaneous heating and cooling can be provided by the introduction of a heating coil in the terminal device, and the supply fan pressure and delivery temperature are either constant or reset by zone demands. Usually, a minimum flow for single-duct VAV terminals is predetermined by the ventilation requirements (Fig. 2.5b). Single-duct systems vary in the type of terminal devices used. Individual zones can be served directly by a diffuser that varies the flow of supply air that is directly introduced into the space, or the supply air can be mixed by a fan or induction effect with return/space air prior to delivery to the conditioned zone. Generally, fan-powered terminal devices are either series or parallel configuration. With the parallel fan arrangement, the fan draws in only the secondary or plenum air to mix with the supply

---

Packaged VAV System

+ Single source of responsibility
+ Limited design engineering required
+ Can be lower cost in smaller sizes
+ Easy to obtain parts and training
+ Performance proven in the factory
− Limited to under 50,000 cfm
− Flexibility limited to most popular configurations
− Performance is not optimized for application.

Built-Up VAV System

+ Can use best components available
+ Size is virtually unlimited
+ Configuration and flexibility is almost unlimited
+ Can be lowest-cost alternative for large applications
+ Can be customized with special features
+ Can be fit better into space available
− Contractor rather than vendor assumes responsibility
− Training and parts could become a problem
− No factory testing or proven performance
− Requires extensive engineering

Figure 2.6   Packaged/pre-engineered VAV compared to built-up VAV system.

or primary air. Parallel fan-powered terminals are often used in perimeter zones with reheat coils to supply supplemental heating when required by climate. With the series fan arrangement, both the primary and recirculation air is introduced through the fan. This arrangement is well suited to handle low-temperature primary air.

The VAV concept is implemented primarily with the single-duct distribution system that is connected to pressure-independent terminal devices. Lower-cost electronic controls and higher reliability permit packaged systems to be installed with relatively little system-specific engineering. Therefore, it is beneficial to distinguish these pre-engineered general-purpose packaged systems from custom-built systems that are engineered for a specific application. (See Fig. 2.6.) Each of these two types of systems has its own advantages and drawbacks which should be understood and explored.

2.5   Induction

A common form of VAV system found in hospital patient rooms is the VAV induction system (Fig. 2.7) that uses an induction/reheat terminal device containing an induction nozzle(s) and a reheat coil. This system usually mixes about one-third cold supply air at 55°F with two-thirds of induced room air. If the cooling load drops, the primary cold air is reduced and the warm secondary/induced air is increased to maintain the space temperature. If supplemental heating is required, the reheat

MIXED AIR
TO CONDITIONED SPACE

COOLING/REHEAT COIL
ELECTRIC, CHILLED, HOT WATER OR STEAM

CASING

INDUCTION NOZZLE(S)

SECONDARY
AIR

CONTROL DAMPER

PRIMARY/SUPPLY
AIR

FROM
AHU

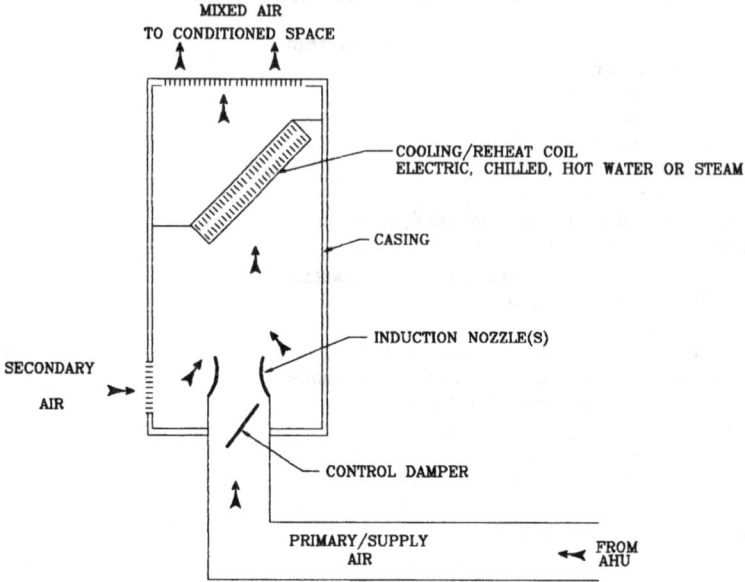

Figure 2.7 Induction VAV terminal.

coil is utilized to raise the mixed air and space temperatures. The system is capable of using the induction effect instead of a fan to maintain air motion and circulation within the space.

2.6 Packaged

VAV systems can be obtained as complete pre-engineered packages with the refrigeration system consisting of a direct expansion refrigeration coil that is supplied by integral or remote compressors. An intrinsic control problem with variable airflow through the refrigerant coils is the danger of the coil freezing up at low load and airflows. The refrigeration capacity control method used in these systems and the anticipated lowest load become very critical. If the latent loads are high compared to the sensible loads and/or the total load falls below the unloading capability of the compressor, coil freeze-up may occur. Some systems use multicircuited coils with multiple refrigeration compressors that are staged; others use multispeed or variable-speed compressors and supply air bypass ducts; still others use heat pipes and hot gas bypass refrigerant systems on the compressors to resolve low-load operation. All of these variations in design, though, may cause the system to lose control of space humidity if the supply air temperature is allowed to rise too high during low-load conditions. The majority of these systems use a centrifugal fan of the backward-curved or airfoil type, and the larger systems include an integral return or exhaust fan. Dampers for outdoor air intake, exhaust, and return are typically integral with the unit. Housings are typically lined for sound and thermal considerations, and controls are often the electronic type, with smaller systems using analog control and the larger systems digital control. If heating is required, it can be supplied as either an electric, direct-fired heat exchanger, hot water, or steam preheat coil. Most systems use a draw-through design with the filters and coils placed in the airstream ahead of the fan. This design feature is common because it permits a uniform distribution of air across the coil face in the smallest package.

Packaged chilled water air-handling systems do not have the low-load coil freezing problems of direct refrigerant systems, and they operate well under a wide range of load conditions. The small- to medium-sized packaged chilled water systems typically use electronic controls that manage the discharge temperature and down-duct static pressure well. Like the direct refrigerant systems, most packaged systems use centrifugal fans of the backward-curved or airfoil type, and integral control dampers in a draw-through design.

2.7 VAV Exhaust and Ventilation Systems

The last type of VAV system is the exhaust/ventilation system, used not so much for temperature control but for removing or displacing air

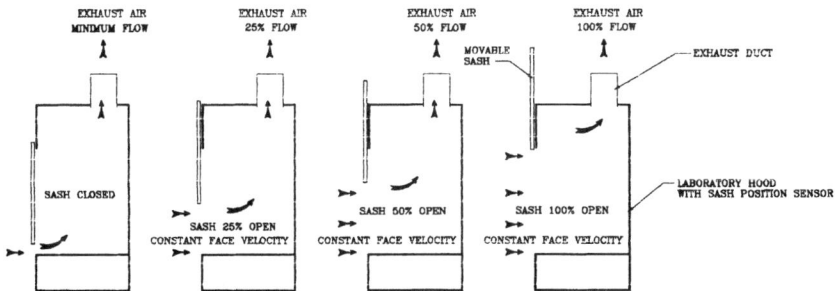

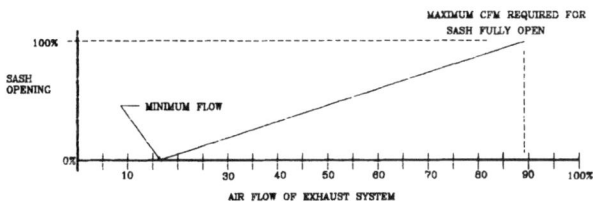

Figure 2.8 Typical VAV laboratory hood exhaust system.

from the space. Laboratory ventilation systems have been energy-intensive due to the relatively high air change requirements and ventilation standards. VAV technology has been adopted to the laboratory to provide sophisticated exhaust and supply tracking systems that dramatically lower energy consumption while meeting all ventilation requirements. VAV exhaust and ventilation systems are designed to vary the flow of air as the need for the airflow changes. For example, a laboratory hood (Fig. 2.8) has a movable sash and the containment of the potentially harmful fumes is maintained by providing constant air velocity through the sash opening. If the sash is fully open, more air is required to maintain the minimum safe velocity than if the sash is closed. The VAV laboratory hood exhaust system varies the airflow in response to sash position to maintain the correct velocity of air through the sash. This process saves energy and maintains safety. The laboratory exhaust system is also interconnected with a tracking VAV supply and return system that maintains the laboratory space pressure and ensures adequate makeup air for the hood under all design conditions.

Other common variable volume ventilation systems are used with cooking hoods, paint booths, medical isolation rooms, and similar activities that require more ventilation when the space is in use and less ventilation when the space is not in use.

Size of VAV Systems

The size of the VAV system is sometimes controlled by the size of the project, the size of specific areas to be conditioned, or the preferences of the designer, contractor, or owner. Some very large projects have been done with one air-handling unit that serves 30 or more floors, and many smaller projects have used VAV units that serve areas no larger than a few hundred square feet. This range in system size and application illustrates the flexibility that these systems have. However, there are several generalizations that should be considered.

The larger air-handling units can be made more sophisticated and the cost of controls per cfm delivered declines as the unit grows larger. The piping, electrical, and control costs associated with installing units make units below 5000 cfm very expensive on a dollar per cfm basis. Therefore, using larger units can be cost-effective. The next consideration is space available. Smaller units on every floor or those scattered throughout the occupied space consume valuable floor space. Large units placed in a basement or rooftop equipment room save valuable floor space. Larger units have more ductwork cost associated with their installation and there may be more fan energy consumed per cfm if the main distribution ducts are extensive. The physical space available for installing the ducts may limit the size of the unit that can be used. Load diversity favors larger systems, and fewer installed cfm are required to serve a given building load in most cases when large units are used. Larger units are also constructed from components that are more durable, and the large built-up VAV systems tend to have much longer service lives than the small packaged systems. Large systems can be interconnected, with one serving as a backup to another. When the one unit fails, load diversity usually allows the backup unit to adequately handle the increased load.

In contrast to the factors that favor using large systems, smaller systems can be packaged and delivered at lower cost per cfm by the vendor. Smaller systems have less complex duct design. Smaller systems affect fewer people and smaller areas when they fail. If the building is a one-story configuration, small rooftop units are generally the lowest-cost system.

3.1 Small Packaged

Systems are generally considered small when they deliver less than 5000 cfm. (See Fig. 3.1.) Small packaged systems can be found on any size project, and their chief advantage is low cost to procure and install.

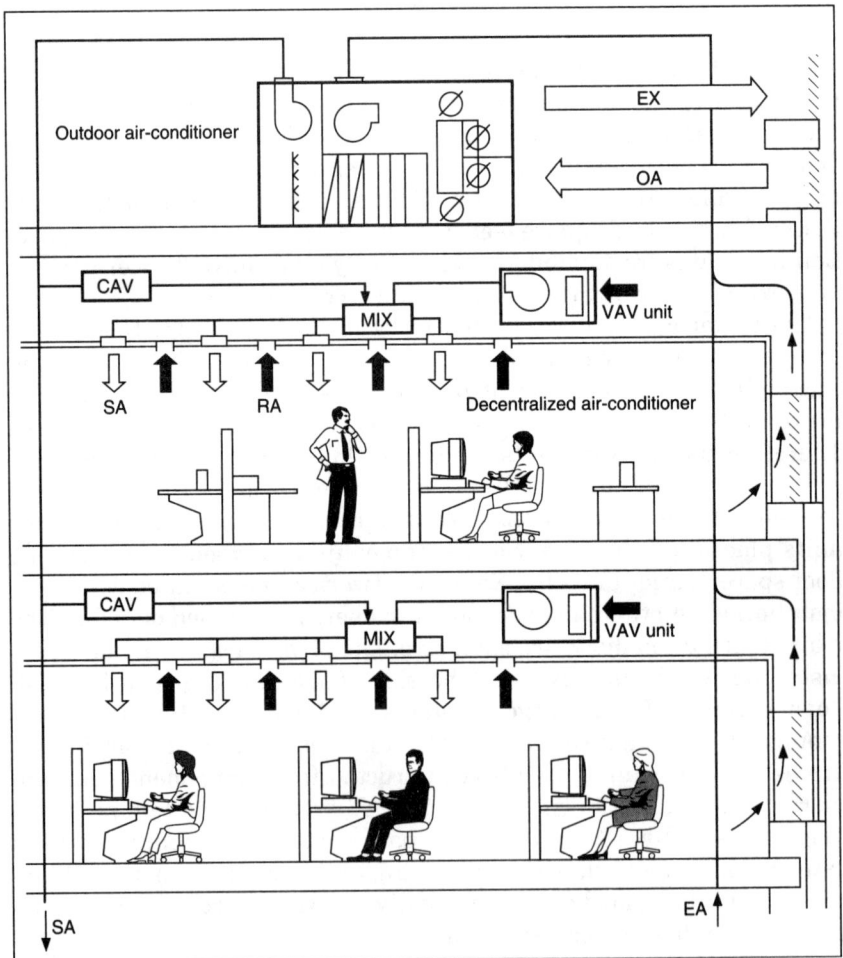

Figure 3.1   Small packaged VAV units for a large office building. (*Source: ASHRAE Journal, January 1994. Reprinted with permission.*)

Many designers prefer small systems because there is less to go wrong with these systems. The ductwork design is very simple and the controls are usually packaged with the system. These units seldom require a return or exhaust fan, and the one supply fan is almost always a centrifugal-type fan with vortex dampers or VSD. The cooling source is generally direct expansion when only a few units are used, and chilled water is common when many units can be connected to a larger chiller plant.

3.2 Medium Packaged

Medium-sized packaged units range from 5000 to about 40,000 cfm. (See Fig. 3.2.) These units are more rugged and sophisticated than the small systems, and they may have return or exhaust fans due to the more extensive ductwork used. The controls can be packaged with

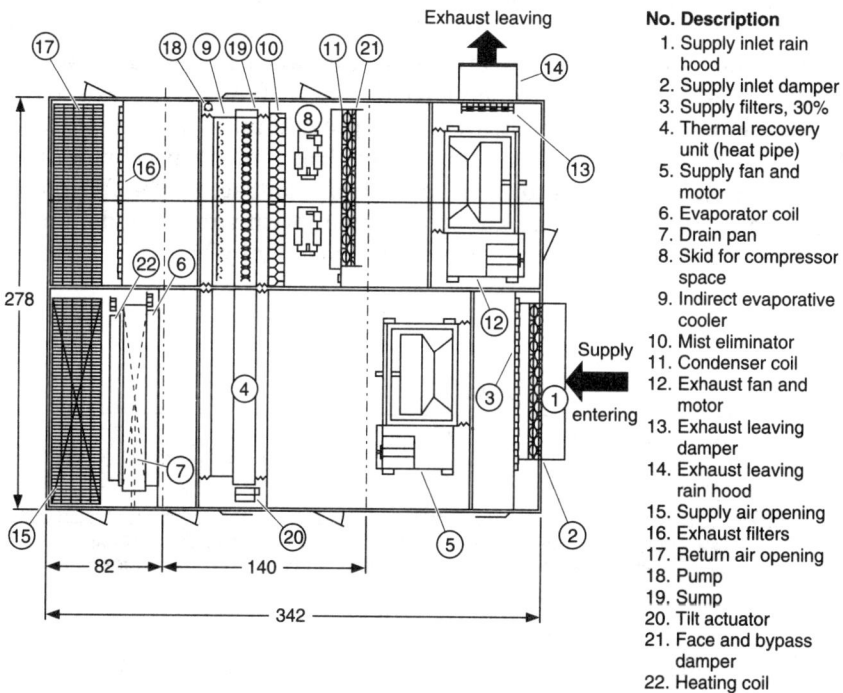

No. Description
1. Supply inlet rain hood
2. Supply inlet damper
3. Supply filters, 30%
4. Thermal recovery unit (heat pipe)
5. Supply fan and motor
6. Evaporator coil
7. Drain pan
8. Skid for compressor space
9. Indirect evaporative cooler
10. Mist eliminator
11. Condenser coil
12. Exhaust fan and motor
13. Exhaust leaving damper
14. Exhaust leaving rain hood
15. Supply air opening
16. Exhaust filters
17. Return air opening
18. Pump
19. Sump
20. Tilt actuator
21. Face and bypass damper
22. Heating coil

Notes: All dimensions are in inches; unit height is 139 in.; unit ships in three sections.

1 A 40,000 cfm supply and 36,000 cfm exhaust California VAV outside air machine showing the location of heat exchanger and refrigeration components to optimize energy savings. Simplicity of design and ease of maintenance are key requirements for classroom air conditioning equipment.

Figure 3.2 An example of VAV components packaged for classroom applications. *(Source: HPAC, January 1994. Reprinted with permission.)*

the unit or furnished by a controls contractor. Medium-sized units can be rooftop-mounted in an exterior package or installed in a mechanical equipment room. The smaller units in this range may be direct expansion cooled, while the large sizes are generally for use with chilled water. Medium-size packages offer the lower cost of factory assembly with the improved performance and efficiency their size offers. The fans, drives, and control devices used in these units are typically much more efficient than the smaller packaged systems.

3.3 Large Built-Up

Large built-up units should be considered when the cfm requirements exceed 50,000 cfm. (See Fig. 3.3.) These units offer the highest fan efficiency and use very durable components. Duct designs and space requirements typically are a major design issue. These systems have to be built in to the overall building plan, and they can be exterior or internally mounted using prefabricated panels. Because of their size and diversity effects, they can be built with redundant components that provide a limited but effective fault tolerance. Large built-up systems almost always require exhaust or return fans due to extensive ductwork, and the control strategies to provide many energy-savings features can become very complex. The controls are always customized for the application and project by a controls contractor. The fan types can be either airfoil centrifugal or vaneaxial, with vaneaxial dominating the larger sizes. Capacity control is typically by VSD or variable-pitch fans.

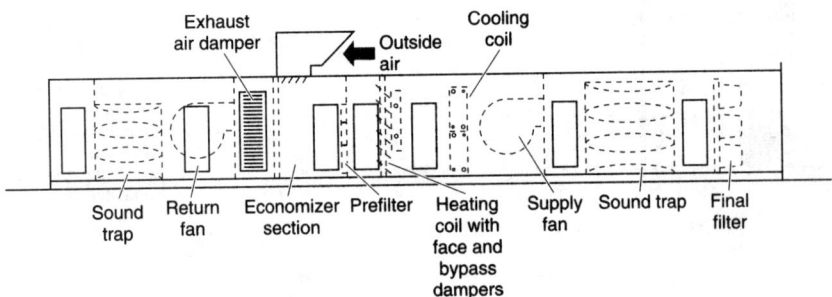

Figure 3.3   An example of large built-up VAV system. *(Courtesy of Industrial Acoustics Co.)*

References

1. *ASHRAE Journal,* January 1994.
2. *Heating/Piping/Air Conditioning,* January 1994.

VAV System Configurations

4.1 By Fan Arrangement

4.1.1 Draw-through air-handling units

For cooling applications, a chilled or DX coil that serves to both dehumidify and cool the air is placed either ahead of or behind the VAV fan. As shown in Fig. 4.1, the fan placement after the coil is identified as a draw-through unit. The typical draw-through air-handling unit raises the discharge air temperature above the cooling coil leaving conditions due to the addition of the supply fan energy. The draw-through is the most common air-handling-unit configuration, and it has been preferred by most packaged system designers for its lower cost of construction and compact size. Reheat coils are sometimes installed after the cooling coil for dehumidification purposes, but most applications today avoid using reheat, if at all possible, to save energy. This configuration is easy to package because the filter and coil sections typically require the air to move at the same low uniform velocity, from 350 to 550 fpm, and these sections can be stacked together since they have a uniform cross section. The draw-through unit has been packaged in both horizontal and vertical configurations, and most major vendors offer fan, filter, coil, and mixing box modules that can be assembled into a package that can fit a wide variety of physical space limitations.

4.1.2 Blow-through air-handling units

As shown in Fig. 4.2, another common configuration is the blow-through unit with the coils located after the fan section. Sometimes the blow-through air-handling unit is used because of its ability to deliver colder supply air. Because the blow-through design places the fan ahead of the cooling coil, the fan heat is absorbed by the coil, and the discharge air is therefore colder. Colder air is preferred in the VAV sys-

Coil location

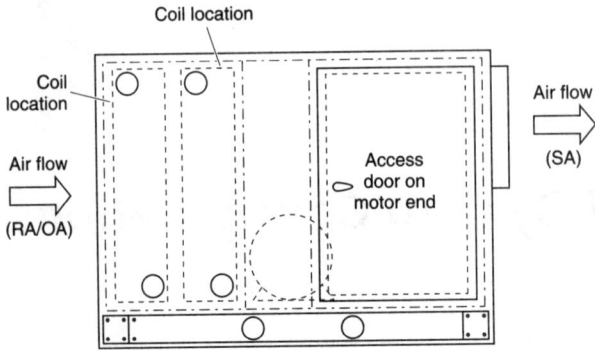

Figure 4.1 Draw-through air-handling unit. *(Courtesy of Carrier Corp.)*

tem because it can result in smaller ducts and less airflow to meet the space cooling loads. The blow-through design has the discharge air leaving the cooling coil with no source of reheat. Any water that blows off the coil in the form of drift most likely will not evaporate because the leaving air is generally 95 to 97 percent saturated. Furthermore, if air bypasses through or around the coil, it will not be dehumidified by the coil and will carry moisture through to the discharge duct, where it may condense due to the colder temperatures there. The cooling of the bypass air can produce condensation in the duct.

Another problem is the deviation in supply air temperature and coil temperatures. With a blow-through system, the air leaving the unit is so close to saturation that several conditions can produce saturation and condensation. The first condition is a momentary loss of cooling. If the discharge duct surface is cold and the cooling coil begins to discharge warmer air due to a problem such as a rise in chilled water tem-

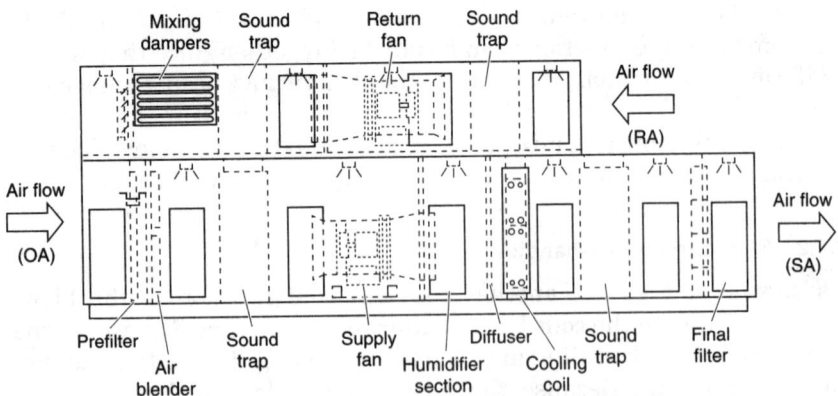

Figure 4.2 Blow-through air-handling unit. *(Courtesy of Buffalo Air Handling Co.)*

perature caused by a power failure or refrigeration problem, the higher dew point of the supply air can cause condensation on the interior surface of the duct. Systems with internal liners and insulation are particularly sensitive to this problem. Even typical control valve hysteresis and the cycling of compressors in a DX system could result in the formation of condensation. At partial load, the cooling coil is oversized in comparison to the amount of air flowing through the coil. At these conditions, it is relatively easy for the coil to generate an airstream that is saturated. The coil valve and capacity control means also becomes oversized, and the potential for the fluctuation of the coil temperature due to control deviations increases. At low loads, the coil surfaces may also display an erratic temperature range because the control system is measuring the average discharge temperature. Some of the air leaving the coil can be warmer and colder, with mixing occurring downstream of the coil. Mixing warmer and cooler air is one way condensation and saturation can occur.

Another source of moisture from condensation is rapid pressure loss such as occurs in discharge filters or branch fittings. Pockets of moisture can form and collect within the system. Higher moisture leads to the ideal environment for the growth of bacteria, mold, and fungi. Ensuring the dryness of the duct interior is a key means of limiting potential sources of contamination, and the blow-through unit is incapable of providing dryness under all conditions. If DX cooling is employed, the control of the coil temperatures will generally be less effective. Any effort to drive down the discharge temperature and raise the saturation level of the air increases the potential for air quality problems due to excessive moisture in the supply ducts. Draw-through units, because of the intrinsic reheat produced by the supply fan, have far fewer problems.

The fan energy savings of producing colder air with the blow-through design may not be as great as intuitively believed. The draw-through fan configuration moves air through the fan that has been cooled and has a higher density than the warmer, more moist air that moves through the fan in the blow-through design. For example, the condensate that flows off the cooling coil is water vapor that does not have to be moved by the fan. Some systems employ terminal reheat to maintain space temperatures under light loads, and the higher air-handling-unit discharge temperatures produce less need for reheat. In the draw-through design some of the fan heat will be exhausted before it reaches the cooling coil, while in the blow-through design, all of the fan heat reaches the cooling coil. VAV systems often operate at part load, and the savings that could result at full load is diminished by the load factor. The anticipated fan energy savings of using a blow-through unit may be cut in half if the load factor is 75 percent or less.

Another blow-through unit consideration is the need to distribute the nonuniform, high-velocity, small airstream from the supply fan in a manner that permits moving the air though the cooling coil at lower uniform velocities. After leaving the coil, the air is typically increased in velocity to pass through the supply ducts. In contrast, the draw-through unit can accelerate the slow-moving air from the cooling coil and discharge it at high velocity directly into the supply duct. This difference results in the draw-through unit generally being physically smaller and less costly to install. It also saves some energy because of the loss produced by slowing down and speeding up the air.

Because of IAQ considerations, the use of the draw-through fan configuration is recommended due to less likelihood of supply duct condensation that leads to the potential growth of harmful organisms in the duct. Blow-through units are a potential cause of poor indoor air quality and contamination due to the condensation that can occur in the supply duct. The intuitive supply fan energy savings of the blow-through design compared to the draw-through design is generally not as great as believed, due to greater fan efficiency, less turbulence, and fewer losses in the draw-through unit. A draw-through unit should require less space and be less expensive to acquire and install.

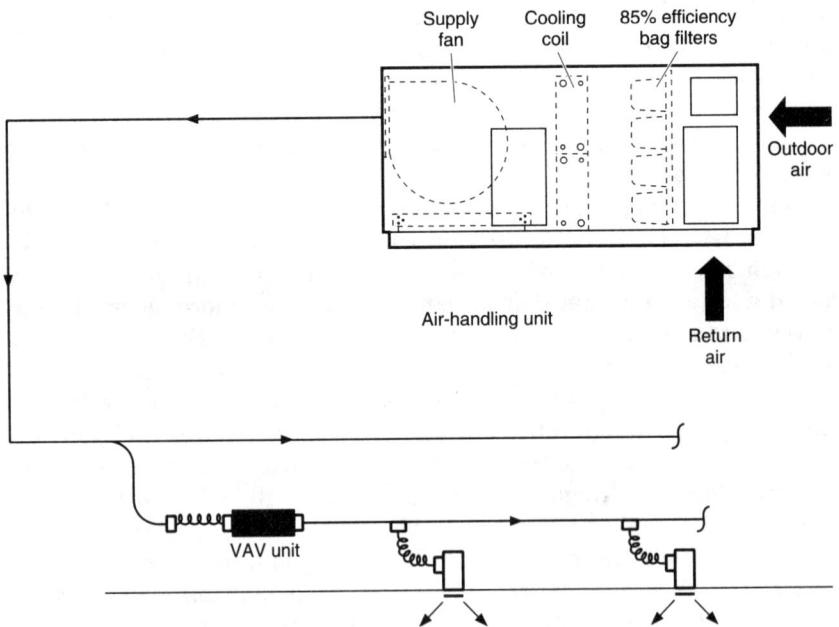

Figure 4.3 Air-handling unit with integrated outdoor air intake.

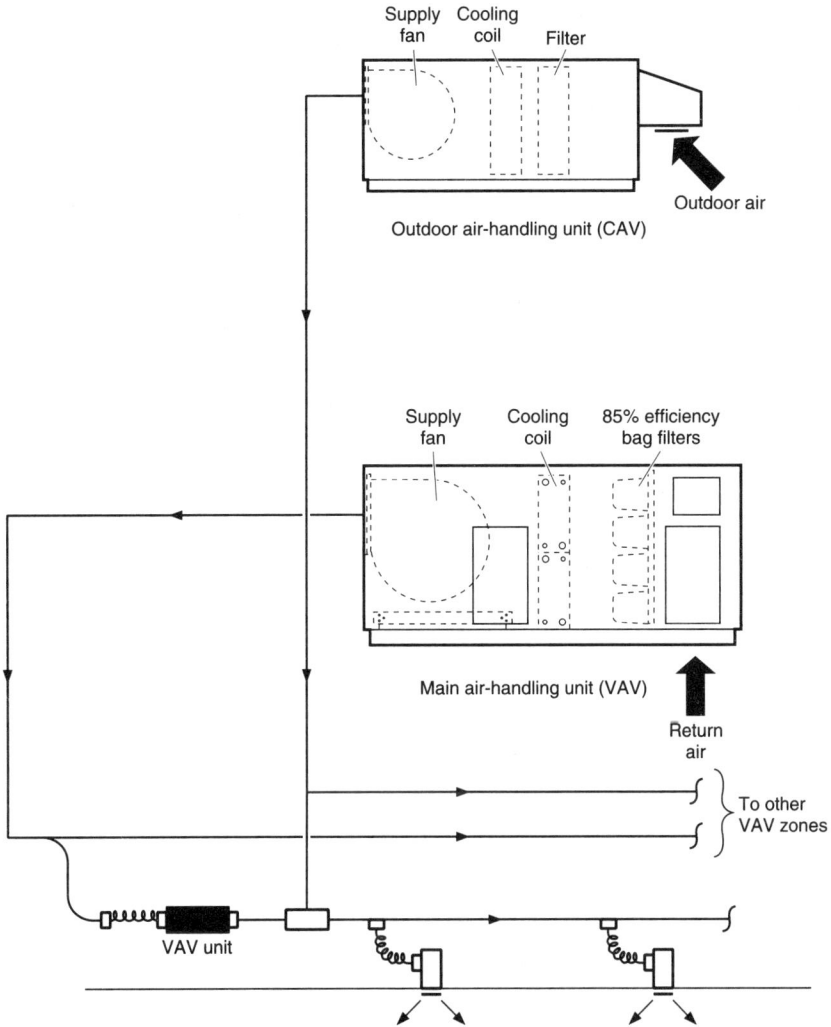

Figure 4.4 VAV system with a separate outdoor air-handling unit.

4.2 By Outdoor Air Introduction

The VAV system designer can select two means of outdoor air introduction. The outdoor makeup air can mix internally with the main VAV air-handling-unit return air (Fig. 4.3) or it can be delivered to the space by a separate air-handling unit (AHU) and duct system (Fig. 4.4). Because the ventilation requirements are independent of thermal load, many engineers favor a separate system that provides ventilation in a manner that is not affected by thermal load. Other designers have favored the integrated approach due to its low cost. ASHRAE has published requirements for outdoor ventilation air in a quantitative manner, but the distribution of the mandated air quantities have remained ambiguous. A VAV terminal that restricts the flow of supply air to a specific space can reduce the ventilation effectiveness to that area even though the total cfm outdoor air serving the air-handling system is at or above present standards. Therefore, based on IAQ concerns for the specific space air quality, a separate constant volume outdoor air system could deliver the proper amount of air to the space that would not be possible with the integrated concept.

A potential solution for designers who want to integrate ventilation with VAV systems is the use of a terminal device with a minimum flow that provides adequate air changes with an air-handling unit designed to bring in proportionately more outdoor air as the total flow diminishes. This type of system can compensate for the reduced ventilation at low thermal loads. At minimum, the integrated system should have a constant volume of outdoor air introduced regardless of thermal load, and total AHU minimum flow should never be allowed to fall below the ventilation requirements.

4.3   By Terminal Type

The description of VAV systems often refers to the type of terminals used with the air-handling system, though terminal types are easily mixed on a system. The most common descriptions used are:

- Pressure-dependent single duct
- Pressure-independent single duct
- Induction
- Fan powered
- Double/dual duct

5

VAV System Components

All VAV systems share the same basic components, though the size, design, and packaging of these components vary widely. (See Fig. 5.1.) All VAV systems use a fan with some type of capacity/volume control, a terminal device for controlling flow to a zone and a diffuser or grille for distributing the airflow within the zone. Other components such as dampers, ductwork, filters, coils, sound attenuators, and controls are included as needed to create a system that meets the specific HVAC requirements of the application. Each of these components must be selected and applied properly in order to provide a system that meets all HVAC requirements. The following discussion illustrates the basic features and functions of each of these components.

5.1 Fans

Many of the same type of fans that are used in other HVAC systems are used with VAV systems, and all the same basic fan laws and application limitations apply. However, there are specific requirements imposed by a VAV system that limit fan selection. First, the fan must be capable of having its volume delivery modulated in an energy-efficient manner that does not create excessive noise or erratic operation. Second, the fan must produce adequate static pressure to permit the terminal devices to operate properly. Finally, the fan characteristics must be compatible with other fans in the system. For example, many VAV systems use multiple fans that work together, such as supply, return, or exhaust/ventilation fans. Attention to the fan characteristics is critical to developing a system that will deliver the air in a quiet and energy-efficient manner.

Component	Configuration

- Fans
  - (1) Supply air fan
  - (2) Return air fan
  - (3) Exhaust fan
  - (4) Outdoor air fan

- Ductwork and terminals
  - (1) Single duct with VAV terminals
  - (2) Single duct with bypass terminals
  - (3) Single duct with induction terminals
  - (4) Single duct with fan-powered units
  - (5) Dual conduit with CAV and VAV terminals
  - (6) Dual duct with VAV terminals

- Outlets
  - (1) Self-regulating
  - (2) Non-self-regulating

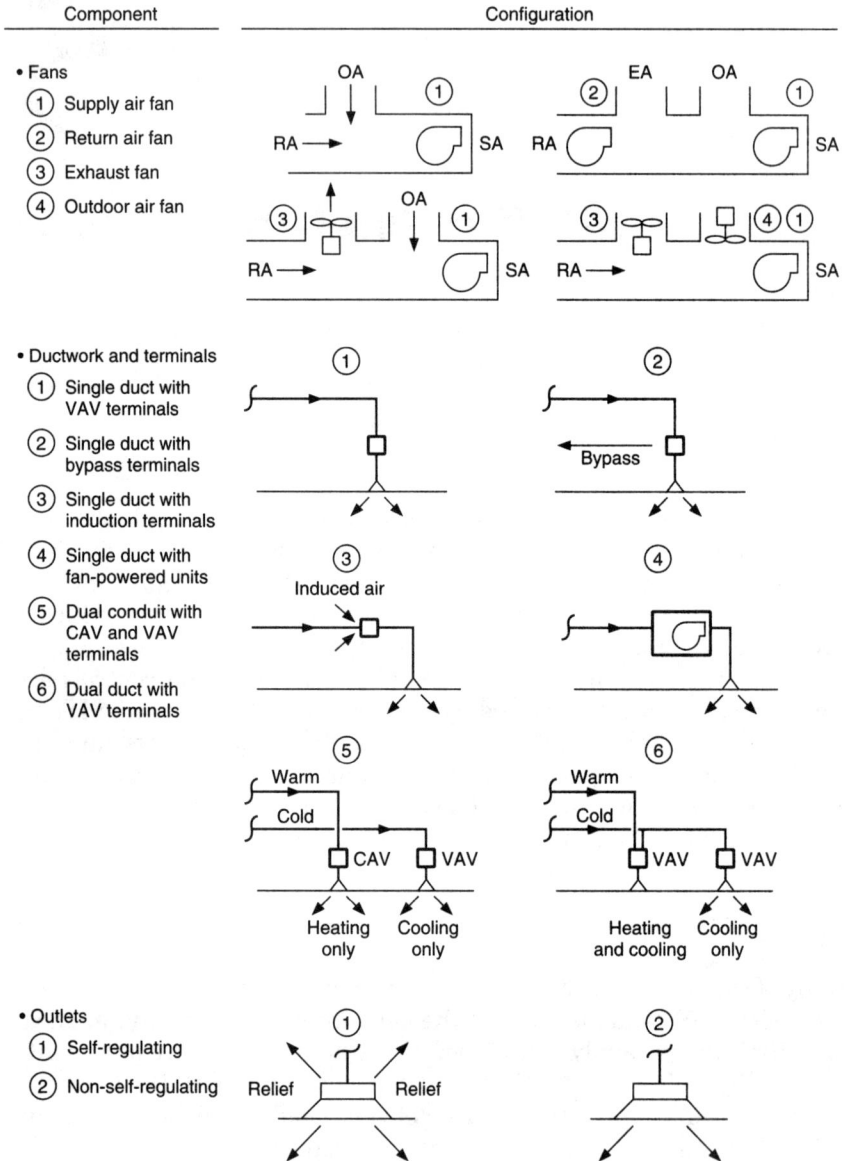

Figure 5.1   Major components of a VAV air-handling system.

5.1.1 Types

The following fan types are used in VAV systems (see Fig. 5.2):

- *Centrifugal,* with four possible wheel types

  Backward inclined (very common)

  Forward curve (limited to small system applications)

Airfoil (very common, especially higher pressures)

Radial paddle wheel (generally exhaust applications only)

■ *Tubular centrifugal* (common)

■ *Vane axial* (common for large systems and multiple fans)

■ *Propeller* (used mainly in exhaust and very low pressure applications)

The centrifugal versions are one of the more popular selections for the system supply, return, outdoor air, and exhaust fans. These fans typically have an air intake that is located at a right angle to the air discharge. The backward-inclined fan is favored for its nonoverloading

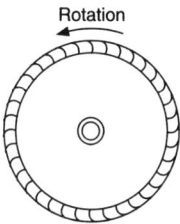

Rotation

Forward curved wheel.
Many small blades used on
smaller systems at low pressures.

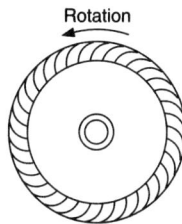

Rotation

Backward inclined wheel.
Fewer and larger blades than forward curved.
Wide range of flow and pressure uses.

Rotation

Radial paddle wheel.
A few straight large blades
used on exhaust for applications
where the fan could be clogged by debris.

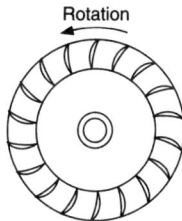

Rotation

Airfoil (backward inclined).
Large thick blades that are shaped
like aircraft wing. Higher pressures and flows.

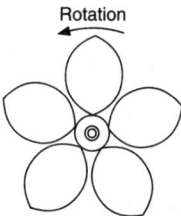

Rotation

Propeller fan.
3–8 large blades typically.
Very low pressure and high flow.
Air flows parallel to shaft.
Blades may be flat or airfoil shaped.

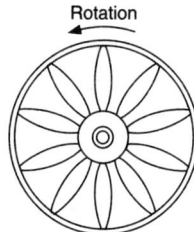

Rotation

Vane axial.
Blades are typically airfoil shaped.
Blades may be fixed or movable pitch.
Used on high pressures and flows.
Air flow is parallel to shaft.

Figure 5.2   Major fan types used on VAV systems.

characteristics and stable air delivery. These fans have higher efficiency than the forward-curve wheels, but they produce more noise due to the higher operational speeds. Typical application pressures range from 0.5 to as high as 8.0 inches w.g. The airfoil wheel is a form of backward-inclined fan that has the blades shaped like an airfoil or aircraft wing to improve the aerodynamic performance. These fans feature slightly more efficiency and higher operating speed than the backward-inclined fans. Their efficiency and stability at high delivery volumes have made them a very popular fan choice.

One form of centrifugal fan that is gaining popularity in both constant volume and VAV is the plug fan that uses a compact design that fits into a small space. These fans have no scroll, and their pressure capabilities are generally limited to below 5.0 inches w.g. The plug fans are often integrated into packaged systems where the package housing encloses the fan.

The forward-curved wheels are primarily used in small, low-pressure systems that are applied to light commercial and residential applications. These fans have the main advantage of moving a relatively large amount of air at low operating speeds and noise.

The radial blade fan is used primarily in exhaust applications where fouling of the fan wheel is a consideration. They can deliver high pressures and volumes at the expense of noise and efficiency, but their self-cleaning feature and ruggedness allow them to operate in an environment that would damage other types of fans.

The tubular centrifugal fans are typically centrifugal fans that have a special housing design that places the fan discharge in line with the fan intake. The housing typically contains vanes or a scroll that redirects the fan discharge so that the fan can be placed in line with the entry and discharge ducts. The vaneaxial fan is a propeller or turbine type of fan that accelerates the air by generally using numerous airfoil-shaped blades arranged around a hub. The vaneaxial fan can vary flow either by varying the pitch of the blades or the speed of the wheel. Vaneaxial fans work well in VAV applications and they have the advantage of being very stable over a broad range of delivery pressure and flow, with less tendency to stall than airfoil centrifugal fans. The vaneaxial fan excels in multiple-fan installations and very large capacities in excess of 20,000 cfm.

All fan characteristics are classified by AMCA, and centrifugal fans have been divided into four classes based on the outlet velocity and static pressure range. Each class from I to IV has progressively higher ratings for pressure and velocity. Using the proper class of fan ensures that the metal gauge, bearings, and component speeds are within accepted ranges.

Propeller fans are typically used for very low pressure, high-volume applications in the system. Small- to medium-sized packaged air-handling equipment sometimes uses a propeller fan as the exhaust fan. Larger systems seldom use propeller fans due to their low-pressure capability.

The packaging of VAV components may integrate the fans within an air-handling-unit housing, or the fans can be connected to the ducts and other components on an individual basis. Packaging of fans plays a large role in controlling noise and providing efficiency. For example, a fan placed inside an acoustically treated air-handling-unit enclosure may generate less noise than the same fan connected outside by duct-work alone. Putting the electric motor and drive components inside the enclosure may protect them from exposure to harmful elements, but it will also add the motor heat and drive slippage energy to the air passing through the fan. It is very common to mix equipment configurations on one VAV system, with the supply fan inside a housing and the return fan attached to the ductwork. (See Fig. 5.3.) The fan type selection for VAV applications is affected by many factors. Figure 5.4 lists key factors for fan type selection and special considerations required for proper VAV applications.

5.1.2 Capacity control

There are a number of methods of controlling both the pressure and the delivery volume of fans for VAV applications. Smaller systems can merely let the fan ride its natural fan curve (Fig. 5.5) when it encounters increasing or decreasing resistance to flow. A simple damper in the outlet can close to increase resistance and diminish the flow. However, this method is not generally energy efficient, and it can produce excessive noise if the velocities through the damper become excessive (above 3000 fpm). Small fan coils and packaged systems often use multispeed motors that change the fan delivery capacity in discrete steps. Several-step types of capacity control are the two-speed motor or two motors

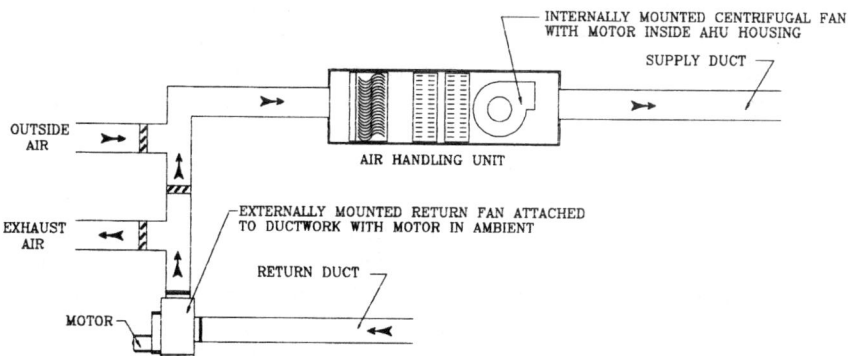

Figure 5.3 Example of internally and externally mounted VAV fans.

VAV Application Fan Type Selection is Based On:

1. Pressure

2. Airflow

3. Efficiency

4. Stability over operating range (fan curve)

5. Noise

6. Physical space available and ductwork configuration

7. Relative cost of installation

8. Maintenance requirements

9. Special requirements (dust or exhaust use, etc.)

10. Availability from suppliers in configuration required

VAV Special Requirements That Must Be Considered Include:

1. Stability and efficiency over full flow and pressure range

2. Compatibility with flow modulation method

3. Compatibility with other fans in system if auxilliary fan

4. Starting/stalling if used in parallel application

Figure 5.4   VAV fan type selection criteria.

connected to one fan that are selected to provide different operating speeds.

Centrifugal fan capacities were previously controlled by vortex or inlet dampers that pre-rotated and limited the amount of air entering the fan scroll. This vortex damper control was popular before the advent of low-cost electronic speed controllers. A major drawback to vortex control was a limited range of control and poor mechanical efficiency. Another means that was previously common but has been virtually discontinued is the mechanical speed control that used variable pulley diameters to change the fan speed. These devices proved to be mechanically unreliable and expensive to operate compared to electronic speed control. A previously popular type of electronic drive is the eddy current drive which has lost acceptance due to the superior energy efficiency of the inverter/VFD type of control.

Most of the larger fans are best controlled by the modulation of the fan speed using an electronic speed control that is called a *variable frequency drive* (VFD), *variable speed drive* (VSD), or *inverter.* These devices modulate the power going to the AC induction electric motor so that the motor speed changes in response to the changing frequency of the power produced by the drive electronics. (See Fig. 5.6.) For systems requiring more than 2 horsepower, the choice of a three-phase AC

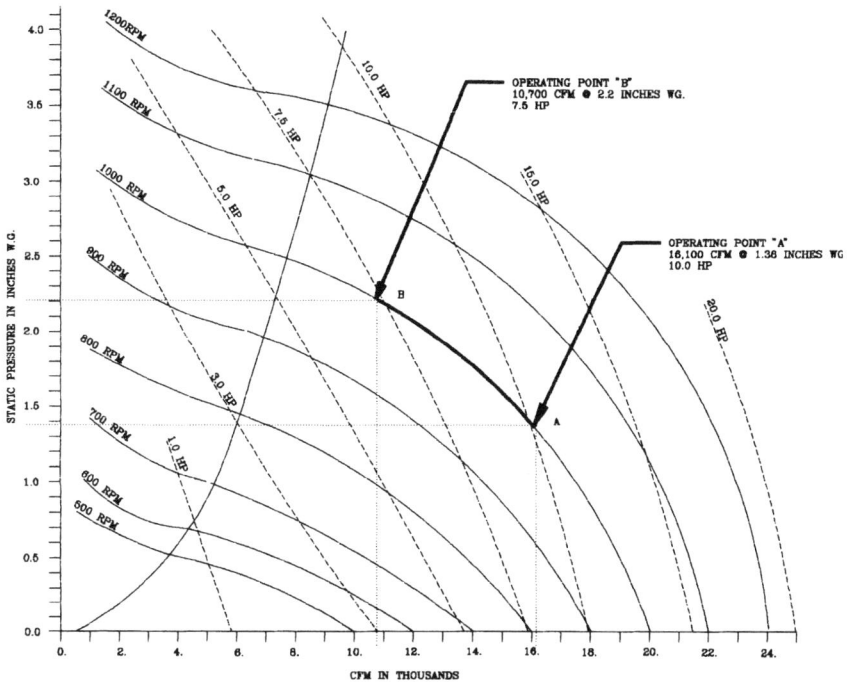

Figure 5.5 Fan curve for forward-curved fan riding the curve at 1000 rpm. Power drops 25%, cfm falls 34% and static pressure rises 70%.

induction motor is the most common, with smaller systems using the DC motor selection. DC devices change the RMS voltage being supplied to a permanent-magnet, brushless DC motor. Sizes of available drives range from fractional horsepower up to 500 horsepower.

The electronic speed control device cannot only vary the fan delivery volume and pressure, but also it can save energy, reduce the electric surge produced by starting the motor, protect the motor from overload, and protect the drive belts and bearings from the shock produced by sudden starts. The electronic speed control is also capable of the widest range of control. Drawbacks of the VSD/VFD include the potential for electronic noise or interference produced by the drive, acoustic noise produced by the motor which has harmonics with VFDs/VSDs that are not found with across-the-line operation, and the cost of the VFD/VSD.

The last method of capacity control that is employed is the incremental use of multiple fans of identical capacity and design. The airflow or pressure is typically measured, and multiple fans are started and stopped to maintain the correct range. A major problem with multiple fans, though, is backdraft through the idle fans and starting the idle

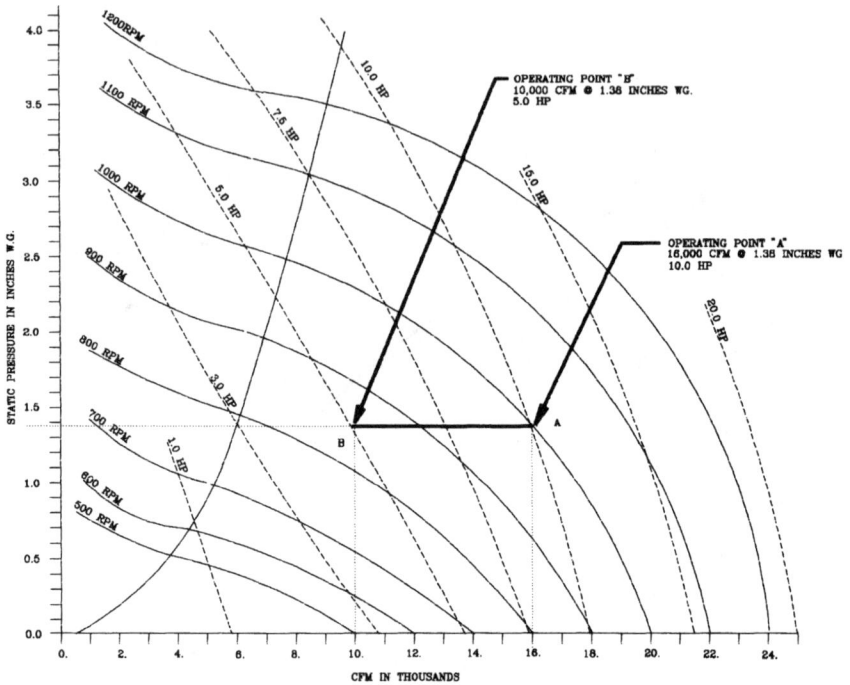

Figure 5.6   Fan curve for forward-curved fan with VSD from 1000 to 850 rpm. Power drops 50%, cfm falls 38% and static pressure is held constant.

fans without the fans stalling. Often, elaborate controls must be developed to control fan stability and starting. Centrifugal fans are difficult to operate in parallel, but the vaneaxial variety can operate very successfully without problems. Multiple fans have the advantage of introducing redundancy and protection from the failure of a single unit. Multiple fans can also broaden the airflow range of the system, given that they are carefully selected within the proper operating range.

5.1.3 Pressure control

Most of the larger VAV systems use some method to control fan static pressure. A sensor typically measures the static pressure in the duct and adjusts the fan capacity to maintain the desired pressure. Important features of pressure control involve where the pressure is measured, how it is measured, and how it is controlled. The accepted industry standard for measuring pressure has been a static pressure probe placed two-thirds of the distance down the supply duct from the fan. This method is stable and it has the feature of being independent of the air terminal devices. Some control vendors today are advocating elimination of pressure control in favor of using the worst-case terminal device flow as an indicator of the need for more or less fan capacity.

In this type of control strategy, the fan capacity is increased until the worst-case air terminal is delivering its required airflow with the integral terminal damper completely open. Theoretically, all other terminals would have their control dampers slightly throttled. There are many variations of this scheme, but the one drawback is the supply fan capacity being directly controlled from the air terminals. If the air terminal reading fails or is in error, then the whole VAV system is affected. This technique can also mask a poor duct design by raising duct static pressures higher than required to feed one terminal that is encountering a flow problem due to a duct restriction.

Pressure control in VAV systems takes another form, and that is safety pressure control that measures and limits the fan performance to a value that is not harmful to the duct or fan system. (See Fig. 5.7.) Many VAV fans are capable of literally blowing apart the duct system if a damper were to be incorrectly positioned. These pressure controls can relieve excessive pressure through mechanical means such as a safety valve, or they can stop the fan by means of interrupting the control circuit to the motor controller. Many code jurisdictions require this type of safety system to protect the system and personnel. Safety pressure control should be examined based on the capacity of the fans and the potential for items such as fire or smoke dampers causing an excessive pressure buildup. It is also possible to build in a control sequence that limits down-duct static pressure to allow the ductwork to be constructed of lighter-gauge materials, thus saving construction cost.

5.1.4 Pressure range

The energy consumed and the noise produced by a VAV system are directly related to the system pressure range. Wherever possible, mea-

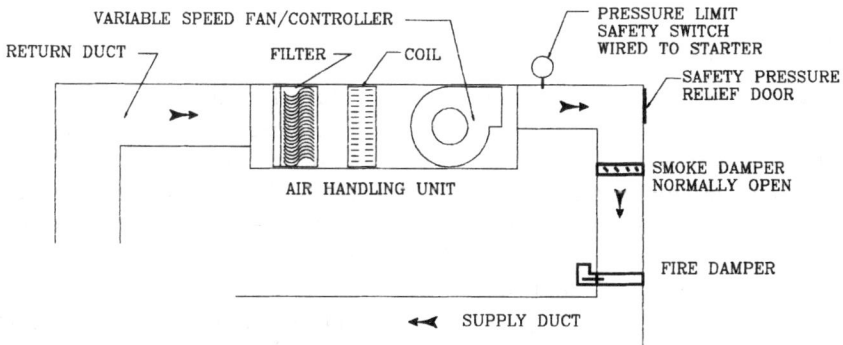

NOTE: CLOSURE OF SMOKE OR FIRE DAMPER CAN CAUSE EXCESSIVE DUCT PRESSURE.

Figure 5.7 Two methods of providing safe control of duct pressure. (1) Pressure limit switch wired to starter shuts down fan. (2) Safety door built into duct relieves excessive pressure.

sures should be taken to reduce the required operating pressures. However, due to the need to reduce first (i.e., initial) cost and fit components within physical space limitations, higher duct velocities and pressures are often required. A VAV system should be designed for a specific pressure range as designated by AMCA and SMACNA. If possible, supply duct pressures at the fan discharge should be reduced to below 5 inches w.g.

5.1.5   Return, relief, and outdoor air injection fans

The size and complexity of many VAV systems demand the supply fan to be assisted by one or more auxiliary fans that serve specific purposes. The return fan is used to assist the supply fan in overcoming the pressure losses of the return system and in expelling air when an economizer cycle is used. Exhaust fans are used to generally provide removal of air from the system, and air injection fans assist in moving outdoor air into the space. These additional fans should be used only when necessary, and the overall configuration of the air-handling system should be kept as simple as possible. A good application of these fans is their use with specialized items, such as air-to-air heat wheels, plate-type heat exchangers, HEPA filters in exhaust, and other specialized applications. Too often, designers use one configuration of air-handling system without examining the need or consequences of using a configuration with auxiliary fans.

Using auxiliary fans demands the control means must be provided to control the auxiliary fan operation so that it works in harmony with the main supply fan. For example, a return fan must track or have its capacity modulated to work properly with the supply fan.

5.2   Dampers

HVAC dampers are simple devices that vary the area through which the air passes. They can modulate open or closed to control either/or flow and pressure, or they can be open/closed devices that provide on-off flow control. Though these devices appear simple in function and appearance, they can prove to be difficult to select and apply in a VAV system. Dampers are rated and specified by leakage, by application (such as smoke control or fire dampers), by style/construction (such as butterfly, opposed blade, parallel blade, clamshell, and guillotine), and by control (either automatic or manual). In a VAV system, the proper selection becomes more difficult because of the variable flow encountered in the system.

A flow-modulating damper that is sized for full design airflow may become grossly oversized at the normal or average operating range. A damper that is sized for an adequate pressure drop at full flow may create virtually no pressure drop at low flow rates. To obtain proper damper operation in modulating applications, it is advisable to divide the damper area into smaller modules that can be incrementally controlled. This permits extending the throttling range of the damper without limiting full flow performance. Any VAV system that is moving 5000 cfm or more may need multiple-section modulating dampers to provide adequate flow control over typical VAV operating ranges.

Common damper design and selection errors include:

1. The assumption that an incorrectly sized/selected damper can be corrected by using a sophisticated control system such as DDC.

2. The assumption that the damper leakage will not be significant.

3. The assumption that a damper of a given construction will have linear or even proportional control over the entire airflow range in a VAV system. With wide variations of flow and sizing constraints, a damper may exhibit linear characteristics over only a small operating range. Flow characteristics in an actual system may differ considerably from the theoretical or flow "bench" performance. As a damper in a VAV systems closes, the pressure could be changing, and the proportional movement of the damper may not generate a proportional change in flow. For example, in Fig. 5.8, an increase in damper opening from 12 to 28 percent produces an airflow increase from 40 to 64 percent.

5.2.1 Damper selection

Automatic dampers are available in a variety of construction configurations, with the most popular being the parallel- and opposed-blade dampers (Fig. 5.9). Traditional information provided for damper use describes differences in the opposed-blade and parallel-blade dampers along with the effects of sizing and series resistance. In VAV applications, the flow variance makes the sizing calculation and series resistance value for design full flow virtually meaningless. As the flow drops, the series resistance drops, the damper becomes oversized, and the static pressure may be rising. This is one reason many engineers in the past have not had success in obtaining the desired linear control characteristics and stability with dampers. The design of automatic dampers in a VAV application must include:

1. Leakage at lowest flow and highest pressure

2. Pressure drop and damper position at full design flow

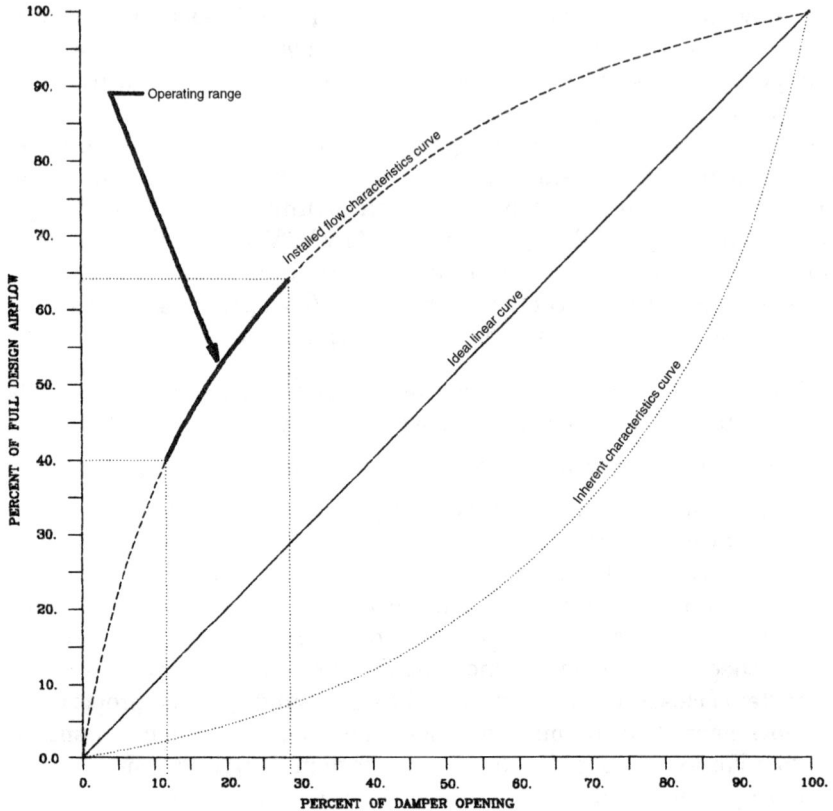

Figure 5.8 Typical VAV damper performance. Damper modulates over 16% of opening range for opposed blade damper properly sized for 100% flow conditions.

3. Pressure drop and damper position at 50 percent design flow

4. Pressure drop and damper position at lowest design flow

5. Maximum differential pressure encountered

The actual damper construction and type are often less important than the understanding of the functional requirements. An opposed-blade or parallel-blade damper choice is generally less significant than the leakage and sizing.

5.2.2 Damper operators

The previous pneumatic systems used linear piston-powered operators that were very reliable and low cost. Today, many DDC systems retain pneumatic operators because of the cost and reliability of these devices. However, as electronic systems have improved, so too have the

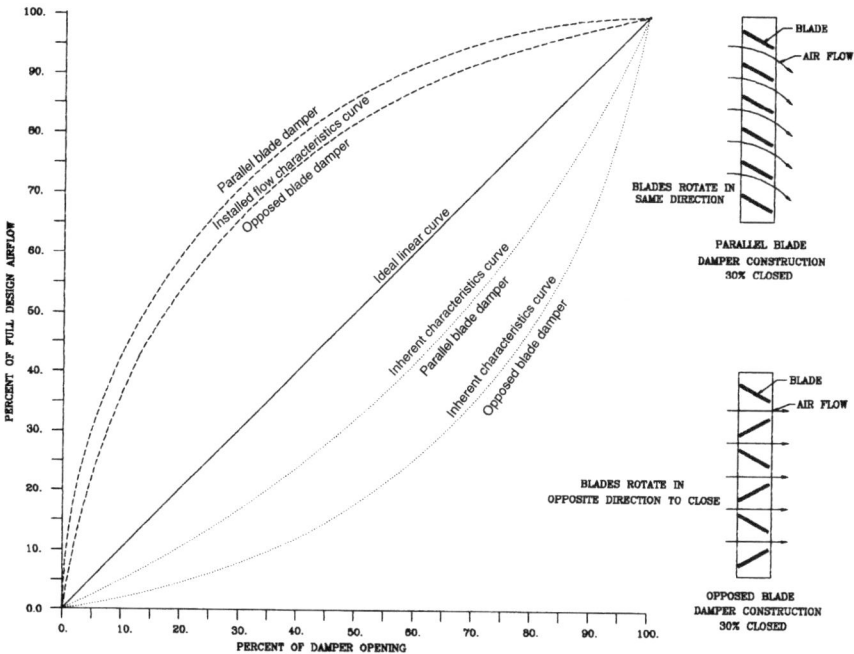

Figure 5.9 Performance of parallel- and opposed-blade damper construction in a typical VAV system.

electric actuators and damper motors. Low-cost electric operators with good reliability are now widely available. In selecting operators, there are several factors that need to be addressed:

1. *Control stability.* Hunting and rapid changes in position can burn out damper operators. Duty cycle and sizing of the operator must be carefully considered. The stability of the control loop can directly affect the life of the damper operator.

2. *Adequate power source.* Electric operators are sensitive to having an adequate power supply and can fail if not properly served with the correct size of power supply and wiring.

3. *Adequate size.* A pneumatic motor that is too small will generally just fail to position properly while an electric-motor-operated device could burn out or fail completely.

4. *Accurate positioning.* Bell cranks that slip or those that are poorly aligned/positioned and flexing of the motor mounting can lead to unstable control operation (Fig. 5.10).

5. *Protection from electrical surges caused by power transients and lightning.* Rooftop systems are especially vulnerable to lightning effects.

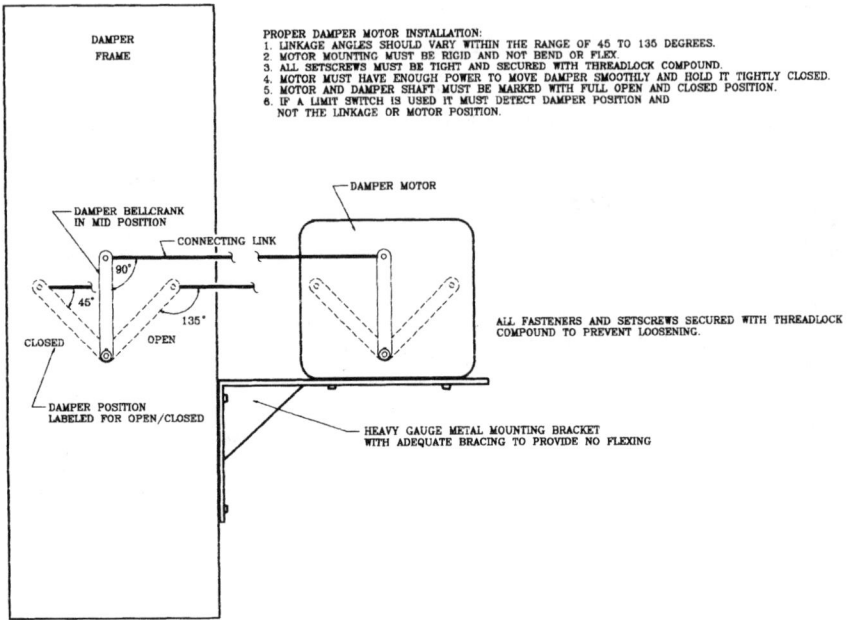

PROPER DAMPER MOTOR INSTALLATION:
1. LINKAGE ANGLES SHOULD VARY WITHIN THE RANGE OF 45 TO 135 DEGREES.
2. MOTOR MOUNTING MUST BE RIGID AND NOT BEND OR FLEX.
3. ALL SETSCREWS MUST BE TIGHT AND SECURED WITH THREADLOCK COMPOUND.
4. MOTOR MUST HAVE ENOUGH POWER TO MOVE DAMPER SMOOTHLY AND HOLD IT TIGHTLY CLOSED.
5. MOTOR AND DAMPER SHAFT MUST BE MARKED WITH FULL OPEN AND CLOSED POSITION.
6. IF A LIMIT SWITCH IS USED IT MUST DETECT DAMPER POSITION AND NOT THE LINKAGE OR MOTOR POSITION.

Figure 5.10   Proper damper motor installation for accurate damper positioning.

5.2.3 Manual dampers

In a constant volume system, manual dampers are often used on main and branch ducts to control/balance the flow of air. In a VAV system, all manual dampers should be eliminated throughout the system except for proportioning dampers that are installed on the discharge side of VAV terminals. (See Fig. 5.11.) Manual dampers add series resistance to the flow, which changes with the flow and becomes negligible at low flows. Thus, balancing with manual dampers at full design flow is not effective for lower flow rates. Even installing manual dampers on a diffuser connected to a pressure-independent terminal is incorrect, because the terminal will modulate its damper to overcome the effects of the diffuser damper. When multiple diffusers are connected to a single VAV terminal, manual dampers in the ductwork can proportion the flow of air among the diffusers, but it cannot change the total flow. A common problem is a user closing a diffuser damper which then diverts flow to other areas and creates overcooling. (See Fig. 5.12.) Manual damper settings should not be easily changed by persons who may not understand the effects on the system operation.

5.3 Terminal Equipment and Outlets

The VAV air-handling system can be connected to a wide variety of terminal devices that include:

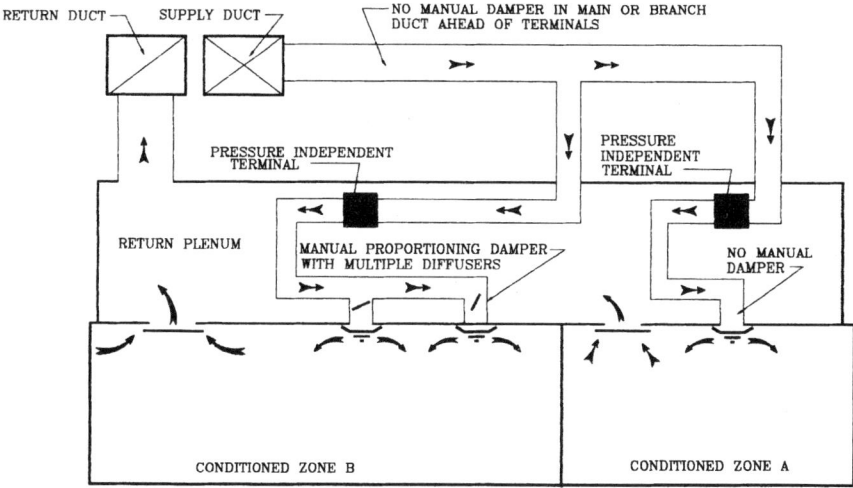

Figure 5.11   Manual dampers on the discharge side of VAV terminals.

- Variable flow pressure-dependent terminals and outlets
- Variable flow terminals that are pressure independent
- Constant flow terminals and outlets
- Terminals that are integrated with the outlets
- Induction terminals
- Serial or parallel fan-powered terminals connected to an outlet network
- Dual-duct terminals that can be both constant or variable flow

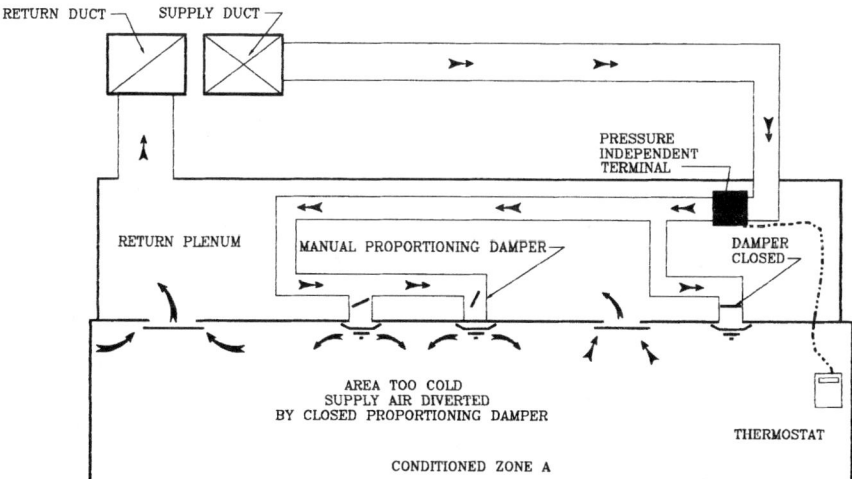

Figure 5.12   Diverted cold air creating overcooling.

It must be noted that a VAV system can use one or more or all of these devices based on specific space conditions. These terminal devices can be mixed with no loss of performance. Typical VAV terminal design features include one or more of the following:

- A volume control, regulation, and sensing means
- Measurement of the supply air or mixed air temperature
- Sound attenuation liners and construction to limit air-produced noise
- A means of mixing and/or inducing recirculation air

5.3.1 VAV terminal devices

Pressure-dependent variable volume terminals. The most elementary VAV terminal device is a control damper inserted into the supply duct that responds to a control signal generated by a wall thermostat. (See Fig. 5.13.) When the space needs more cooling, the signal from the thermostat opens the damper; when the space needs less cooling, the thermostat closes the damper. Though simple in concept, this air terminal design does have several limitations. The first is the variance of the supply pressure serving the terminal, resulting in the air delivery changing independently of the space thermal requirements. This pressure-dependent control of the airflow leads to an instability in the space temperature that is often unacceptable.

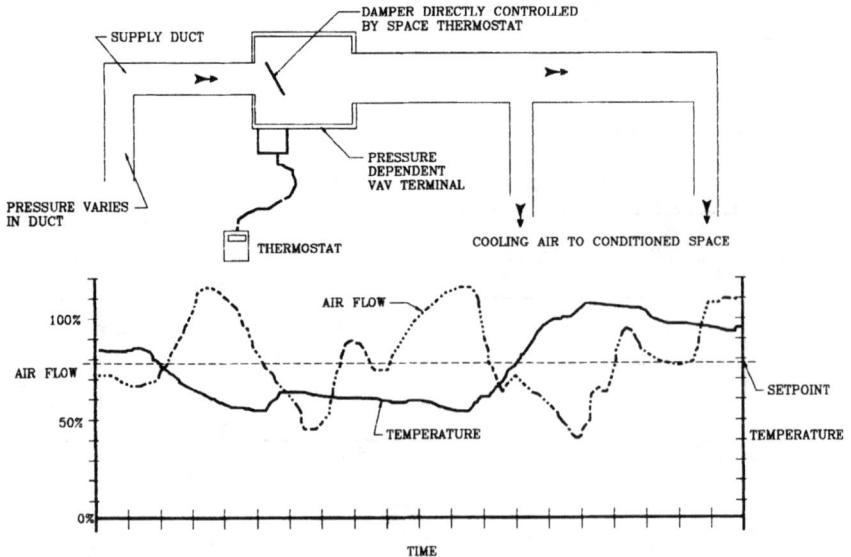

Figure 5.13 Diagram of airflow and temperature for steady-state load with pressure-dependent VAV terminal.

Pressure-independent variable volume terminals. A more sophisticated and stable control of space temperature can be obtained by introducing a flow measurement device within the terminal. (See Fig. 5.14.) The space thermostat again measures the need for cooling, but the control signal now does not directly control the damper in the terminal but instead establishes an airflow setpoint that is changed in response to the need for cooling. The terminal damper is positioned to maintain an airflow quantity that is determined by the setpoint. If the pressure in the system changes, the damper is repositioned to maintain the airflow established by the control setpoint. The supply of air to the space is now made independent of the supply duct pressure changes. These pressure-independent terminals are the most popular type, based on the stability of operation offered.

Pressure-independent constant volume terminals. The VAV system can have zones that are of constant volume. To provide a constant volume zone, a pressure-independent terminal is configured to deliver a constant volume of air that is set by the integral flow-measuring sensor. This terminal then is usually used with a reheat coil that is controlled by the space thermostat to maintain the space temperature. Using a constant volume reheat terminal (Fig. 5.15) in this manner is common with zones that must maintain a specific airflow or number of air changes or pressure relationship with other spaces. Constant volume

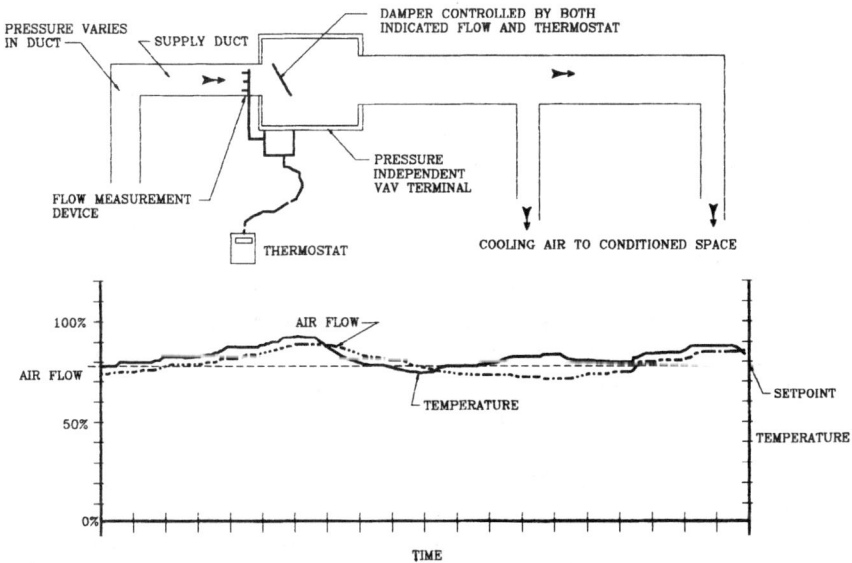

Figure 5.14   Diagram of airflow and temperature for steady-state load with pressure-independent VAV terminal.

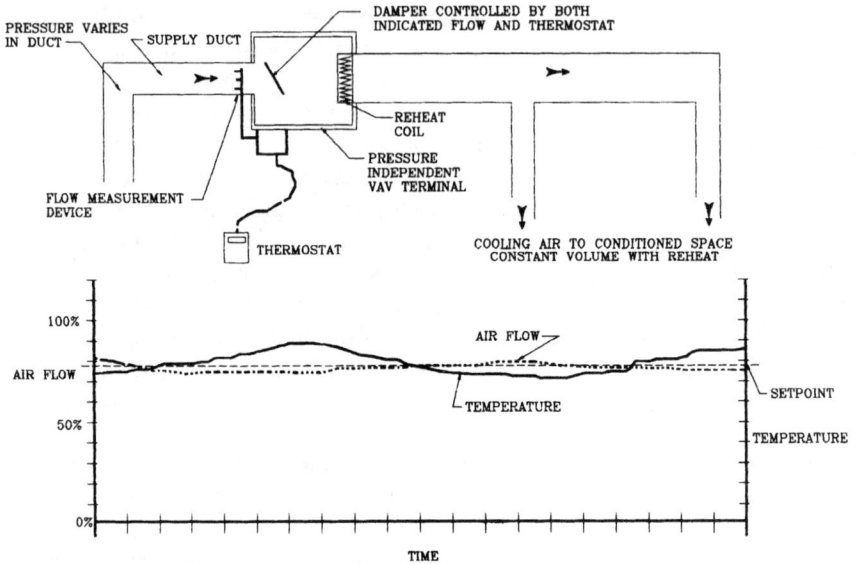

Figure 5.15 Diagram of airflow and temperature for steady-state load with pressure-independent VAV terminal operating in CAV reheat mode.

control may provide makeup air for exhaust fans or provide an adequate flow of air to pressurize the space.

The flexibility of today's air terminal controls makes possible a multi-mode air terminal that can function as both a constant volume terminal and as a variable volume terminal. An example of such an application would be a hospital room that had the terminal operate in constant volume mode when the room was occupied to maintain air changes and pressure, and then operated in variable volume mode when the room was unoccupied. In IAQ applications, the air terminal may have a constant volume minimum flow, with additional flow provided to meet high cooling load conditions. Minimum flows may be set so high that the terminal is a constant flow device 60 to 80 percent of the time.

Whenever a constant volume control mode is used with a VAV air-handling system that has a discharge temperature controlled independently of the zone temperature, some form of reheat must be provided to prevent overcooling the space when the cooling load air volume demand is below the constant-volume delivery.

Integrated terminals and diffusers.   The air terminal can be separate from the diffuser, being connected by a duct, or it can be integrated with the diffuser. The advantage of using an integrated terminal is the feature of having the damper and controls pre-engineered to work well with the diffuser. Sizing, noise, dumping, and other selection problems are potentially reduced. However, a disadvantage is the increased cost associated with giving each diffuser control damper and associated controls. Two varieties of devices have been used. The first is a smart terminal that uses DDC controls that can report zone conditions, remotely controlled and operating in an overall control network. The second is a dumb terminal that uses controls that are electrical, mechanical, or pneumatic to control temperature without any communications means. These dumb terminals are popular for their low cost, but many do not provide the level of control needed. They can malfunction and operate poorly, and troubleshooting is difficult. Devices that sense temperature at the terminal depend on air motion produced by the terminal to provide an accurate sample of the room temperature. However, if the air-handling system is turned off, these systems will sense only stagnant air at the ceiling, which may prove to be grossly inaccurate. Many integrated designs are purposely designed at a reduced cost to compete with the more expensive DDC air terminals, with the result that quality and accuracy may be diminished in favor of a lower price.

Induction terminals.   The supply air leaving the VAV air-handling unit is often too cool to introduce directly into the space. To warm up the air leaving the diffusers, the VAV air terminal is designed with an integral induction nozzle or nozzles (Fig. 5.16) that blend air from the return plenum or space with the supply air. This induction effect warms the discharge air, raises the air volume delivered and the number of air

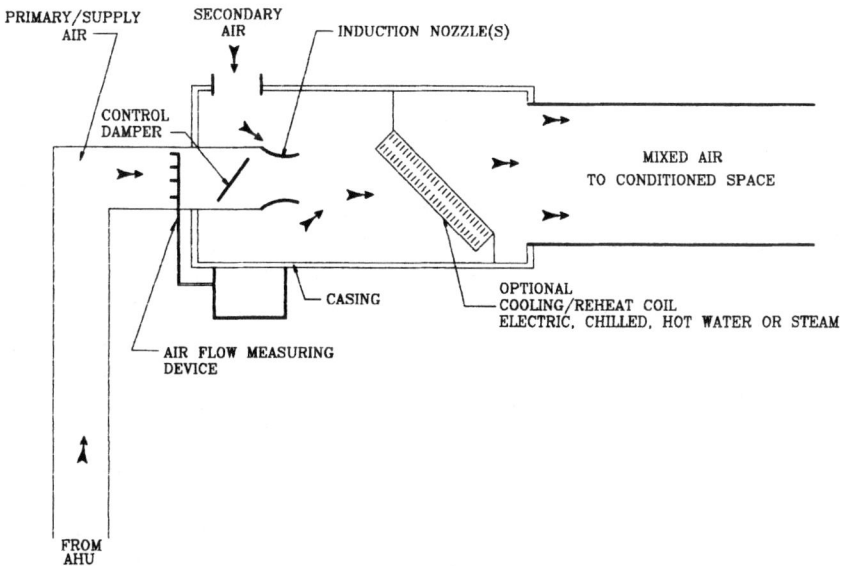

Figure 5.16   Pressure-independent duct-mounted VAV induction terminal.

changes, and provides potentially better air distribution and comfort. The induction effect can also be used to move the mixed supply and space air through a reheat coil to provide heating. Induction terminals typically require a relatively high minimum static pressure for operation, and they are often confined to specific applications such as hospital patient rooms and perimeter zones that require heating.

Pressure-independent fan-powered terminals.   Fan-powered terminal devices are configured in one of two configurations. A series fan-powered terminal (Fig. 5.17a) typically has a small centrifugal fan placed in series with the supply air and recirculation air. The fan usually runs all the time, delivering constant volume through the air diffusers and, as the amount of primary air is reduced, the fan pulls in more recirculation air and blends it to effectively raise the temperature of the air being delivered to the space. Thus, a series fan terminal converts the VAV air-handling unit's variable supply into a constant volume at the terminal. The series fan has adequate static pressure to deliver air through a short duct from the terminal and an optional reheat coil. Series fans typically operate at static pressures from 0.3 to 0.6 inches w.g., 100 to 2500 cfm, from 3000 to more than 6000 hours a year and are shut down only when the space is not occupied. They have relatively low mechanical efficiency compared to the main air-handling unit, and their cumulative air-moving energy consumption may be as great or even greater than the main supply fan.

Parallel fan-powered terminals (Fig. 5.17b) have a fan installed in the recirculation airstream that typically is turned on when the primary airflow falls below some set minimum. The fan operation blends the recirculation air with the cool supply air and raises its temperature. If more heating is needed an integral heating coil can be energized to provide heating. The parallel fan operates only to maintain minimum air circulation or the heating, and it consumes less energy than the series type of terminal. Though the parallel fan saves energy by operating less often, the change in airflow rates and noise produced by the fan can be very noticeable and objectionable in some applications. The parallel fan is sized for the minimum space airflow requirement, or typically 50 to 80 percent of the terminal design cfm and not over 0.7 inches w.g. static pressure. About 500 to a maximum of 2500 fan operating hours per year occur in most applications.

Regardless of series or parallel configuration, fan-powered terminals are generally less energy efficient than the other terminal types due to the relatively poor efficiency of the small internal fans. The fan-powered feature, though, is a means of providing constant-volume operation in the space, increasing the number of air changes, and in delivering warmer air. Fan-powered terminals have often been used where perime-

INSULATED TERMINAL
CASING
MIXING BAFFLE
CONTROL DAMPER
SECONDARY
RETURN/PLENUM
AIR
FAN
MIXED AIR
TO CONDITIONED SPACE
PRIMARY/SUPPLY
AIR
AIR FLOW MEASUREMENT
SENSOR
CONTROL DAMPER
REHEAT COIL
ELECTRIC, HOT WATER OR STEAM
DISCHARGE DUCT
CONNECTION
FROM
(a)

INSULATED TERMINAL
CASING
BACKDRAFT DAMPER
CONTROL DAMPER
SECONDARY
RETURN/PLENUM
AIR
FAN
MIXED AIR
TO CONDITIONED SPACE
PRIMARY/SUPPLY
AIR
AIR FLOW MEASUREMENT
SENSOR
CONTROL DAMPER
REHEAT COIL
ELECTRIC, HOT WATER OR STEAM
DISCHARGE DUCT
CONNECTION
FROM
AHU
(b)

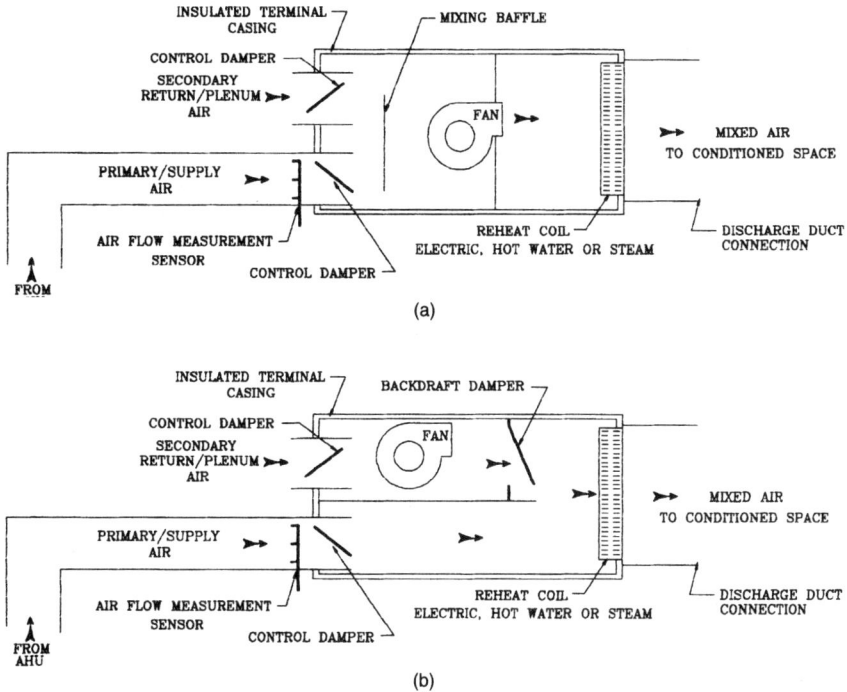

Figure 5.17   (a) Series fan-powered VAV terminal. (b) Parallel fan-powered VAV terminal.

ter heating is required or where cold supply air (below 55°F) is used, demanding substantial blending with recirculated air.

Double dual-duct terminals.   The double- or dual-duct terminal (Fig. 5.18) has two primary air connections. One is cool air and the other is warm air. The two supplies are mixed in varying proportions to provide the desired amount of cooling or heating as determined by a space thermostat. Because each supply is a variable volume system, the terminal can operate in a constant volume mode where the sum of the two supplies is kept constant, or the sum can be varied to provide variable volume control. Furthermore, each terminal can provide heating and cooling on a simultaneous basis without the need for a switchover mode. In some applications, where the dual-duct terminals are used for perimeter heating, a separate heating VAV air handler is provided that operates on a seasonal basis. When perimeter heating is needed, the relatively small heating air handler provides heated air to the perimeter terminals, and this unit can be turned off when the weather is mild.

Dual-duct terminals are relatively expensive because they use two sets of supply ducts. However, generally only the cool duct system needs to be insulated. Dual-duct systems can serve a wide range of

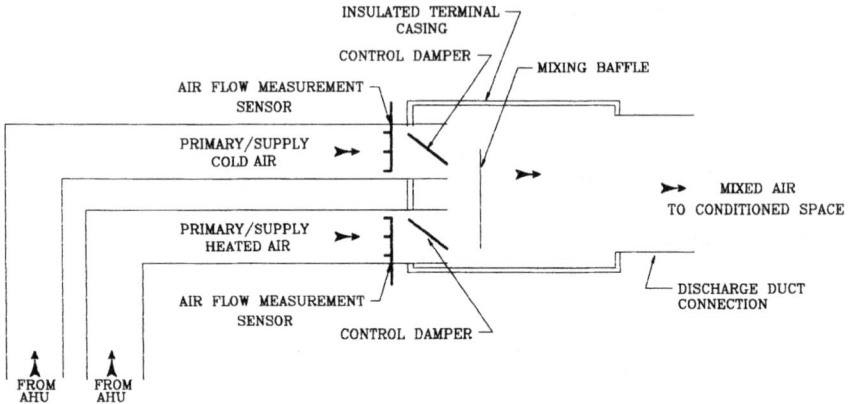

Figure 5.18   Dual-duct VAV/CAV terminal.

loads with a minimal need for reheat. They do not use the small, inefficient fans found in fan-powered terminals, and they can provide both constant and variable flow on a programmable basis. In many respects, the dual-duct terminal provides the highest-quality environment without sacrificing operating energy or IAQ.

5.3.2   VAV supply outlets

The VAV system can be used with the two basic HVAC air distribution concepts. In the well-mixed concept, the supply and return diffusers and grilles are typically located high within the conditioned space. (See Fig. 5.19.) Typically, ceiling-mounted or high-sidewall grilles, diffusers, or nozzles are installed to vigorously mix the supply air from the air-handling unit with the space air. This mixing effect is desirable, as is a minimum air motion throughout the space to prevent stagnation. Mixing or induction and the resulting air velocity and temperature are carefully selected to provide a uniform condition throughout the space without hot or cold drafts. VAV systems generally have difficulty maintaining adequate throw and mixing at low supply volumes when the velocity and mass of the supply air exiting the diffuser loses adequate kinetic energy to mix the air. This phenomenon results in the cold air flowing unmixed from the diffuser and is commonly known as *dumping*. To prevent this effect or reduce the point at which it begins, various vendors have created special diffuser designs that are tailored to the wide flow range of VAV distribution. Some designs are more successful than others, and the terminal mixing and throw characteristics need to be examined carefully for each application.

   Another VAV terminal problem caused by the ceiling-mounted or high-sidewall diffuser location is the difficulty in distributing heated

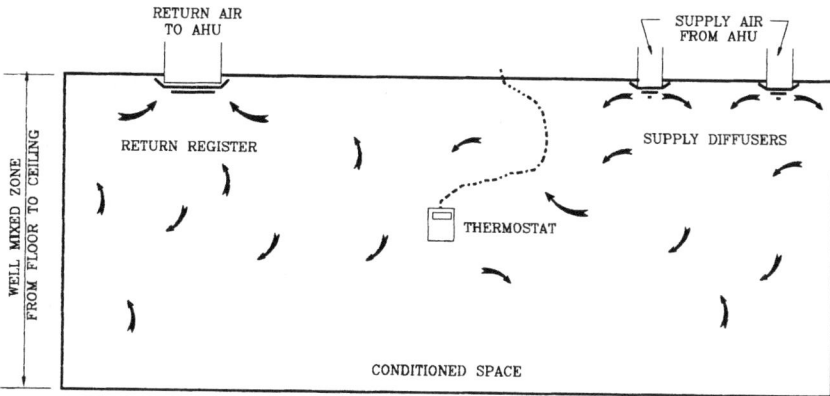

Figure 5.19 Typical well-mixed air distribution system with ceiling-mounted supply and return.

air from the system. (See Fig. 5.20.) Heated air tends to hug the ceiling, leaving drafts at the floor and poor mixing. Some vendors have developed switchover diffusers that change the airflow pattern when the device must provide heated air. VAV systems typically lower the airflow in the heating mode, and heating season problems are not uncommon unless countermeasures are taken in the design process.

Well-mixed ceiling/high-sidewall supply/return distribution. The typical VAV system can use the same style of supply outlets or diffusers as does a constant volume system if care is used in the selection. The VAV process varies the mass of air exiting the outlet, but dilemmas arise when selecting the outlet. If the outlet is selected to provide good induction and air mixing at low flows, it may have an excessive throw

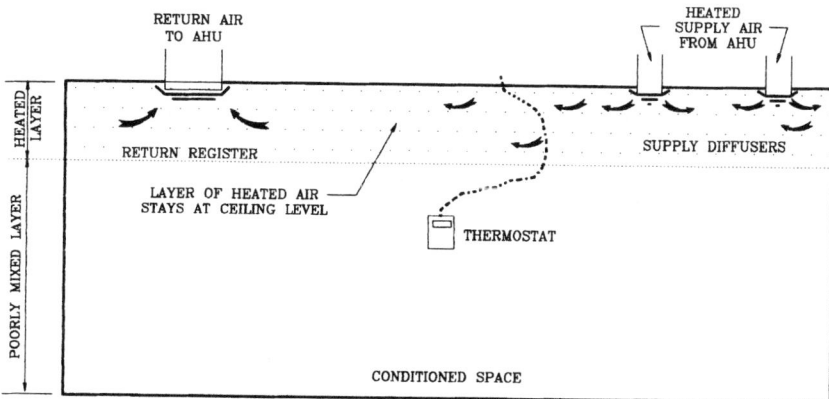

Figure 5.20 Typical well-mixed air distribution system with ceiling-mounted supply and return showing poor mixing with heating mode.

and be noisy at full design flow. If an outlet is selected to provide good induction, air mixing, and throw at full design flow it may cause dumping and drafts at minimum flow. Because of these problems, most outlet selections are a compromise, and the slot type of diffuser is favored in VAV because of its wide range of flow without losing its ability to mix the air and keep it from falling as a cold draft at low flows. Perforated grilles and diffusers with multiple induction nozzles, slots, or deflectors are also common in VAV applications.

One of the newer air distribution concepts for VAV applications involves using a diffuser that can vary its outlet area in response to the airflow volume to produce a constant velocity, variable volume effect. This class of device offers many advantages in terms of throw, mixing, and acoustics, but it involves additional components in the diffuser that raise its cost.

Displacement supply distribution. The displacement concept has been applied to commercial and industrial projects, both large and small. (See Fig. 5.21 *a* through *d*.) This technology has seen limited application

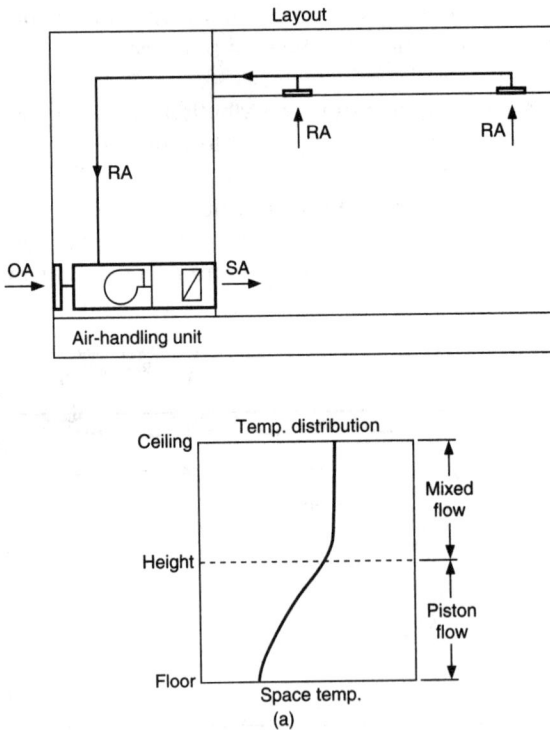

Figure 5.21 Displacement air distribution. (*a*) Near floor displacement air distribution.

Layout

Supply air volume versus temp. distribution

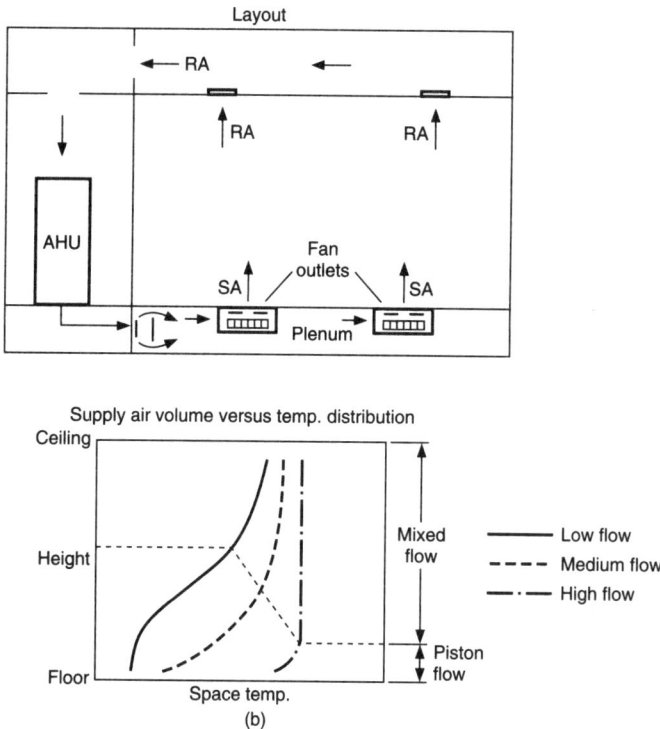

Figure 5.21 (*Continued*) (*b*) Under floor air distribution with variable speed fan outlets.

in the past but the IAQ and lower energy consumption potential of this system may cause it to gain in popularity. When VAV is used with displacement systems (Fig. 5.21*d*), the design of the floor diffuser becomes critical. Both the exiting air temperature and velocity must be maintained within a narrow range to provide mixing in the well-mixed zone within 6 feet of the floor, while not disturbing the stagnation layer above. The displacement concept demands a warmer air supply that can range from 60 to 70°F, compared to 45 to 55°F for the traditional systems. Furthermore, if the air terminal leaks cold air and it is not mixed with the room air, a cold draft can develop at the floor. Too little or too much diffuser induction effect can also produce discomfort.

5.4 Sound-Attenuation Devices

Ideally, the fans and other components are selected in a manner that limits the HVAC system sound to acceptable levels. However, cost and space constraints often raise fan speeds, air velocities, and turbulence produced in the system to a point that another means must be used to reduce the sound produced. These devices typically take the form of spe-

Layout

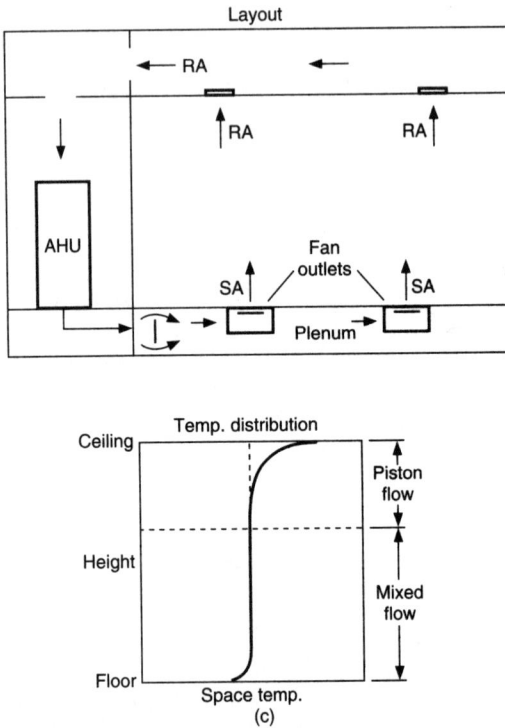

Figure 5.21 (*Continued*) (*c*) Pressurized floor plenum distribution with manually adjusted outlets.

cialized duct construction, housing construction, and devices inserted in the airstream to reduce noise.

Most sound attenuators use a mechanical means to cancel the sound by interference or by absorption. The typical sound attenuator uses perforated, metal-lined chambers that are filled with glass fibers or other media that can absorb the sound. Because the attenuator must be inserted in the system, the attenuator can raise static pressure losses and create noise of its own. A common mistake in selecting attenuators is the oversight of the noise caused by the increase in static pressure and the air noise caused by the attenuator. Less-than-anticipated noise reduction can result from poor device selection.

5.4.1 Passive sound attenuators

Passive sound attenuators generally take the form of some device that is inserted into the HVAC system duct. (See Fig. 5.22.) As the air passes through the device, the sound energy is absorbed in media or canceled by passages that reflect the sound out of phase. These devices must be selected based on the range of frequencies to be attenuated, the amount of attenuation required, and the acceptable pressure drop

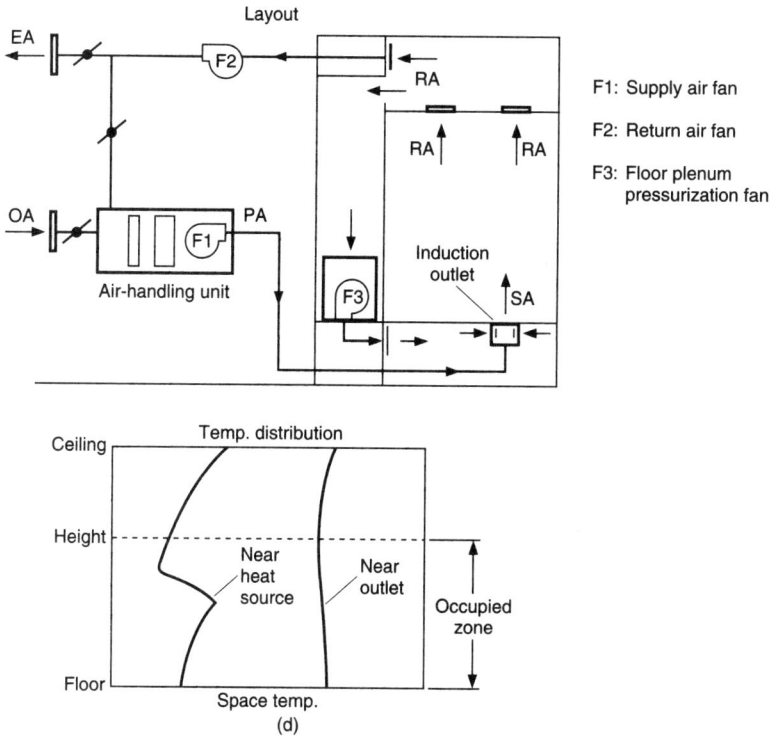

Figure 5.21 (Continued) (d) Under floor ducted air distribution with VAV primary air induction outlets.

caused by the device. These devices tend to be large and expensive, and if they create a substantial pressure drop, their operating cost is very significant.

The air-handling unit casing, the ductwork itself, and the air terminal devices all represent potential passive sound attenuation devices that can reduce the fan-generated noise. In many instances, slight changes in the fan selection, the ductwork configuration, duct lining, and the air terminals can eliminate the need for installing sound attenuators.

5.4.2 Active sound attenuators

The development of low-cost electronic devices that can sense, analyze, and process acoustic information in real time has led to the development of active sound attenuators. (See Fig. 5.23.) Typically with this type of attenuation, small microphones are placed in or near the HVAC duct, and a speaker is connected to the duct that transmits sound that is of virtually the same frequency but out of phase with the sound produced by the system. This sound that is out of phase can theoretically cancel the original sound; depending on the sophistication of the sys-

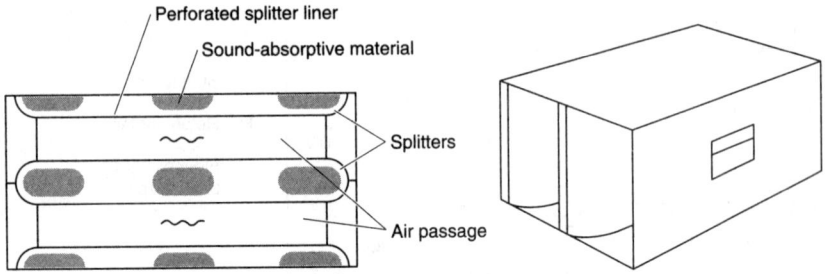

Typical performance (5 ft)
(+ 1000 fpm face velocity)

Octave bands

63	125	250	500	1000	2000	4000	8000

Dynamic insertion loss (DIL) in decibels

8	18	24	40	45	46	41	26

Self-noise power levels, dB re 10^{-12} watts

55	49	49	47	46	49	42	32

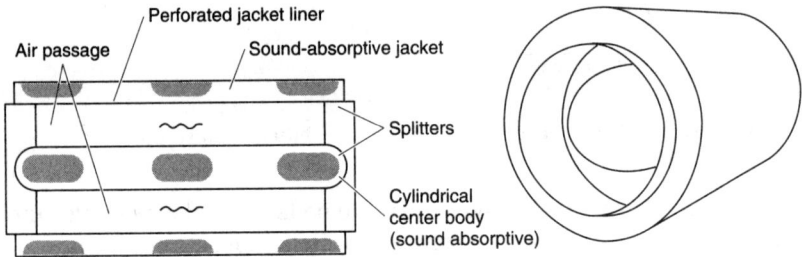

Typical performance (48 in. dia.)
(+ 2000 fpm face velocity)

Octave bands

63	125	250	500	1000	2000	4000	8000

Dynamic insertion loss (DIL) in decibels

10	18	33	34	35	24	20	16

Self-noise power levels, dB re 10^{-12} watts

56	54	57	57	57	57	53	47

Figure 5.22 Passive sound attenuators (silencers). *(Courtesy of Industrial Acoustics Co.)*

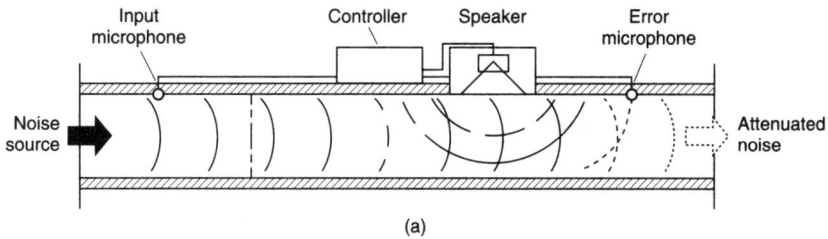

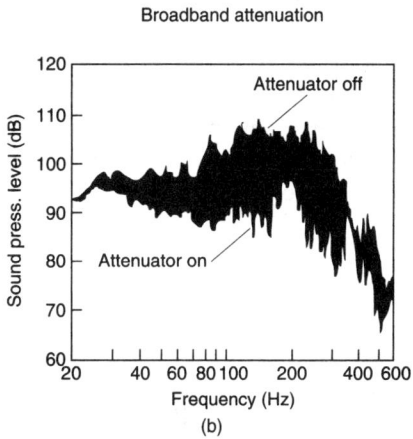

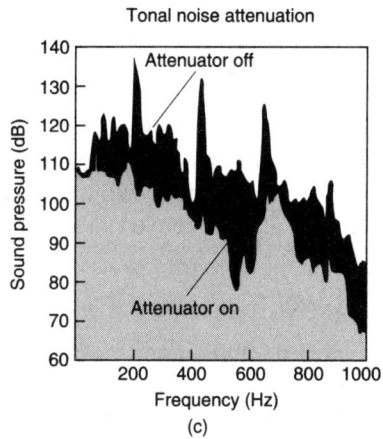

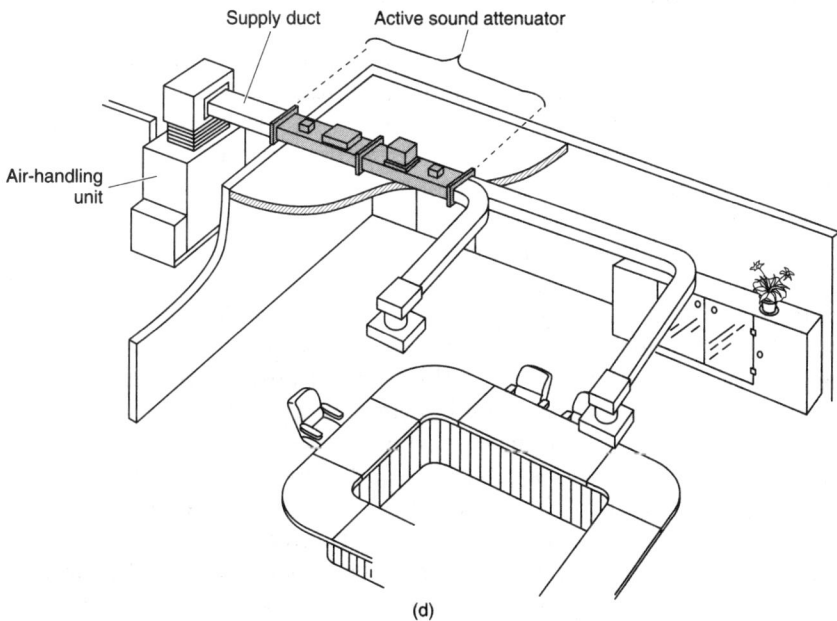

Figure 5.23 Active duct sound attenuator. (*a*) System component, (*b*) broad band attenuation, (*c*) tonal noise attenuation, (*d*) an example of application. *(Courtesy of Mitsubishi Electric Co. Ltd.)*

tem, the combined sound level is significantly reduced. Active sound attenuation can be tuned to reduce only a portion of the sound spectrum or a broad band, depending on the system type and need.

The advantages of this system are its real-time ability to adapt to changing sound caused by changing fan speeds, airflow, pressures, and so on. Also, this system can be used without requiring large space or restriction in the airflow. These are very significant and worthy features, but there are drawbacks. If the active sound system fails, the operation of the system becomes very noisy. This system does not have the reliability of a passive system, which seldom breaks or fails. Furthermore, this technology is not fully proven or developed and it remains somewhat experimental and expensive to use. Museum and theater applications where very low noise is required may justify the expense and increased maintenance.

5.5   VAV Controls

Until the 1990s, pneumatic controls were regarded as the most widely used controls on most HVAC systems. Smaller packaged VAV system controls were converted to electronic control in the 1980s, but the larger systems were often pneumatic-based. However, pneumatic, mechanical, and analog electronic controls today are obsolete, inflexible, and provide poor performance when compared to direct digital controls (DDC). DDC control is based on programmable digital logic that provides superior flexibility, diagnostics, and efficiency of operation. Continued improved performance with lowered costs have virtually eliminated any advantages of the previous pneumatic/electric control systems. Today, the air-handling system and air terminals can be integrated in a powerful DDC network that provides the user with diagnostic and programming capabilities that are almost unlimited. Zone temperatures can be set, monitored, and alarmed prior to the space occupants noticing a problem. The electronic communications capabilities of these systems even permit the control and monitoring to be remotely distributed around the world if desired. Self-diagnostics and the abundance of operating information available provide quicker and easier problem resolution. Today's service technicians are well acquainted with digital electronics, and repair and maintenance of these systems is not the problem it was when technicians were mechanically rather than electronically oriented.

New industry standards, open protocols, object-oriented programming, and many other features have driven down the cost of installation, expansion, and repair of these systems. The number and possible configuration of the control programs within a system is generally limited only by the imagination of the design engineer. Control techniques that were virtually impossible or cost-prohibitive with the previous technology are now available at low cost.

5.5.1 Supply air temperature and humidity control

The VAV air-handling unit is typically a constant temperature, variable volume device, and with previously limited function controls, the supply discharge air temperature is set at a fixed value (e.g., 55°F). However, there are times when it is advantageous to abandon this control practice. For example, if the system has an economizer feature, the discharge air temperature may be raised or lowered to take advantage of the outside air condition. By raising the supply air temperature from 55 to 58°F, it might be possible to meet the space cooling load without the need for mechanical cooling. If the outside air conditions provide 50°F air, then further lowering the permissible discharge temperature may save energy heating the air and in reducing the volume of air required and the fan energy, too. With a higher supply air temperature, the VAV air-handling unit would supply more air and use more fan energy, but this may be more efficient than operating a refrigeration system at an inefficient low-load level. In a similar manner, it may be advantageous to raise the supply air setpoint to move more air in the system and provide a greater number of air changes and ventilation. Control programs are available that examine the outdoor air conditions and reheat and airflow needs, and then select a supply air temperature that minimizes both cooling, reheat, and fan energy.

A common energy savings practice is to raise the chilled water supply temperature. If the cooling load and space humidity levels permit, raising the chilled water supply temperature reduces mechanical refrigeration energy consumption. However, with VAV systems, this refrigeration savings may be offset by the VAV fan energy required to move a greater volume of the warmer discharge air. A DDC system can compute the mechanical refrigeration savings and the VAV fan energy penalty and develop dynamic setpoints for the VAV discharge air and chilled water that are optimized for energy reduction. Furthermore, the airside controls can monitor the higher relative humidity caused by the warmer chilled water temperature and adjust the chilled water temperature to prevent loss of humidity control.

In many applications (such as static sensitive electronic assembly/ repair, laboratories, museums, and hospitals) it is desirable to maintain a high indoor relative humidity above 40 percent. If the cooling load is high, the fixed supply air temperature may remove too much or too little moisture from the air. By adjusting the discharge air temperature, it is possible to control the amount of moisture removed and the overall relative humidity. Humidity controls can be incorporated into the air-handling controls design to override the energy-conserving features when the humidity reaches an undesirable level.

5.5.2 Static pressure control

The traditional VAV air-handling system has a fixed supply duct static pressure that is typically measured by a sensor located two-thirds of the way down the duct system from the fan. This sensor location generally produces both reliable and stable duct static pressure. It has been proven to work with pneumatic, electric, and electronic controls. However, with DDC technology there are other alternatives. One control scheme examines the damper positions of the air terminals and adjusts the fan discharge pressure so that the worst-case terminal has its damper fully open, with the other dampers slightly throttled. Theoretically, this control scheme will reduce the duct static pressure below the fixed static pressure setpoint and thus save fan energy. However, this scheme may not work as well in practice as theory would indicate. If one terminal is always demanding full volume, the system may cause the fan static to be high all the time. Poor duct design, a malfunctioning terminal, or an incorrect measurement of the airflow at the terminal can eliminate the potential energy savings this sequence offers, and it should be backed up with the traditional duct pressure control.

Another potential control scheme for fan static involves purposely reducing the discharge static below the air terminal control range so that all terminals open fully but may not maintain the load. This feature is an alternative to shutting down the VAV system completely during unoccupied periods. It permits a small quantity of supply air to reach each terminal and it is useful for maintaining limited ventilation during unoccupied periods. This form of pressure control can be used to provide an IAQ cycle alternative to the traditional system shutdown during periods of limited occupancy.

5.5.3 Outdoor air control

From the time constant-volume systems became popular, designers have thought of percentage of outdoor air. The percentage concept has little value with VAV technology. Today, VAV systems have a wide throttling range, and most building exhaust systems are constant volume systems. Also, IAQ standards define ventilation in terms of the number of occupants or space area. Therefore, the amount of outdoor air is fixed by building occupancy, exhaust requirements, pressurization needs, and gross area—and not by the thermal load that VAV systems respond to. This implies that the percentage of outdoor air must vary, and it will most likely not vary with the thermal load. The traditional fixed control damper settings or proportional damper settings will not maintain a correct amount of outdoor air.

Generally, a control scheme should be established that maintains a relative constant volume of outdoor air that is independent of the VAV supply air quantity. This can be achieved by several methods. The first involves direct measurement of the flow of outdoor air and varying the outdoor air injection fan or damper setpoint to maintain the desired amount of air. The second involves using a fan-tracking methodology that introduces outdoor air independently of the air-handler supply volume.

5.5.4 VAV terminal unit control

The VAV terminal typically has a unitary controller that functions in a stand-alone manner to maintain the space temperature at a specific setpoint. As control systems have become more sophisticated, this controller has been able to provide control that is based on many factors beyond the wall thermostat setpoint. VAV terminal control features today include:

- User-selectable temperature setpoints that are in an operator-selectable range. Turning a dial on the thermostat may provide from 5 to 15°F (1 to 5°C) of adjustment.

- Occupancy sensors to adjust the space temperature setpoint and amount of outdoor air required based on occupancy.

- Time-clock functions that automatically set up or set back the space temperature setpoint based on planned occupancy.

- Auxiliary control of lights or integration with lighting control to save energy when the space is not occupied.

- Measurement and reporting of space temperature, air supply temperature, damper positions, and occupancy for purposes of energy metering and diagnosis.

- Tracking control of exhaust and supply terminal systems to maintain space pressure relationships.

- Auxiliary control of baseboard heaters or other devices to integrate the heating/cooling equipment functions within the space.

- Feedback of load conditions to the central control system for adjustment of the supply fan, chilled water temperature, or other global parameter.

- Emergency or life safety functions such as those associated with smoke control.

A VAV terminal control feature that is often overlooked is the failure mode or normal position of the dampers. Terminals can be obtained with dampers that fail open, fail closed, or fail in the last position. Each failure mode has its advantages and disadvantages. Local codes and life safety implications of the failure mode should be carefully investigated before the failure or normal mode is selected. Generally, a damper that fails open is most desirable since the system can then deliver air to the space with the terminal logic or power inoperative.

5.5.5 Warm-up control

The practice of using night setback to save energy during unoccupied periods, or the complete shutdown of HVAC systems, requires the HVAC to regain space temperature control when the space is scheduled for occupancy. To accelerate this process, a warm-up control cycle can be employed that raises the air-handling-unit discharge temperature to a point above the desired setpoints, and the air terminal devices open their dampers to the design setpoint flow volumes. This warm-up cycle quickly raises the space temperature and allows the VAV system to function normally when the space is returned to control during occupancy. Without warm-up control, several zones could remain too cold prior to the occupied period due to a lack of load.

Warm-up control is generally a means of recovering from a shutdown period during the heating season, and it is sometimes applied in the cooling season as a cooldown control in which the air-handling system delivers cool air with the air terminals fully open to quickly recover from a shutdown or temperature setup period.

5.5.6 Supply and return fan tracking

One of the least understood, troublesome, and controversial control issues involving VAV applications over the years has been supply and return fan tracking. Many design solutions have been published which have later proven inadequate. There have been literally hundreds of schemes developed, but very few have worked reliably over the long term. These historical problems are the result of:

1. Inadequate design analysis, resulting in design assumptions which are incorrect

2. Failure of available controls to work with the required accuracy and reliability

3. Incorrect information that is passed on as correct although it is flawed

4. Emphasis on thermal comfort rather than IAQ and ventilation

Purpose of fan tracking. Fan tracking is the method of using either a return, outdoor air, or exhaust fan with the air-handling-unit supply fan to move air to and from the conditioned space in a manner that provides control over the conditioned space pressurization, and the proper introduction of outdoor air for ventilation. Space pressurization is critical to energy conservation and indoor air quality to prevent excessive infiltration or exfiltration. Pressurization problems can also produce unwanted "wind" between conditioned spaces. With constant volume systems, fan tracking can be achieved much more easily than with the variable volume system. The VAV systems require special control sequences to maintain the accuracy of the fan tracking over the broad operating range of the air-handling system.

Figure 5.24 illustrates a basic air-handling unit with no outdoor air intake/exhaust openings and dampers serving the conditioned space. This unit describes the process that occurs in moving air to and from the conditioned space, and it will be modified as the described system becomes more complex. Starting with this simplified system allows an opportunity to build up an understanding of the complex VAV air-handling-system process. Throughout the discussion, the symbols and pressure designations will not be changed.

The pressure E represents the external or atmospheric pressure. Pressure S_1 is the space pressure at the discharge of the supply terminal device. Pressure S_2 is the space pressure at the return grille. Pressure S_1 is greater than S_2, and air flows within the space according to this pressure difference. Pressure A is the fan discharge pressure, and A minus S_1 is the combined loss of the supply duct system and the terminal device. Pressure difference S_2 minus D is the return duct and

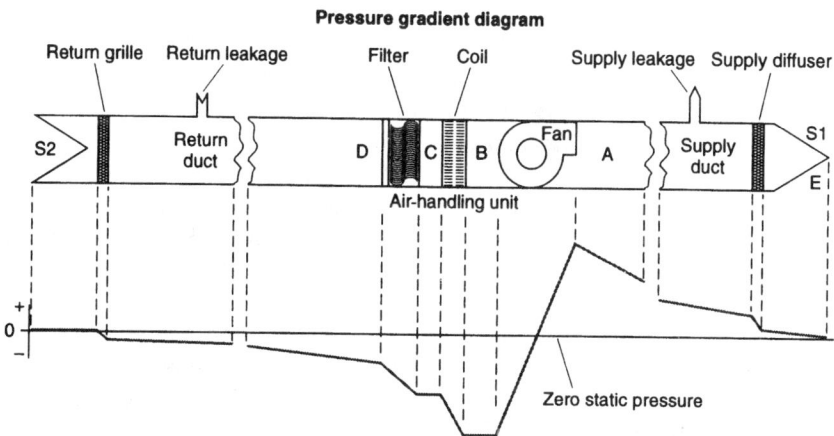

Figure 5.24 Basic air-handling unit.

grille combined pressure loss. Pressure difference A minus B is the total fan pressure produced. Pressure difference C minus B is the coil pressure drop, and D minus C is the filter pressure drop.

The arrow exiting the supply duct before it enters the conditioned space represents duct leakage from the system, and the arrow entering the return duct after the conditioned space represents duct leakage into the system. The bidirectional arrow between the conditioned space and the external space represents exfiltration and infiltration produced by the pressure difference of S_1 or S_2 and E.

Intuitive views of this system would lead to the conclusion that this air-handling system would not affect the conditioned space pressure. However, this is not true. Leakage in the return duct will cause a rise in space pressure. This return airflow must equal the supply airflow, and any leakage will result in less air being drawn from the conditioned space. Leakage in the supply duct produces the opposite effect. Loss of supply air causes a fall in conditioned space pressure because less air is delivered to the space than is removed.

If very precise airflow monitoring stations were placed in both the supply and return ducts, another phenomenon would be observed. If the unit were providing cooling to the space, the supply air volume would be less than the return air volume due to the difference in the air density caused by the difference in the supply and return air temperature and humidity. Some compression would also occur in the supply duct due to the pressure difference between the two airstreams. Therefore, the air volume is not the same in the closed system, but the air mass will be the same. The supply/return air volume difference that exists between a supply and return condition of 43/42 and 78/55 db/wb°F can be substantial. The psychrometric chart indicates the air density for the 43/42 supply condition is about 12.78 cubic feet per pound, while the return condition of 78/55 is about 13.65 cubic feet per pound, for a net difference of 0.87 cubic feet per pound or 6.8 percent. Even a 5.0 percent change in air density causes significant errors in using airflow monitoring equipment without compensation for these temperature effects. Using the conventional approach of airflow monitoring and returning less air to the fan system could lead to excessive overpressure of the conditioned space! In a variable air volume system, all the noted pressures and pressure differences will change as the fan system delivers a variable volume of air. With increased airflow the pressure differences and pressures will increase. Having explored this simplified example, it is now possible to examine the more complex and common system that includes dampers at the supply fan inlet. The air-handling return section is divided into two sections: an exhaust/return plenum and an outdoor/mixed air plenum. One damper and duct serves as an inlet control for entering fresh outdoor air. The second damper and duct serve as an outlet for leaving exhaust air. The third damper, the return damper, develops a pressure difference between the positive pressure exhaust and negative pressure outside air intake sections. Figure 5.25 illustrates this system. The pressures in the system remain as defined previously except for the pressure D_2, which is the mixed air plenum pressure.

For the system in Fig. 5.25 to function properly, pressure D must always be greater than pressure E so that the system can push the air out the exhaust damper, duct, and louver. Pressure D must be positive and pressure S_2 must even be more positive to overcome the return duct loss. Thus, the supply fan pushes air out of the conditioned space, and such a system will always require a positive space pressure to function. Without a positive pressure, the exhaust function will not work!

For the intake of outdoor air, the pressure D_2 must be low enough to overcome the outdoor air louver, duct, and damper losses represented by the pressure difference E minus D_2. The resistance and sizing of the return damper are very important in this system.

The return damper will generally have a negative pressure on the outdoor mixed air plenum side and a positive pressure on the exhaust/return plenum side. If the resistance of the outdoor air and exhaust

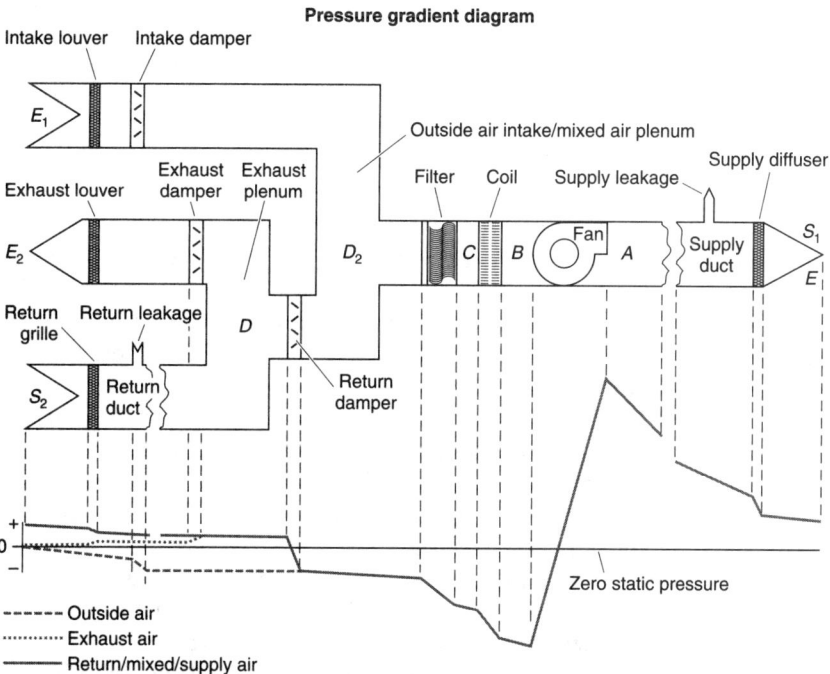

Figure 5.25 Air-handling unit with outdoor air and exhaust ducts.

ducts/louvers is high, this damper may not develop an adequate pressure drop under all flow conditions. The return damper position that develops an adequate pressure drop at full system flow may prove inadequate at partial flow. When the return damper pressure drop is inadequate, the system will not exhaust and intake the anticipated/ required amount of ventilation air. The result may be poor ventilation and poor indoor air quality. The traditional damper-positioning percentages and the control methods that tie the operation of these three dampers together from a single control output lead to ventilation problems in many installations.

Another common limitation with this design is the fact the supply fan must push the return air back to the air-handling unit. The conditioned space must act as a duct, and it may leak excessively. Many buildings that use this simple system can be detected by observing the outside doors standing open when the system is drawing in outside air. The return duct and exhaust path pressure loss is too great! Therefore, this design should be used only with the lowest possible return duct and exhaust losses, generally in the range below 0.05 inches w.g.

The described limitations of using only a supply fan have resulted in the common practice of adding an exhaust or return fan to the air-handling system. The next two figures illustrate how the previous limitations are resolved. In Fig. 5.26, a return fan has been added to the system.

The configuration in Fig. 5.26 provides a return fan that now can pull back the air to the supply fan. It can also help build the positive pressure required to exhaust the air out the exhaust damper, duct, and louver. In a constant volume system this fan would be easy to add to the system, but with the wide range of flow rates and pressure losses in the VAV system, a form of fan-tracking control will always be required. This return fan must provide the correct flow and pressure relationships to keep the conditioned space pressure within the required range while providing the correct pressures for the exhaust and intake of air. The damper positions will, of course, affect the performance of this return fan.

Many methods over the years have been tried for providing correct fan tracking. Unfortunately, most prove inadequate and fail to work as intended. Even some of the most notable books published by control companies include incorrect control arrangements. Investigation of failed systems has generally revealed an absence of the control of plenum pressures in the mixed air plenum and the exhaust plenum. Usually, the plenum pressures vary widely and unpredictably due to the control damper sizing and control methods that generally fail to address the variable volume characteristics of the system. Inability to control these plenum pressures leads to an incorrect balance between

Pressure gradient diagram

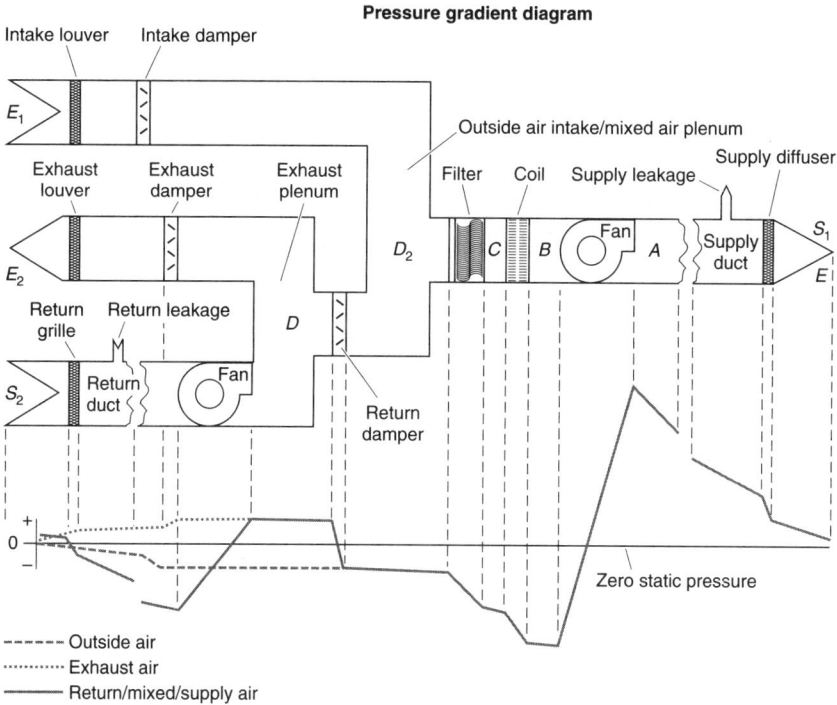

Figure 5.26 Air-handling unit with return fan.

the exhaust and outdoor air intake, which defeats the fan tracking control system.

Another issue that affects fan tracking is the nature of building exhaust systems that often must work with the VAV system. Toilet exhausts, kitchen exhausts, and other exhaust systems are generally constant volume systems, and the VAV systems are often introducing outside air on a supply fan flow percentage basis. This implies that the outside air volume will vary depending on the thermal loads. Therefore, at times the exhaust flow may be greater than or less than the outside airflow. Also, ASHRAE recommends that the outdoor ventilation air be based on occupancy quantities, or carbon dioxide concentrations, but these systems lack independent control over the outdoor airflow.

The following are the most common fan-tracking methods that have been used.

Slave return fan control. With this control design, the volume/pressure control signal that is used to control the supply fan is slaved or duplicated for the return fan. As the supply fan speeds up, or as its vortex dampers or pitch control change, the return fan changes its speed, dampers, or pitch proportionately. In the rarest of fan and duct configurations this simplistic scheme would work. It generally fails because the supply fan is being controlled by the duct static pressure that is set to a constant value. The system curve and fan curve for the supply fan are usually very different from those of the return fan system, resulting in incorrect tracking. Poor tracking leads to excessive conditioned space pressure, too little space pressure, and/or the inability to control the amount of outdoor air that is moving through the intake system.

Space pressure control. The most intuitive solution suggests that the conditioned space pressure could be used to regulate the return fan performance. If the space pressure becomes too high, you speed up the return fan, and if the pressure is too low, then you slow down the return fan. Despite the use of variable-speed fans, vortex dampers, or variable-pitch fans, this system often proves impractical. The main problems are the space leakage and the very low control range of the conditioned space pressure. Even with expensive high-accuracy pressure sensors and DDC control, the typical 0.01 to 0.03 inch w.g. pressure is too low to measure accurately. Furthermore, many VAV system control zones lack walls to contain the pressure. Direct space pressure control alone is seldom a practical fan-tracking-control solution.

A variation of the space pressure control scheme locates a pressure sensor in the return duct to maintain some constant pressure. This change in pressure-sensing location, though, does not resolve the problems identified.

Airflow monitors. Most commonly accepted control schemes previously published use an airflow monitor in the supply duct and the return duct to measure the supply and return fan airflow. These monitors first used pneumatic devices that proved expensive and difficult to calibrate, and the newer versions use solid-state DDC control. Though published accuracies are very high (some vendors claim plus or minus 1 to 2 percent), field investigation proves these devices to be too inaccurate to perform the fan-tracking function (plus or minus 20 percent is not uncommon). As discussed previously, air density compensation is seldom or never used. The often required "ideal" straight section of duct and nonturbulent airflow is seldom found in the field. The transmitter errors, airflow turbulence, conversion errors, and a host of other control system limitations lead to flow accuracies that typically are no better than 15 to 20 percent for most field systems. This control accuracy is not acceptable. The premise of accurate flow measurement that is the basis of this tracking methodology is difficult to attain in the field.

The air monitor tracking method is based on an assumption that the control of the airflow in the supply duct and the return duct are the parameters that need to be controlled. This assumption is not totally correct, though intuition implies it should work. The problem with this assumption is the lack of control over the pressures D and D_2. If the fan-tracking system does not have control over these pressures, the fan-tracking control will fail. If these pressures are not controlled, the amount of air that is drawn through the outside air path and discharged through the exhaust path will not have the correct ratio to maintain the correct space pressure. This pressure problem will fight against the control sequence that tries to maintain the flow tracking. If pressure D is too low, not enough air will be exhausted and the conditioned space pressure will rise, or the air will exit as excessive exfiltration. If pressure D is too high, too much air will be exhausted, and the conditioned space pressure will be too low, leading to infiltration. If pressure D_2 is too low, then too much outdoor air will come into the system, and the space pressure will be too high or too much exfiltration will occur. If D_2 is too high, then the required amount of outdoor air will not be brought in and the space pressure will be too low.

Another problem that arises with this scheme is the changing pressures that occur throughout the air-handling system flow range. The damper settings for full flow must be changed for partial flow, but the control system employed does not compensate for this. Also, the return damper is generally oversized and the correct pressure drop across it becomes impossible to achieve unless it is fully closed.

Another problem with this system is the intrinsic control hysteresis. The return fan is in series with the supply fan, and what the return fan fails to pull may be pushed by the supply fan. Thus, if the conditioned space leaks very little, the conditioned space pressure can rise to the point where a return fan will move the correct airflow. Some systems have been observed where the return fan is turned off by the control system, and the space pressure alone pushes the air back!

The airflow monitors have been located in the exhaust and outside air intake ducts with much better success. Though some inaccuracies remain, placing the airflow monitoring at these locations has proven to be one of the more successful methods. Unfortunately, airflow monitors are expensive, and it is often difficult to find a section of ductwork with limited turbulence that allows them to operate correctly.

New approach: plenum pressure control method. One limitation of the previous methods was the basis used to control the return, exhaust, and outside air dampers. These dampers are often sized incorrectly for the airflow, and the control system does not consider the plenum pressures that affect the flow through these dampers. The ultimate result has been poor fan tracking and problems with indoor air quality and space pressure. Dampers can be sized correctly for the maximum flow of the VAV system, but they will ultimately become oversized for the airflow when the system is not operating at full volume. Too often, these dampers are the full size of the duct or plenum, or set by the space available. The result is an inability to maintain the correct airflow and pressure.

The flow of air through a duct, damper, or orifice is controlled by the resistance/area of the passage and the pressure difference across the passage. This relationship provides a basis for constructing an accurate fan-tracking system. There are two versions of this method that have been proven. The first is a constant-plenum-pressure, variable-damper method, and the second is a variable-plenum-pressure, fixed-damper method. Each of these methods can produce good return fan tracking, control of space pressure, and most important, accurate control of outdoor air ventilation rates for providing indoor air quality.

Fixed-plenum-pressure method.   Referring to Fig. 5.26, a pressure transmitter is installed in the return/exhaust air plenum to measure pressure D. The return fan performance is modulated in response to this pressure to provide an essentially constant pressure. The pressure D setpoint should be selected by air balance testing to permit the full required exhaust flow to exit the exhaust louver. Normally, this pressure is set as low as possible using PID control, but in the event this pressure is too low to measure accurately, the exhaust damper can be partially closed and the pressure raised to provide a usable control range. The pressure at D_2 is measured by another pressure transmitter, and this outdoor air intake plenum pressure is controlled by the return air damper setting also at a constant value. The D_2 pressure should be set according to air balance measurements at a value that allows the full required outdoor air to flow through the outdoor air damper when it is fully open. The test-and-balance procedure should provide a pressure and airflow curve for the system using traverse readings or other accurate flow measurements for at least four damper positions (25, 50, 75, and 100 percent). If this pressure is too low to measure accurately, the control range can be increased by partially closing the outdoor air damper and raising this pressure. The D_2 pressure should be used with a PID control loop to modulate the return damper position. To maintain the correct flow of both exhaust air and outdoor air, the control system modulates the positions of the outdoor and exhaust dampers. Due to nonlinear conditions, these positions may be different, and an offset value is often used to provide an outdoor air quantity that is greater due to the need for slight space pressurization, losses from other exhaust sources, or other need. The air balance testing is critical to establish the relationship of the damper position to the quantity of air being admitted and exhausted.

The fixed-plenum-pressure method allows the control system to modulate the intake and exhaust of air using the outdoor air and exhaust damper positions independently of the air-handling-unit flow rate. It also provides a means of control that uses low-cost and accurate pressure transmitters in place of the less accurate and more expensive airflow monitors. This method works the same and costs essentially the same to install on any air-handling unit regardless of size. Return fan tracking, space pressure, and indoor air quality are controlled and managed with this technique. The control system can be embellished with economizer, air quality override, smoke control, and other sequences such as mixed air control without affecting the fan-tracking function.

The fixed-plenum-pressure method has two drawbacks. The first is the potential waste of energy caused by maintaining a fixed plenum pressure that is higher than what may be needed to meet process conditions, and the second is the need to coordinate the control setpoints with the air balance readings.

Fixed-damper method. To overcome the energy waste of the fixed-plenum-pressure method, the fixed-damper method has been developed. With this method, the exhaust and outdoor air dampers are generally set fully open when the air-handling unit is in operation and the flow of outdoor air and exhaust air is controlled by controlling the plenum pressures D and D_2. The return fan performance is modulated in response to a pressure transmitter that measures pressure D. Pressure D is raised to exhaust more air and lowered to exhaust less air. The actual pressures are set with a PID control sequence with the setpoints selected by air balance readings. The test-and-balance procedure for these readings requires an accurate traverse of the duct or another type of air volume measurement to construct a curve that relates the plenum pressure to the actual airflow. The exhaust damper can be set partially closed to raise the measured pressure to a reasonable range for better accuracy. The return damper is modulated in response to the D_2 pressure measured by a pressure transmitter. A PID control sequence varies the measured pressure and damper position according to the required outdoor airflow. Again, the control setpoints are established by air balance data.

This method uses the same hardware as the constant-plenum-pressure method, and it too produces control of the outdoor air and exhaust airflow that is relatively independent of the air-handling-unit flow rate. The fan tracking, conditioned space pressure, and air quality can be controlled well with this method, and it saves fan energy compared to the fixed-plenum-pressure method. This system allows the air-handling unit to move a constant volume of outdoor air rather than a percentage of the supply air volume. When the thermal requirements fall, the rate of ventilation air does not change, and the outdoor air volume can vary up to the total supply fan volume if the ductwork permits it. This independence of the outdoor air volume control allows the intake of outdoor air to occur at a fixed rate, implying the percentage of outdoor air increases when the thermal load falls. Maintaining the flow of outdoor air despite thermal requirements is the key to controlling indoor air quality. Fan energy is saved compared to the fixed-plenum method because the total fan pressure of the system is less under partial flow conditions. The supply fan and return fan do not have to maintain the full flow pressures when not required.

Two disadvantages of this method are the fact that an operator may be concerned about the dampers being fully open regardless of flow. The control process is not intuitive when the air-handling unit is physically observed. Also, the air balance data is again critical to the proper operation of the system. Obtaining accurate pressure and flow readings during the test-and-balance operations are essential to correct control operation.

Both methods offer a means of overcoming nonlinear characteristics in the systems. Oversized/undersized dampers and excessive pressure losses through ducts and louvers can be addressed. The former problems of poor tracking and lack of control over ventilation are eliminated with an inexpensive solution. Too often in the past, fan tracking has been compromised due to cost of implementation, and indoor air quality has been sacrificed. This method overcomes those problems. Control device accuracy and stability are better than with the airflow monitor method.

Fan-tracking-control conclusion. Most medium and large VAV air-handling installations need a return fan to maintain proper conditioned space pressures. With most VAV systems, the former fan-tracking methods have proven inadequate, and a major problem is the loss of control over conditioned space pressure, ventilation, and indoor air quality. The recommended control over the exhaust/return plenum and outdoor/mixed air plenum pressures offers a solution that will lead to control of conditioned space pressure, ventilation, and indoor air quality at a reasonable first cost using standard control components of proven accuracy. This method can compensate for many installation limitations, such as oversized/undersized dampers and duct turbulence/sizing problems.

The air balance measurements and setting of the control sequence parameters according to the air balance data remain a critical step in ensuring proper system performance. Though the recommended methods can produce accurate fan tracking and stable control operation, the ultimate system performance is dependent on the test-and-balance accuracy.

5.5.7 Exhaust fan control

The configuration of VAV air-handling systems has evolved over the years. Many traditional configurations have operational limitations that are not well publicized. For example, the use of energy conservation devices (e.g., heat pipes, plate-type heat exchangers, and heat wheels) has resulted in the addition of auxiliary fans to the basic air-handling unit. Return fans are often used, though the return fan configuration may not be suitable for the application. Use of a properly designed and selected exhaust fan configuration may be superior to using a return fan configuration. Also, it may function better if the exhaust static losses are high due to the use of heat recovery devices.

Exhaust fan compared to return fan.   The previous fan-tracking discussion described a system that used a return fan. However, though the use of return fans is popular, it may not be the best method or the most efficient method of accomplishing the airflow management objectives. Figure 5.27 illustrates a design configuration that uses no return fan. In place of the return fan, an exhaust fan has been provided. The exhaust fan overcomes the return duct pressure loss (pressure S_2 minus D) and the exhaust duct, damper, and louver loss (pressure E minus D). This fan is no larger than the previously described return fan, but it can be made smaller if the exhaust quantities are not as large as the total return quantities. This feature often reduces the first cost and size of this fan compared to a return fan configuration.

Another feature of this design is the relative pressure difference between D and D_2. With the return fan design, the pressure D was positive compared to E and D_2. With this design, the pressure D is negative compared to E. Pressure difference D_2 minus D is now very small compared to the previous return fan design, and the selection of the return damper size is much less critical. The lower pressure drop across the return damper causes this design to work better and be more stable. The supply fan is both pushing the supply air and pulling the return air, and it must be larger to accommodate this larger pressure requirement. In most systems this requirement will add about 0.5 to 1.5 inches w.g. to the supply fan total static pressure.

Though the supply fan may have to work against more static pressure than does the return fan design, the energy consumption will be less with this design. The air flowing in the previous system had to pass through two fans in series, whereas in this design the air passes

Pressure gradient diagram

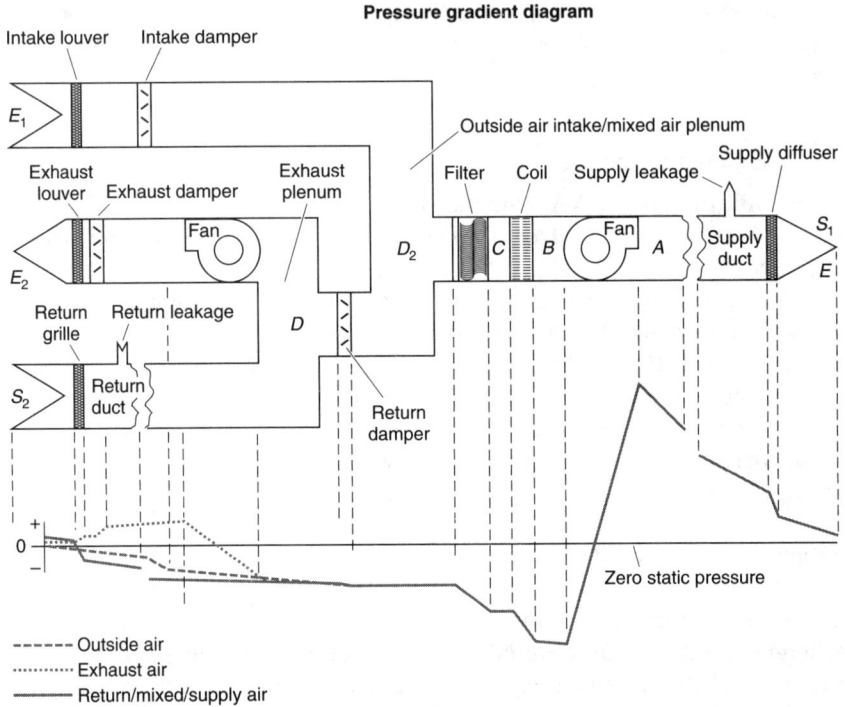

Figure 5.27 Air-handling unit with exhaust fan.

through the fans in parallel, with none of the air passing through both fans. This feature saves the energy that is normally called the *fan entrant loss*. Every time air passes through a fan some energy is lost due to turbulence and the speeding up and slowing down of the air. Over the life of the fan system, this energy savings can be very significant. This design also saves some refrigeration energy as the exhaust fan heat is totally exhausted, whereas some, if not most, of the return fan heat reached the cooling coil.

Exhaust fan tracking methods. The fan-tracking-control methods previously described for the return fan can be applied to the exhaust fan with almost no change. (See Fig. 5.27.) The pressure transmitter that previously measured pressure D should now be located to measure pressure D_2 and the system will operate correctly. Modulating pressure D with the exhaust fan is employed to control the flow of exhaust air, and modulating pressure D_2 by controlling the return damper controls the flow of outside air. The lowered pressure drop across the return damper will yield an energy savings when compared to the return fan design.

If desired, an airflow monitor could be placed in the exhaust duct to measure the exhaust flow and modulate the exhaust fan capacity. However, the limited space that is usually available limits most airflow monitor installations. The use of the exhaust fan as an alternative to the return fan does not affect the use of an economizer cycle, smoke control, or other similar sequence so long as the exhaust fan is properly sized.

Exhaust fan use summary. With most medium-to-large VAV systems, the use of an exhaust fan is preferred to the return fan approach. Potentially, a smaller exhaust fan can be substituted for the return fan. The exhaust fan offers more stable operation and less energy consumption because of a lower pressure drop across the return damper. The exhaust fan design also saves some energy through the elimination of the entrant losses of the second fan. All air now passes through one fan only. Refrigeration energy savings are obtained because all of the exhaust fan heat is exhausted compared to only some of the return fan heat.

The use of the exhaust fan configuration with the fixed-damper/variable-plenum-pressure fan-tracking method is recommended as saving both first cost and energy cost. Proper design and installation following these guidelines will save energy and provide better control of indoor air quality resulting in total system quality. Previous configurations and control methods should be reviewed and compared to this new recommended design. For indoor air quality considerations, the VAV fan-tracking system should provide the introduction of outdoor air in a constant volume that is independent of the thermal system requirements. Exhaust fans can be used effectively with energy-saving heat exchangers without the need of a return fan. This implies that fewer fans can perform the same ventilation and heat recovery function.

5.5.8 Outdoor air fan control

Outdoor air fans are generally used in two configurations. The first is a separate ventilation-only system that is typically used to provide makeup air for the exhaust systems and to ventilate the space. These systems are typically constant volume systems that need only to interlock with the exhaust system so that they operate when needed. The second type is an auxiliary fan that is used as an integral component in a VAV air-handling system to overcome pressure losses in the outside air system. Due to ductwork length or sizing or the use of heat recovery devices, an auxiliary outdoor air fan may be required to deliver ventilation air to the main VAV air-handling unit. This fan system may be constant volume or a VAV unit that is designed to track the VAV system.

If the outdoor air fan is VAV it must track the main supply fan and exhaust fan operation. Typically, control can be implemented by using the plenum-pressure method previously described or an airflow monitor to measure the fan delivery volume.

5.6   VAV Ductwork

VAV ductwork generally is identical to constant volume ductwork except for a few minor exceptions. The first is the elimination of most branch duct balancing dampers at the takeoff from the main duct. These intermediate dampers in the VAV duct system serve little purpose since they create a pressure drop that is meaningful only at the operating point in which the system is balanced. Typically, all intermediate dampers should be avoided.

VAV systems can dynamically adjust the fan performance, and in doing so they can sometimes mask duct design errors. For example, one poorly designed or bad-fitting/branch duct takeoff may cause the entire VAV system to operate at excessive pressure to satisfy the one terminal. Also, VAV terminals have an operating pressure range that must be maintained. If duct dynamic static losses vary too greatly between terminals, some terminals may be operated outside their control range. Generally, the operating range of the terminals sets the maximum permissible pressure loss of the duct system. Therefore, if the terminals have a 2.5-inch-w.g. static pressure range, the duct static pressure loss can be no greater than 2.5 inches w.g. among the terminals.

5.6.1   Duct construction

VAV duct construction should follow SMACNA standards for the operating pressure of the system. However, there are several points in the system that need special attention. The air-handling-unit exhaust, return, mixing, and discharge plenums need special attention due to the varying working pressure and flow. Varying pressures can cause flexing that leads to premature duct failure if additional reinforcing is not provided. For best service, these plenums should be made out of material that is one gauge heavier than SMACNA minimum requirements. The point at which the outdoor air and return air enters a mixing plenum is also another trouble area. Too often, the designer ignores the potential for air stratification, with the result that coils can freeze or temperature sensors can provide a false reading. Airflow may create adequate turbulence at design flow but decrease to laminar, stratified flow at lower rates. Adequate space and configuration for mixing airstreams is a problem that should be carefully addressed when the equipment and ducts are designed.

A unique IAQ problem with VAV duct construction is the accumulation of dirt, dust, and other debris in the duct during periods of low flow that is suddenly purged when the unit reaches design flow. Typically, during the heating season the VAV air delivery may be low, allowing this debris to build up; later, during the cooling season at peak flow it breaks loose. Because of this potential, the VAV supply duct should be free of areas that can collect debris, and more access doors for duct cleaning should be provided.

5.6.2 Duct lining

With today's IAQ concerns over fiberglass particle contamination and the potential for molds and fungi to grow on duct liners, duct lining materials have become controversial. Many applications such as hospital, nursing home, and laboratory systems should not use internal duct lining to provide a smooth as possible duct surface that is easy to clean if needed. Duct lining in other applications can be provided with the new liners that incorporate a membrane that seals the liner, preventing the flaking of small particles. Liner systems that encase the internal duct liner with a perforated metal liner are also superior to bare lining systems.

Traditional construction with internal duct liner has produced less than satisfactory performance in far too many applications. Over time, liners have become brittle, the adhesives have failed, and the liner has separated from the duct wall in either pieces or whole sheets. Lining joints have been improperly sealed and the bending or damage to the duct exterior can often dislodge the interior liner. Liners represent a challenge to effective duct cleaning, and even the best cleaning efforts have proven inadequate. When duct linings become wet they tend to hold the moisture and it is easy for fungi or molds to begin growing within the liner.

At this time, duct lining methods and materials are improving because of IAQ concerns. However, duct lining involves a quality control problem that requires many detailed steps during construction. If at all possible, duct lining should be avoided, and the ductwork should be insulated on the exterior. The acoustic and labor-saving benefits of duct lining may not offset the risks to IAQ.

5.6.3 Flexible ducts

Flexible ducts are often used to connect the VAV terminal devices and the diffusers to the main and branch ducts. Generally, the flexible ducts should be kept as short and straight as possible. Ideally, only a short run of flexible duct is required to provide vibration isolation,

Basic looping

• Constant cross-sectional area
• Low fan static pressure
• Minimum duct pressure fluctuation

Reduced size looping

• Reduced duct sizes
• Duct material savings
• Duct sizing based on both pressure
  drop/100 feet and duct velocities

Looping with crossover ducts

Crossover
duct

Crossover
duct
(typical)

A crossover duct to equalize air flow   Multiple crossover ducts

Small
pressure-
equalizing
duct

Multiple duct runs to reduce main   A small, pressure-equalizing duct when air
duct sizes   movement is limited at the end of distribution

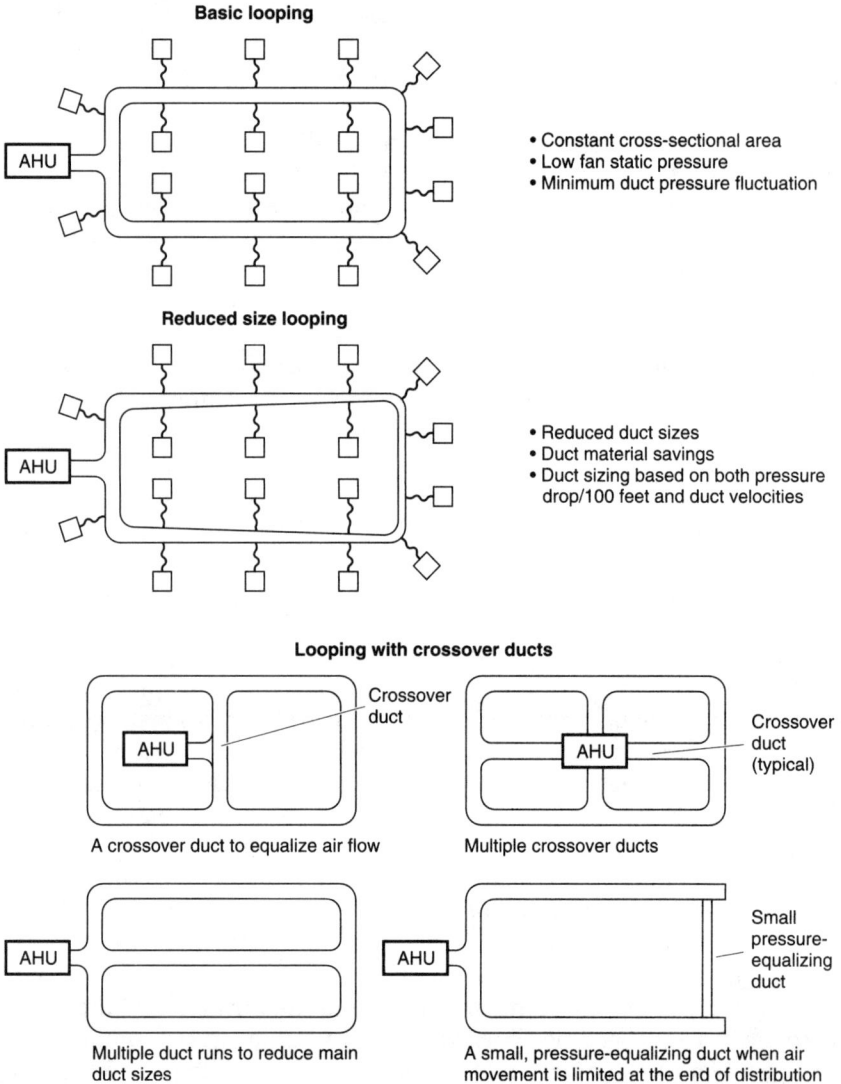

Figure 5.28   Various duct-looping concepts.

room for expansion and contraction, and adjustment for slight duct
misalignment and positioning. Unfortunately, many designers and contractors misapply flexible ducts.

Common misapplication includes using a long, flexible duct to
replace a takeoff duct. Often, 6 to 20 feet of flexible duct is used when
no more than 2 feet should be used. Flexible ducts often contain many
bends, and the duct may be partially collapsed to fit in the space. Using
flexible duct in a deformed state often generates turbulence, noise, and

excessive pressure losses that lead to inadequate terminal performance. A typical problem with VAV terminals is a direct connection of flexible duct to the terminal inlet that produces a right-angle bend at the terminal inlet where the velocity pickup is installed. This bend in the flexible duct causes turbulence that prevents the pickup from providing a correct airflow reading. Noise is another problem with flexible ducts. Flexible ducts typically have limited capability to provide a barrier for transmitting the noise present inside the duct to the exterior. These ducts also may be deformed and bent, creating turbulence that produces excessive noise due to high velocities.

5.6.4 Main supply duct looping

The traditional HVAC ductwork design features radial distribution to/from the air-handling unit and the connected terminal devices. Thus, there is only one path through which air can flow to and from the unit. However, water and electrical systems have used a different type of distribution configuration known as the *loop*, which permits at least two or more paths from the source to the load. This looping architecture can be applied successfully to HVAC duct design, and it has several advantages. Figure 5.28 illustrates various looping concepts. With these configurations, the air can flow from the air-handling unit by means of two or more routes. Looping configurations provide flexibility for future connections, and they take better advantage of diversity. They typically can operate at a lower and more uniform static pressure than radial systems for any given flow. From an acoustic view, they divide the sound power leaving the unit immediately and tend to provide less noise at the first terminal. The major disadvantage is the increased amount of main duct that is required to deliver the design quantity of air, with a corresponding increase in first cost.

Design of VAV Systems

Total Environmental Quality

The concept of total environmental quality (TEQ) implies that a healthy and pleasant indoor environment is created by the design and operation of heating, ventilation, and air-conditioning systems, paying close attention to such environmental factors as thermal comfort, indoor air quality (IAQ), air distribution, acoustics, and building pressurization. In a VAV system, the intention of changing air volumes is to maintain the space temperature for thermal comfort. But by doing so, it may also affect indoor air quality, disturb air distribution, change noise levels, destabilize building pressurization, and even degrade thermal comfort in the occupied space. The result is a compromise of environmental quality. Yet VAV systems are often designed as a simple air volume changing apparatus and the system's impact on environmental quality is ignored.

This chapter explores the fundamentals of five major environmental factors: namely, thermal comfort, IAQ, variable air distribution, acoustics and building pressurization, and their relationships with VAV systems. Thus, the following major headings in Chap. 6 serve as an introduction to the other chapters in Part 2 by emphasizing the importance of designing VAV systems based on total environmental quality:

Thermal comfort (Sec. 6.1)

IAQ and VAV system design (Sec. 6.2)

Variable volume air distribution (Sec. 6.3)

Acoustics for VAV systems (Sec. 6.4)

Building pressurization and VAV systems (Sec. 6.5)

6.1 Thermal Comfort

6.1.1 Conditions for thermal comfort

Thermal comfort is defined as that condition of mind which expresses satisfaction with the thermal environment (Ref. 1). The variables that influence the condition of thermal comfort are:

Activity level (metabolic late)

Thermal resistance of the clothing (clo-value)

Air temperature, humidity, and movement

Nonuniformity of the thermal environment (vertical temperature difference, radiant temperature asymmetry, warm or cold floors, and draft)

A variety of HVAC systems provide thermal comfort through different combinations of these variables. Yet the degree of comfort achieved may not necessarily be the same when different systems are employed. As a result, a VAV system may become less comfortable under certain design and operating conditions when compared with constant volume or other types of HVAC systems. Thus, the VAV system designer should be able to predict the level of thermal comfort that can be provided by this design. Various techniques are available for the prediction of thermal comfort performance of an air-conditioning system.

6.1.2 Prediction of thermal comfort

Fanger comfort equation. Fanger attempted to generalize the physiological basis of comfort and allow comfort for any activity to be predicted analytically in terms of environmental parameters. The result was a series of mathematical expressions described in Ref. 2. The Fanger comfort equation can be used to predict a combination of environmental variables that produce a comfortable environment for a clothed person performing a selected activity.

The generalized comfort charts by Fanger. Comfort charts shown in Figs. 15 through 18 of Ref. 3 depict comfort lines, or various combinations of two variables that define comfort, when all other variables remain constant. The comfort charts are easy to use and preferred by many HVAC system designers. Figure 19 (p. 8.23) of Ref. 3 shows the combined influence of air velocity and ambient temperature on thermal comfort. It is particularly useful for VAV system designers in estimating the impact of VAV operation, or reduction in supply air, which often results in reduced air velocity, and how to compensate by reducing or increasing the ambient air temperature.

Predicted mean vote (PMV). The predicted mean vote (PMV) is an index devised (Ref. 3, p. 8.23) to quantify the degree of discomfort for a large group of people according to the following psychophysical scale:

$$-3 = \text{cold}$$

$$-2 = \text{cool}$$

$$-1 = \text{slightly cool}$$

$$0 = \text{neutral comfort}$$

$$+1 = \text{slightly warm}$$

$$+2 = \text{warm}$$

$$+3 = \text{hot}$$

The PMV is a complex mathematical function of activity, clothing, and the four environmental parameters. It can be determined by the equation in Ref. 2, p. 111, or from a comprehensive table in Ref. 3, an extract of which is shown in Table 6.1.

In a practical design situation, the tabulated PMV values can be used to predict the performance of a VAV system for a combination of variables. For example, for a combination of 1.0 clo clothing, 25°C (77°F) ambient temperature and 0.2 m/s (40 fpm) relative velocity, the PMV value is 0.31 (Table 6.1) which is a value between 0 (neutral comfort) and +1 (slightly warm). However, this value does not indicate what percentage of the people in the group is not happy or is satisfied with the combination of variables. The concept of predicted percentage of dissatisfied (PPD) was developed to relate the PMV value to the percentage of dissatisfied people.

Predicted percentage of dissatisfied (PPD). The PPD predicts the percentage of a large group of people that feel thermally uncomfortable. Once the PMV value is determined, the corresponding PPD value can be calculated from the equation

$$\text{PPD} = 100 - 95 \times e^{-(0.03353 \times \text{PMV}^4 + 0.2179 \times \text{PMV}^2)}$$

or found from Fig. 1 of Ref. 12.

6.1.3 Acceptable thermal environments for comfort

The PMV and PPD indices can be used to predict the percentage of people dissatisfied for a given set of environmental variables. The indices are a useful design tool in predicting or estimating the performance of a

TABLE 6.1   Predicted Mean Vote

Clothing clo[a]	Amb. Temp., °C	Relative Velocity (m/s)			
		<0.1	0.2	0.3	0.4
0.5	23	−1.1	−1.51	−1.78	−1.99
	24	−0.72	−1.11	−1.36	−1.55
	25	−0.34	−0.71	−0.94	−1.11
	26	0.04	−0.31	−0.51	−0.66
	27	0.42	0.09	−0.08	−0.22
	28	0.80	0.49	0.34	0.23
	29	1.17	0.90	0.77	0.68
	30	1.54	1.30	1.20	1.13
0.75	21	−1.11	−1.44	−1.66	−1.82
	22	−0.79	−1.11	−1.31	−1.46
	23	−0.47	−0.78	−0.96	−1.09
	24	−0.15	−0.44	−0.61	−0.73
	25	0.17	−0.11	−0.26	−0.37
	26	0.49	0.23	0.09	0.00
	27	0.81	0.56	0.45	0.36
	28	1.12	0.90	0.80	0.73
1.00	20	−0.85	−1.13	−1.29	−1.41
	21	−0.57	−0.84	−0.99	−1.11
	22	−0.30	−0.55	−0.69	−0.80
	23	−0.02	−0.27	−0.39	−0.49
	24	0.26	0.02	−0.09	−0.18
	25	0.53	0.31	0.21	0.13
	26	0.81	0.60	0.51	0.44
	27	1.08	0.89	0.81	0.75

[a]One clo equals 0.155 m^2·°C/W.
SOURCE: *ASHRAE Handbook 1985, Fundamentals,* Table 9, p. 8.23. Reprinted with permission.

VAV system from a comfort viewpoint. However, the system designer needs to know what percentage of dissatisfaction is acceptable as a good design practice.

ISO 7730 comfort standard recommends that the PPD value of less than 10 percent be used for an HVAC system designed for general human occupancy. This recommendation corresponds to the following criteria for the acceptable PMV range:

$$-0.5 < PMV < +0.5$$

The preceding design standard may be lowered for reasons such as initial cost savings and operating economy. In this case, the thermal comfort standard may be lowered to satisfy less than 80 percent of the occupants, or the PPD value of 20 percent, and the corresponding PMV range would be:

$$-0.8 < PMV < +0.8$$

6.1.4   VAV factors affecting thermal comfort

Variable volume systems are similar to constant volume or any other types of HVAC systems except for one feature: the continuous modulation of supply air to each conditioned space. This affects certain comfort-related variables, such as air movement, temperature distribution (uneven temperature), and humidity.

Air movement.   As stated in Sec. 6.1.1, air movement, or more specifically air velocity, is one of the variables that affects human comfort. However, it is not clearly definable which air velocity limits should be applied to practical HVAC design. Both ASHRAE Standards 55-92 and ISO 7730 (1984) recommend 30 fpm (0.15 m/s) air velocity. Yet a variety of air movement studies have also produced a wide range of air velocity recommendations, depending on variables such as clothing, activity, temperature, and humidity (Ref. 4, p. 26). In practical design situations, the VAV system designer can follow either the ASHRAE or ISO recommendations for normal air-conditioning application, and the air velocity limits can be raised for special applications such as high-temperature and high-humidity environments (Ref. 4, p. 27) or spot cooling in commercial kitchens.

Temperature distribution.   In a conditioned space, the air temperature normally increases with height above the floor. However, the rate of the increase and maximum spread of temperature vary considerably depending on the HVAC system used and its design and operating parameters. Figure 6.1 shows the change in vertical air temperature distribution for a floor-type supply outlet. Both ASHRAE 55-92 and ISO 7730 (1984) limit the vertical temperature spread to $3°C$ ($5.4°F$).

The $3°C$ limit is generally attainable for most overhead supply systems for cooling operation as long as the outlets are properly selected, sized, and distributed. However, the vertical temperature spread may become a problem for certain VAV overhead heating systems. It is recommended that the manufacturer of VAV outlets be consulted for the proper selection, sizing, and layout of VAV outlets for heating applications.

Humidity control.   Two viewpoints exist for humidity control, health, and comfort. Figure 6.2 shows optimum relative humidity ranges for health, while Fig. 6.3 defines relative humidity limits for comfort. The 30 to 60 percent relative humidity range is recommended for most air-conditioning applications. VAV systems in general can maintain this optimum humidity range under normal operating conditions. However, the following design recommendations should be carefully observed to make certain that the space humidity can be maintained within the optimum limits.

Temperature spread

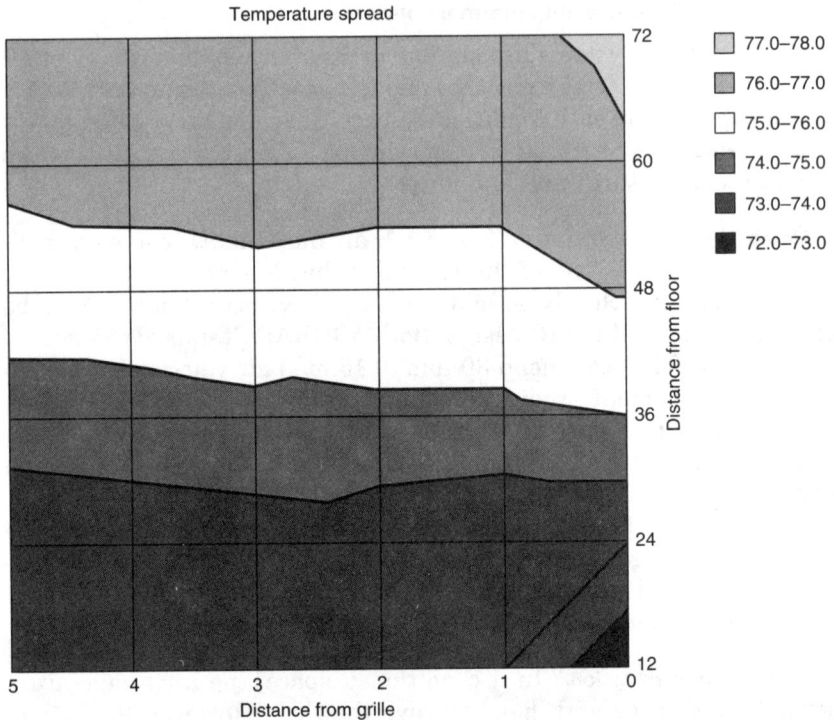

Figure 6.1 Vertical temperature distribution for a floor supply outlet. (*Courtesy of Carrier Corporation. Reprinted with permission.*)

1. When high-occupancy and low-sensible-load areas are connected to a VAV system, the supply air must be maintained at a reasonable dew point temperature to limit the relative humidity in high-occupancy and low-sensible-load areas when such areas need only a fraction of the design supply air quantity because of reduced sensible loads. The VAV system designer must always check probable high humidities at maximum occupancy corresponding to small sensible loads and, if necessary, increase the supply air quantity and apply reheat to limit the rise of space humidity.

2. In cold as well as warm and humid climates, the indoor and outdoor vapor pressure differences may become substantial, and surface condensation and building material damage can occur. To prevent these problems, the VAV system designer may have to conduct a condensation analysis and recommend the installation of vapor barriers.

3. When the building is not properly pressurized and/or moisture protected, the cold and dry outdoor air may be brought into the building by infiltration through the porosity of building materials, wall, ceiling, and door cracks. The amount of infiltration is such that it may cause

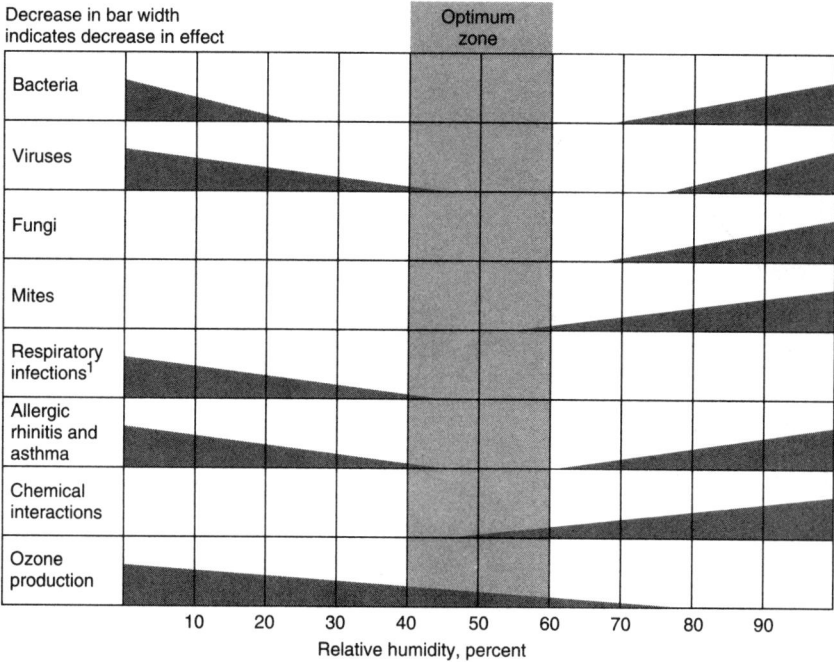

Figure 6.2 Optimum relative humidity range for health. (*Source: ASHRAE Transaction, Vol. 91, Part 1, 1985. Reprinted with permission.*)

a substantial reduction in space humidity levels, even though humidification may still be applied to raise the space humidity levels.

For VAV system design in cold climates, the system designer needs to minimize the loss of moisture through moisture migration and the replacement of warm and humidified space air with cold, dry outdoor air. The former is accomplished by applying vapor barriers to exterior walls, and the latter by properly pressurizing the building and sealing off walls, ceiling, doors, and other building cracks.

6.2   IAQ and VAV System Design

ASHRAE Standard 62-1989 defines acceptable indoor air quality (IAQ) as air in which there are no known contaminants at harmful concentrations as determined by cognizant authorities and with which a substantial majority (80 percent or more) of the people exposed do not express dissatisfaction.

This definition clearly states that the indoor air should be perceived as fresh and pleasant by building occupants. As inadequate ventilation

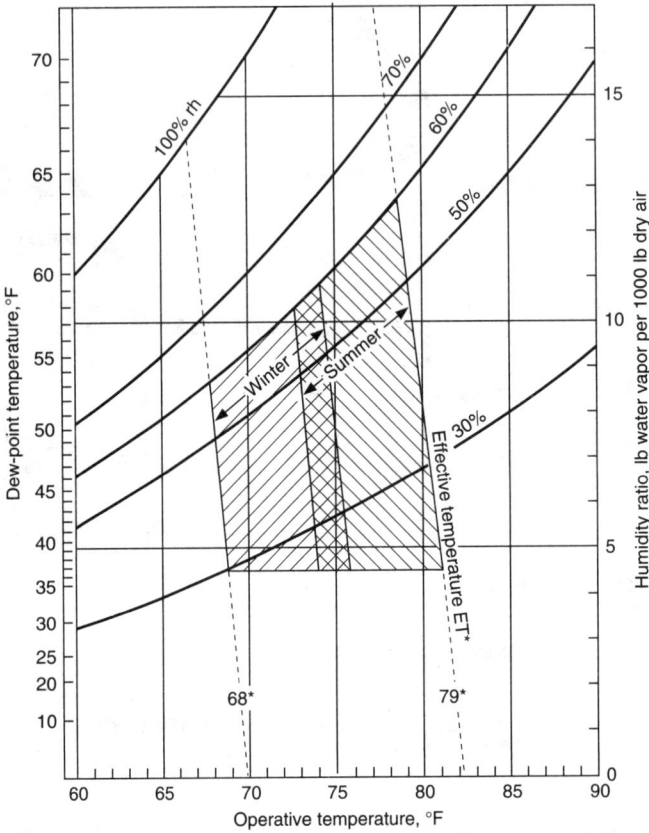

Figure 6.3 Relative humidity limits for comfort. (*Source: ASHRAE Handbook 1993, Fundamentals. Reprinted with permission.*)

is the number one source of IAQ complaints (see Fig. 6.4), HVAC system design needs to address the issue of effective ventilation under all operating conditions, and the conditioned air along with sufficient outdoor air must be delivered to capture, dilute, and filter indoor contaminants.

This poses a unique problem to VAV system designers. VAV systems in general respond only to thermal loads, and not to ventilation effectiveness. Usually, in VAV system design certain outdoor air ventilation rates are selected based simply on type of occupancy and maximum number of people anticipated in each space. Then the ventilation air quantities in each space are summed up to determine the amount of outdoor air to be introduced through the air-handling unit. From there on, the actual amount of ventilation air fluctuates according to the load (and not occupancy) changes in each space. By this design practice, indoor air quality is controlled only indirectly. Sections 6.2.1 through 6.2.3 delineate an IAQ-conscious ventilation design for VAV systems.

6.2.1 Ventilation design

To design a VAV system for good indoor air quality, the designer needs to analyze probable variations in both the outdoor and conditioned air supply to each space under all load conditions, taking into consideration probable changes in occupancy and ventilation requirements.

ASHRAE offers two ventilation procedures to meet the ASHRAE IAQ standard, namely, the ventilation rate (VR) and indoor air quality (IAQ) procedures. The former method allows the system designer to select proper outdoor ventilation rates from Table 6.2 (Table 2 of the ASHRAE 62-1989 standard), and the latter permits the designer to evaluate the ventilation performance of various HVAC systems described in Table E-1, App. E, of the standard.

Still, whichever the designer chooses, the VR or IAQ procedure, the problem of fluctuation in outdoor air supply exists for all the VAV systems supplying air to multiple spaces by a common air-handling system. The problem is one of system origin and must be solved by system design. A variety of system solutions have been suggested, and some of the design solutions are explained in the following sections.

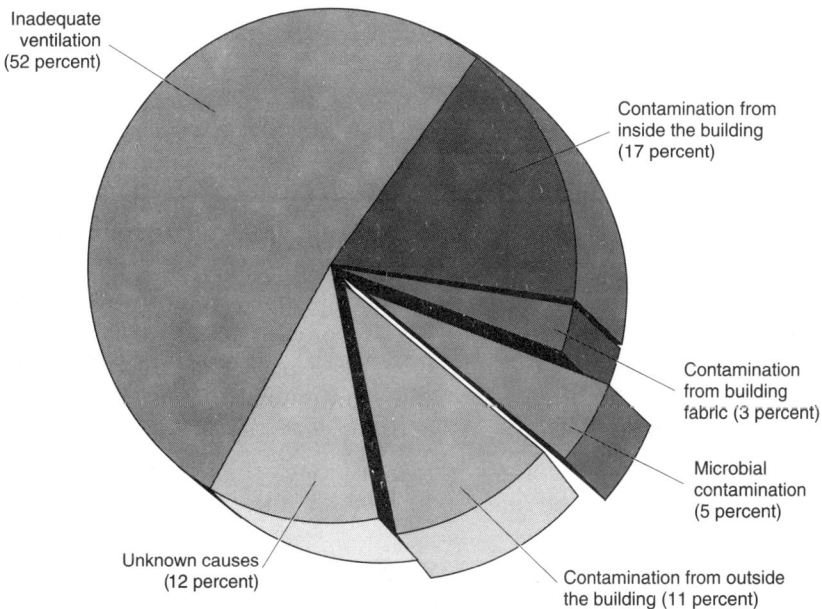

Figure 6.4 Sources of IAQ problems (a NIOSH survey). (*Source: Heating/Piping/Air Conditioning, January 1991. Reprinted with permission.*)

TABLE 6.2   A Partial List of Table 2, ASHRAE 62-1989, Outdoor
Air Requirements for Ventilation

Application	Occupancy ft²/person	Outdoor Air cfm/person	Equivalent OA cfm/ft²
Office spaces	143	20	0.14
Conference rooms	20	20	1.0
Dining rooms	14	20	1.4
Auditoriums	7	15	2.25
Smoking lounges	14	60	4.2
Retail malls	50	(10)	0.2
Retail, basement	33	(10)	0.3
Swimming pools	—	—	0.5

SOURCE: ASHRAE Standard 62-1989. Reprinted with permission.

6.2.2   Design options for VAV system ventilation

The problem of uneven ventilation arises from combining high-occupancy areas with low-occupancy areas and supplying air to all the areas by means of a common VAV air-handling system. The problem, therefore, can be solved by installing separate air-handling systems for areas of different occupancy density, or by providing independent ventilating systems for areas of different occupancy density. Further, the high-occupancy areas should be classified according to their usage.

Design situation 1:   *Large office areas combined with a high-occupancy-density conference room.*   For this situation, several alternative designs can be suggested:

1. Use a separate HVAC system with independent OA introduction (Fig. 6.5, Design 1).

2. Use a common air-handling system with a separate ventilation system supplying properly tempered, filtered, and dehumidified outdoor air to the conference rooms (Fig. 6.5, Design 2).

3. Use an independent ventilation system, usually installed above the conference room ceiling, supplying and/or exhausting air from adjacent, thermally and IAQ neutral areas (usually corridors or hallways) to maintain a reasonable contaminant level inside the conference room (Fig. 6.6, Design 3).

4. Where the peak occupancy duration is less than three hours, the ventilation rate can be reduced to one-half of the maximum (Fig. 6.6, Design 4).

5. Increase the total airflow to a level large enough to supply a sufficient amount of ventilation. Since, in this design option, the supply air quantity is determined solely by the ventilation requirement, it is usually necessary to supply an artificial load, or reheating of the

Design situation: Large, low-occupancy-density office areas combined with a high-occupancy-density conference room
0.14 cfm/ft^2 OA office area,
1.0 cfm/ft^2 OA conference room

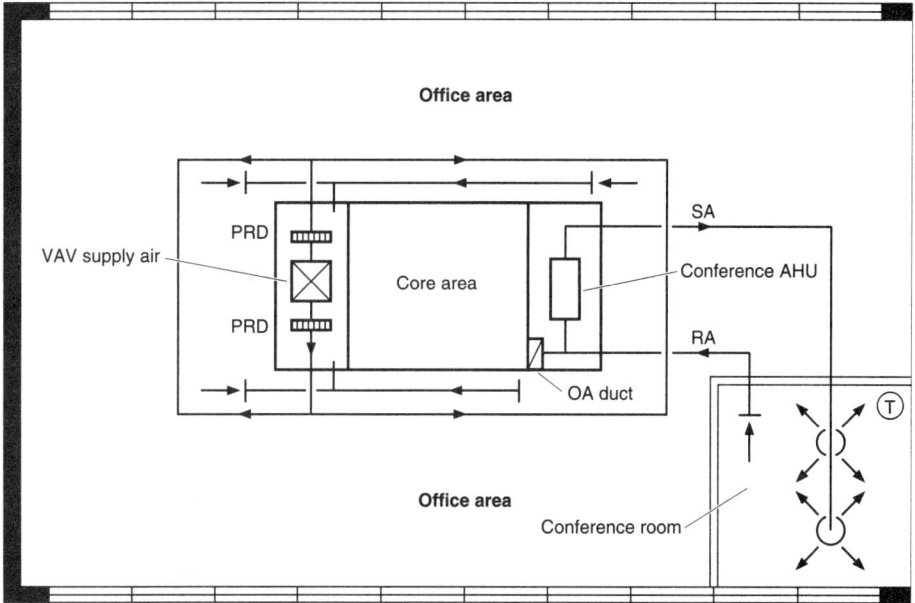

(a) Design 1

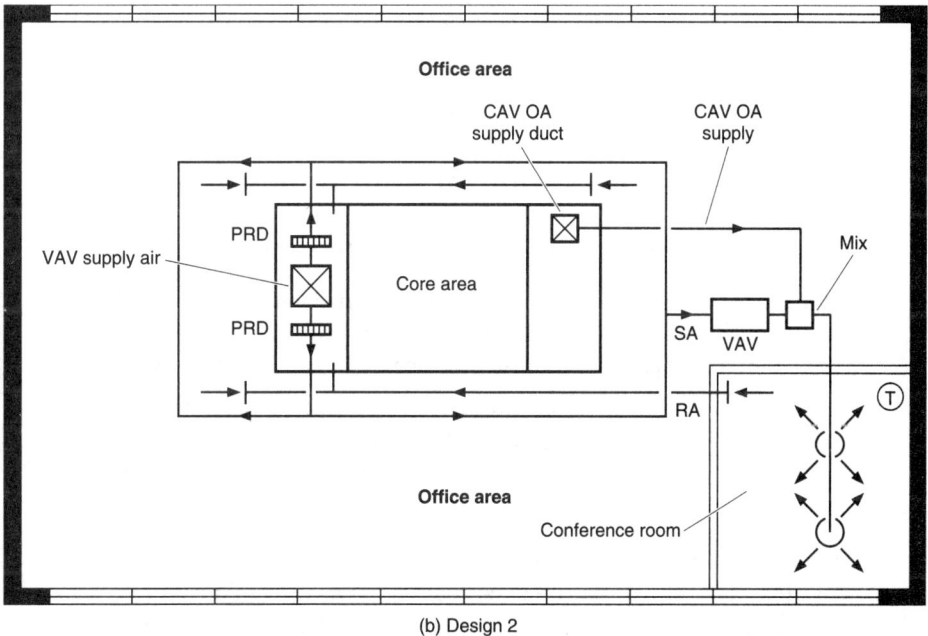

(b) Design 2

Figure 6.5 Design options for VAV system ventilation. (*a*) Design 1: A separate HVAC system for conference room. (PRD: pressure regulating damper.) (*b*) Design 2: A separate OA supply for conference room. (Mix: mixing box.)

supply air (Fig. 6.7, Design 5). Still, this design approach is generally more energy efficient than increasing the ventilation air quantity for the entire system.

Design situation 2: *Private offices mixed with high-load-density work-stations.* In this design situation, several questions need to be answered first.

1. Do the thermal loads in the workstation area vary widely?
2. Is the workstation area much larger than the office area?
3. What are the occupancy densities in these areas? Is the difference substantial?

Design recommendations in this situation would be as follows:

1. A common air-handling system is a viable VAV design in this situation.
2. Construct a system model for the given combination of floor areas, occupancy densities, and load variations, and simulate the probable changes in ventilation air supply to each area.
3. If the calculated minimum is less than the standard set by the design criteria, the reheating and other remedial design options suggested in Figs. 6.5 through 6.7 may be considered.

Design situation 3: *Office area with classrooms.* In Table 2 of the ASHRAE 62 Standard, the occupancy density of classrooms is 7 times that of offices. This is similar to design situation 1 described previously. However, unlike conference rooms, these classrooms may potentially occupy a larger percentage of the total floor plan, resulting in the classroom area receiving much smaller ventilation air, possibly far less than 10 cfm per person.

In this situation, the suggested design approach would be:

1. Set up a ventilation model as suggested in design situation 2, and calculate the variation in ventilation air supply to the classroom areas, using a variety of room-usage scenarios and control strategies.
2. Next, use the findings from the foregoing analysis to determine the feasibility of selecting a separate HVAC system, of either constant or variable air volume type with a separate demand-controlled outdoor air intake.

Unless the classroom is small and its usage infrequent, it is generally not feasible to include the classroom zone as part of a common VAV air-handling system, and a decision to use a common or separate air-handling unit should be made at the early stage of the design process.

Design situation: Large, low-occupancy-density office areas combined with a high-occupancy-density conference room
0.14 cfm/ft² OA office area,
1.0 cfm/ft² conference room

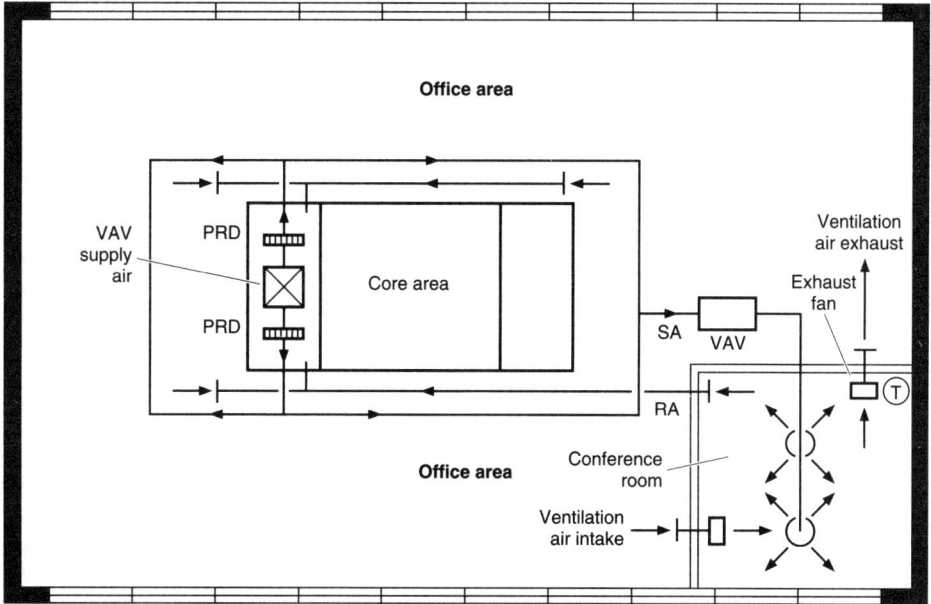

(a) Design 3

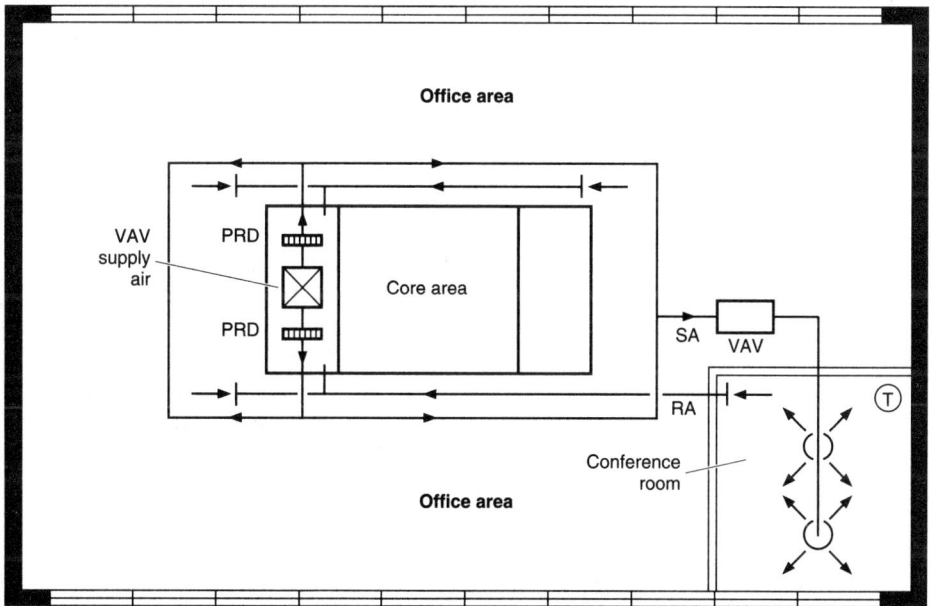

(b) Design 4

Figure 6.6 Design options for VAV system ventilation. (*a*) Design 3: An independent ventilation air supply system for conference room. Additional ventilation air supplied from and exhaust to adjacent thermally and IAQ neutral areas. (*b*) Design 4: Variable ventilation air supply for limited use conference room.

Design situation: Large, low-occupancy-density office areas
combined with a high-occupancy-density
conference room
0.14 cfm/ft^2 OA office area,
1.0 cfm/ft^2 conference room

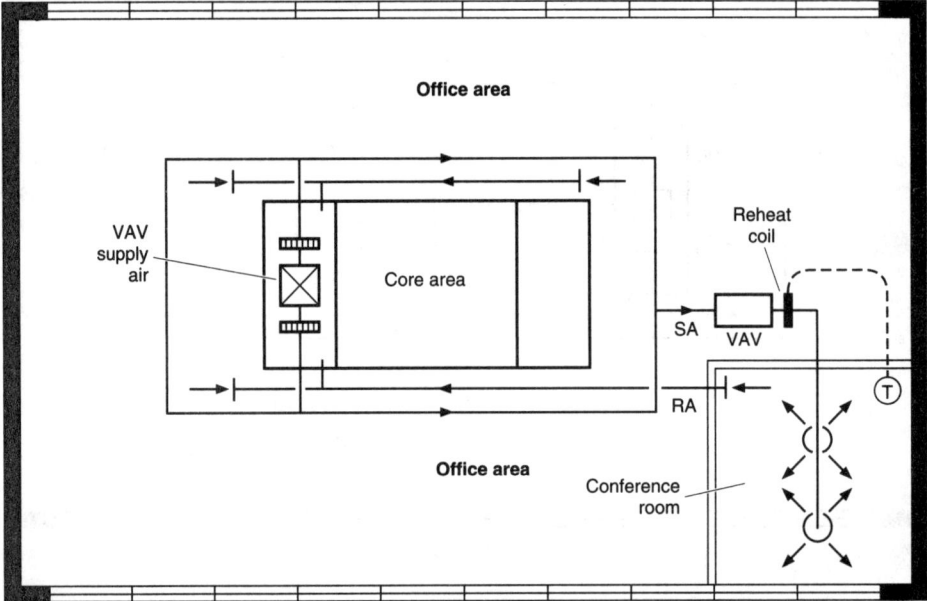

Figure 6.7   Design options for VAV system ventilation. Design 5: VAV supply air to conference room with reheat coil.

6.2.3   Demand-controlled ventilation by design

Overview.   ASHRAE 62-1989 Ventilation Standard for Acceptable Indoor Air Quality calls for substantial increases in outdoor air ventilation rates of as much as 300 percent above that specified in Standard 62-1981. This also means dramatic increases in energy costs for those localities which have hot and humid summers and/or cold and dry winters. For example, the energy required to properly cool and dehumidify 1 cfm of outdoor air can exceed 69 Btu/hr in Taipei, while 56 Btu/hr may be required to heat and humidify 1 cfm of outdoor air on a typical winter day in New York (see Fig. 6.8). Thus, supplying a large quantity of outdoor air regardless of fluctuation in building occupancy is a wasteful practice.

Demand-controlled ventilation offers a solution to this problem by modulating outdoor air quantities according to the space carbon dioxide (CO_2) or volatile organic compound (VOC) levels. Figure 6.9a and b shows a system layout for demand-controlled ventilation and one of the possible control strategies for this layout. Figure 6.9c illustrates the resulting cooling energy savings in several cities in the United States.

Demand-controlled ventilation (DCV) reduces operating costs considerably, especially for large buildings with high density but fluctuating occupancy. Large convention centers, indoor shopping malls, and auditoriums, for example, are the buildings in this category. However, DCV is not a panacea for ventilation air control. It is feasible only for certain design situations. Before deciding to use this control, the system designer needs to go through a design checklist (Table 6.3) to determine the feasibility of applying demand-controlled ventilation under the specified design and operating conditions. In addition, the designer needs to be concerned about the VAV system's tendency to reduce outdoor air supply when the space thermal load decreases.

Thus, unless the outdoor air supply is positively controlled, any attempt to reduce the system outdoor air supply by a CO_2 sensor located in the common return duct only worsens the indoor air quality in the space where the occupancy density remains high.

Energy required to tamper outdoor air

- 68.5 Btuh to coal/dehumidify 1 cfm of OA in Taipei
- 56.1 Btuh to heat/humidify 1 cfm of OA in New York

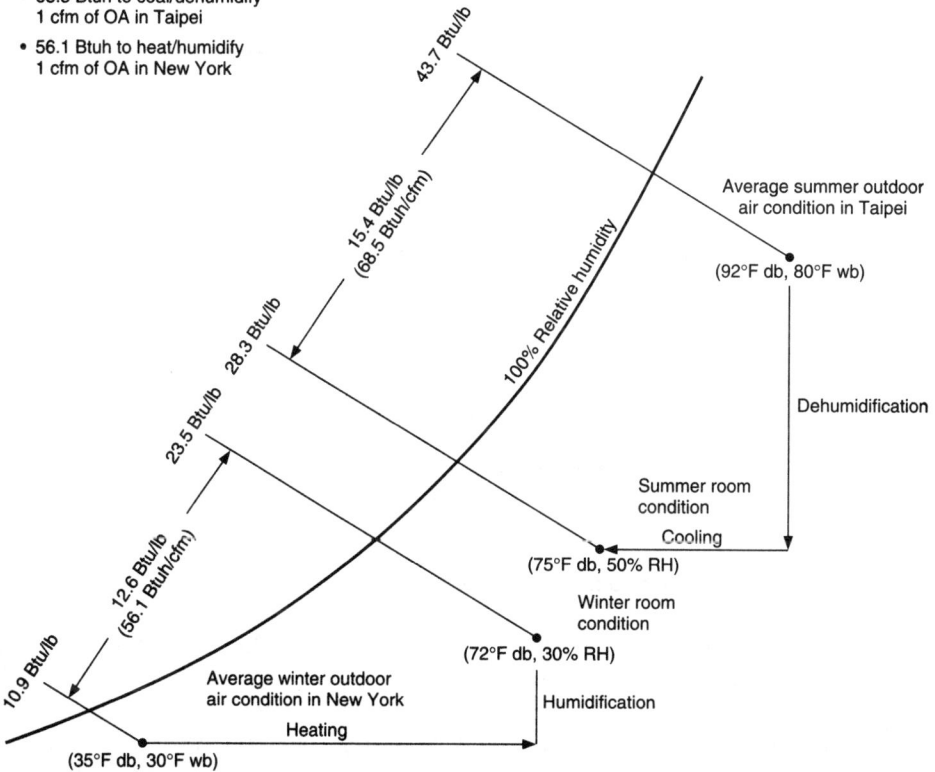

Figure 6.8 Energy required to tamper outdoor air (cool/dehumidify in summer and heat/humidify in winter).

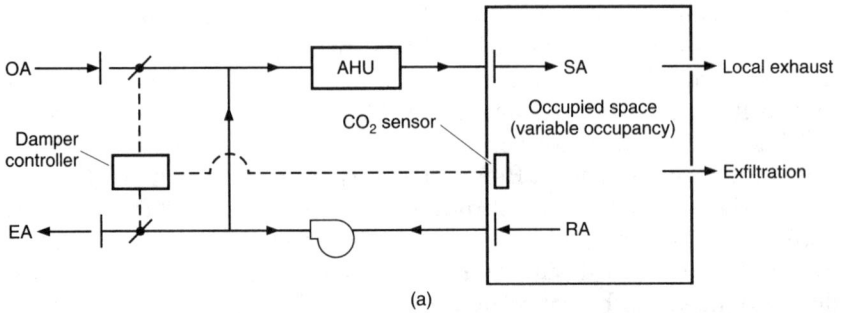

(a)

- On fan start-up, outside air dampers open to 5% minimum outdoor air (this may need to be increased for proper building pressurization).
- If CO_2 concentration in space exceeds 1000 ppm, the outdoor air damper will open an additional 3%. If after a prespecified time interval the CO_2 concentration still exceeds 1000 ppm, the damper will be opened an additional 3%.
- If after another time interval the CO_2 concentration still exceeds 1000 ppm, the damper will continue to open until the outdoor air damper is open to the percentage specified by design or by building operation.
- When the CO_2 concentration drops below 900 ppm, the damper will be closed to the minimum position.

(b)

City	Seasonal OA load (Btu)	Annual cooling season savings	Energy cost
Atlanta	62.7×10^6	$2740	$5.83/10^6$ Btu cooling
Chicago	38.8×10^6	$1700	$5.83/10^6$ Btu cooling
New Orleans	111.5×10^6	$4870	$5.83/10^6$ Btu cooling
Philadelphia	40.7×10^6	$1780	$5.83/10^6$ Btu cooling
Phoenix	69.4×10^6	$3030	$5.83/10^6$ Btu cooling

Notes: 1. Based on average outdoor air saving of 7500 cfm. 2. 10 hours/day, five days/week operation.

(c)

Figure 6.9   Demand-controlled ventilation and cooling energy savings. (a) Schematic layout for demand-controlled ventilation (DCV). (b) A typical sequence of operation. (c) Cooling energy savings.

System implications of demand-controlled ventilation (DCV).   Table 6.3 is based on DCV using CO_2 gas as a sensing medium. When the CO_2 concentration in the breathing zone increases, the ventilation system will introduce more outdoor air to lower the CO_2 concentration. Thus, for this system to be effective, it must be physically practical to introduce more outdoor air to the breathing zone. Otherwise, the system design options described in Figs. 6.5 through 6.7 are probably more desirable than CO_2-based DCV.

TABLE 6.3   Design Checklist for Demand-Controlled Ventilation

Design situations that favor the use of demand-controlled ventilation

1. The air-handling unit serves a relatively large area with a high-design-occupancy density which fluctuates considerably during occupied hours.

2. No major sources of indoor airborne contaminants except people are present.

3. Introduction of large outdoor air quantity up to 100 percent outdoor air is feasible with the HVAC system design.

4. There is no difficulty in finding a location to measure the representative CO_2 level in the space.

5. The ventilation system has the capability to supply the code or ASHRAE standard required outdoor air per person during all the occupied hours.

6. All gases harmful and uncomfortable to occupants can be localized, removed, and exhausted and will not return to the air-handling unit.

7. Multiple spaces conditioned by the common VAV air-handling unit do not cause uneven outdoor air distribution at design and partial load conditions. If this phenomenon exists, the use of DCV only makes outdoor air distribution worse and should not be used.

On the other hand, if a gaseous contaminant–based DCV and a gaseous filter assembly are used, the control of supply and return air circulation becomes the system designer's major concern because more air must go through the gaseous filter assembly for contaminant removal. Thus, the CO_2-based DCV needs to introduce more outdoor air, and the gas contaminant–based DCV requires more air going through the filter assembly. In each case, the system must be designed to satisfy the specific ventilation and air recirculation requirements (Fig. 6.10).

6.2.4   Other design considerations

The preceding ventilation design ensures that the system designer will select proper design options to provide positive ventilation or to supply sufficient outdoor air in the occupied spaces. The outdoor air ventilation is effective in diluting indoor air contaminants, but only when it is introduced to the breathing zone in a way that ensures a reasonable level of ventilation effectiveness. For best results when using outdoor ventilation, the bypass air should be zero, and the supply air is completely mixed with the room air in the breathing zone in Fig. 6.11. Still, there are other considerations that need to be addressed for the effective control of indoor contaminants.

Local exhaust.   The capture of contaminants at their sources is always desirable and generally more economical than providing general ventilation. Local exhaust systems for tobacco smoke, copiers, and plotters are a few examples for local contaminant control (Fig. 6.12).

• Outdoor air V_o increase or decreases according to CO_2 concentration C_s in occupied space

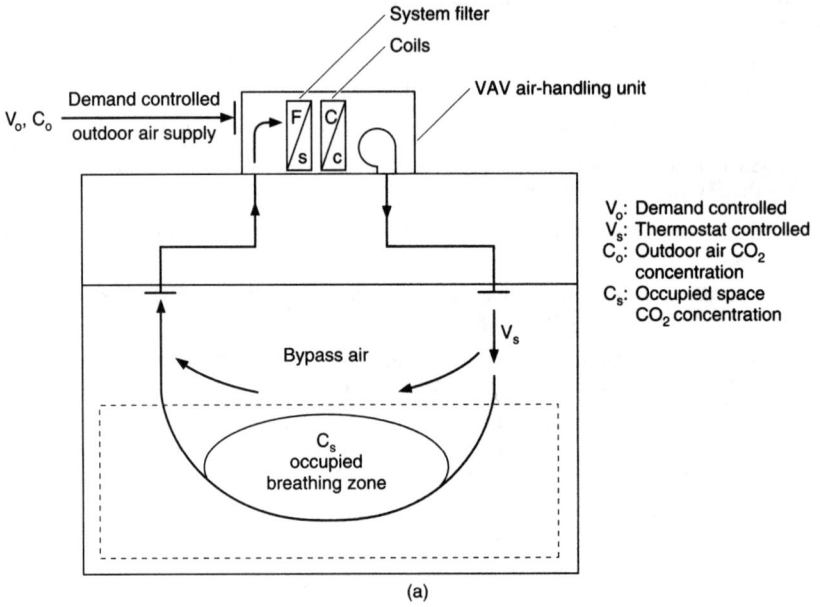

V_o: Demand controlled
V_s: Thermostat controlled
C_o: Outdoor air CO_2
     concentration
C_s: Occupied space
     CO_2 concentration

(a)

• Air circulation through space and filter increased for gaseous containment removal

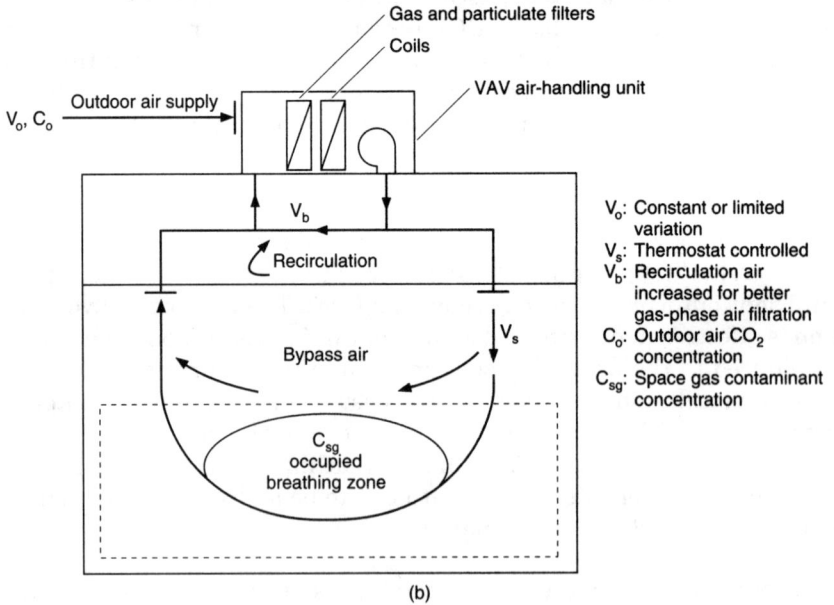

V_o: Constant or limited
     variation
V_s: Thermostat controlled
V_b: Recirculation air
     increased for better
     gas-phase air filtration
C_o: Outdoor air CO_2
     concentration
C_{sg}: Space gas contaminant
     concentration

(b)

Figure 6.10 CO_2-based ventilation versus gas-phase filtration ventilation. (a) CO_2-based demand-controlled ventilation. (b) Gas-phase air filtration ventilation.

Air ventilation for particulate and gaseous contaminant control. Airborne contaminants are classified as vapors or aerosols, solids or particulates, and various gases, and come in a variety of particle and mole sizes (Fig. 6.13). Different technologies ranging from mechanical filters and electronic air cleaners to gas-phase filtration can be used, depending on the type and size of contaminants present.

1. Mechanical filters made of glass fiber and other media are most commonly used for particle removal. However, these filters, with the exception of HEPA (high-efficiency particulate air) filters, are not effective for submicron indoor air contaminants (Fig. 6.14).

2. Electronic air cleaners are more efficient in capturing submicron-size particles. However, these air cleaners produce ozone, which may be harmful to human health if the devices are not maintained and operated properly (Fig. 6.15).

3. Gas-phase filtration is ideal for the removal of gaseous contaminants, tobacco smoke components, formaldehyde, and volatile organic com-

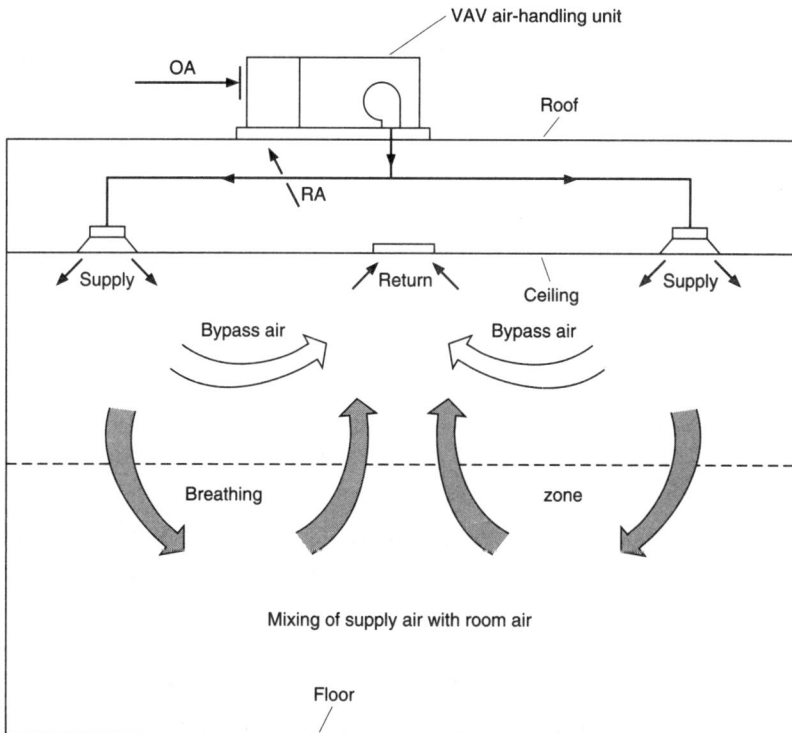

Figure 6.11 Supply air bypass and ventilation effectiveness.

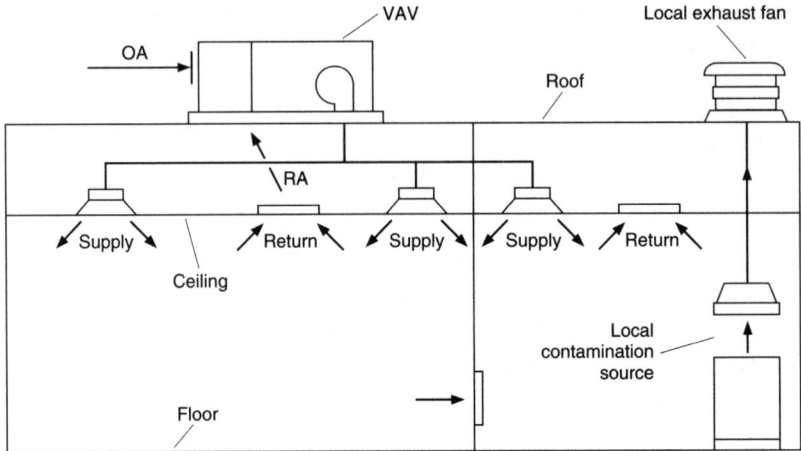

Figure 6.12 An example of local exhaust system.

pounds. However, there are some drawbacks in using gas filters. Carbon monoxide and carbon dioxide are not controlled. When the filter media are spent, they must be replaced. And, above all, there is an increase in operating cost because of higher pressure drops through the gas filters (Fig. 6.16).

6.2.5 IAQ design checklist

In addition to the preceding design considerations, most IAQ problems can be solved during the design stage by going through a checklist of common sources of IAQ problems:

- Do the outdoor air ventilation rates comply with applicable code and standard recommendations and requirements both at design and partial load conditions?

- Is the air distribution method adequate to supply and return and/or exhaust air from the occupied zone with a high degree of ventilation effectiveness?

- Are the outdoor intake openings located away from possible sources of contamination, such as cooling towers, exhaust openings, sanitary vents, and vehicle exhaust?

- Does the system design provide adequate combustion air for combustion appliances installed within the occupied spaces?

- Are building-pressure controls properly designed to ensure the positive pressurization of the building under all operating conditions throughout the year?

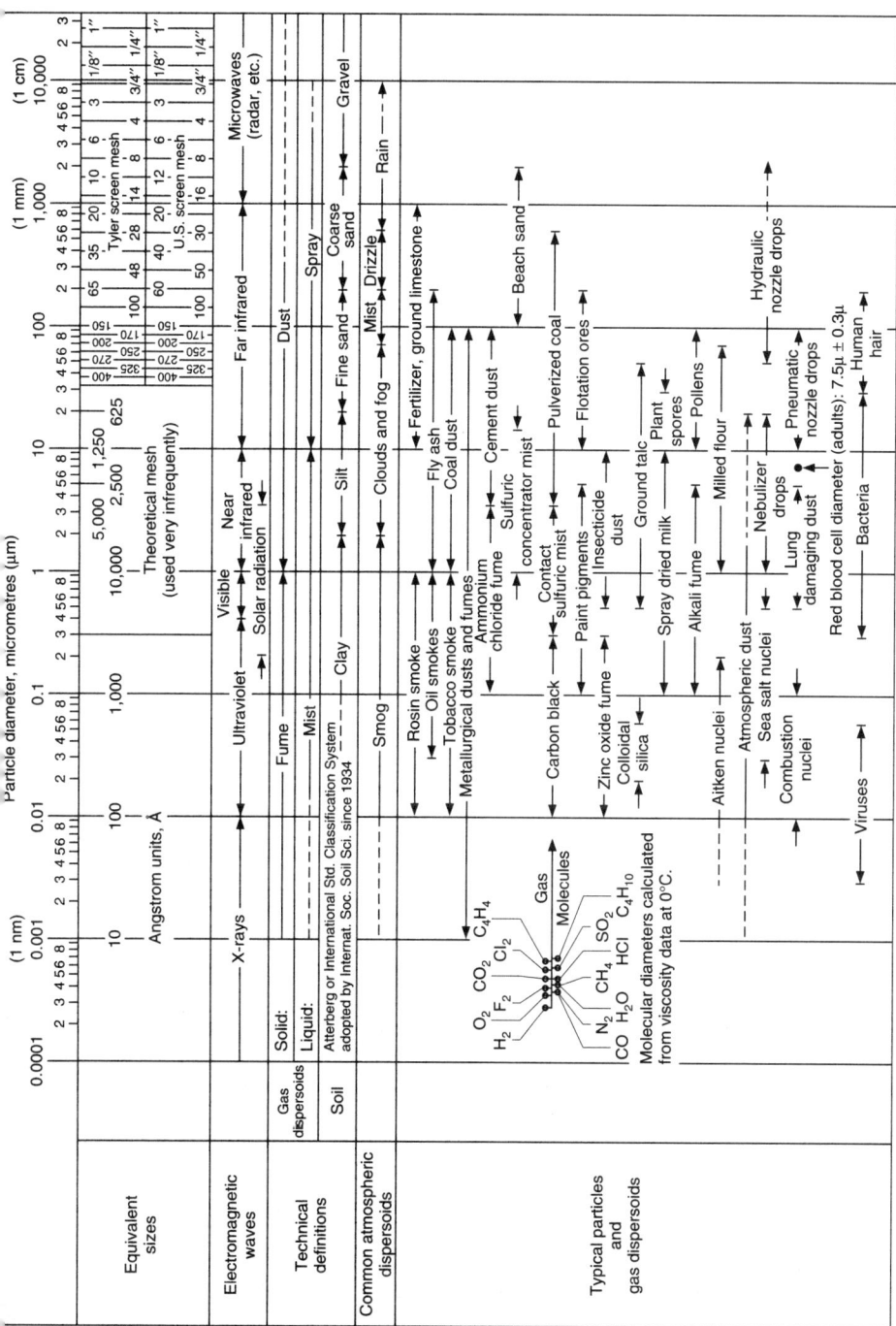

Figure 6.13 Characteristics of particle dispersoids. (*Source: ASHRAE standard 62-1989. Reprinted with permission.*)

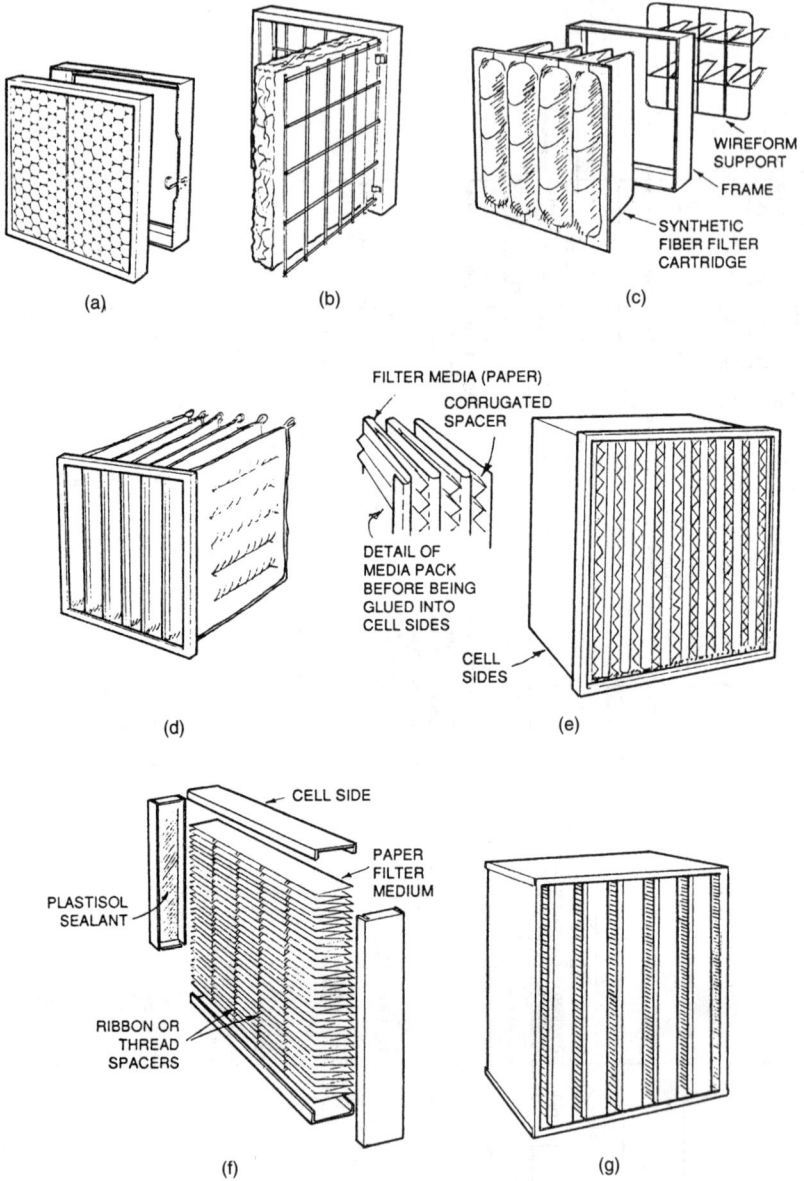

Figure 6.14 Particulate air filter configurations: (*a*) throwaway viscous impingement panel; (*b*) replaceable medium panel; (*c*) replaceable medium extended surface, supported; (*d*) self-supporting "soft" cartridge, extended surface; (*e*) rigid cartridge pleated medium, with separators; (*f*) rigid cartridge pleated medium, no separators; (*g*) zigzag cartridge, no separators. (*Source: Handbook of HVAC Design, McGraw-Hill, Inc. Reprinted with permission.*)

Figure 6.15 Electronic air cleaner. (*Courtesy of Honeywell Inc. Reprinted by permission.*)

Figure 6.16 Side access gas phase filtration modules. (*Courtesy of Purafil, Inc. Reprinted with permission.*)

- Is the building HVAC system designed to flush the building after an extended period of equipment shutdown and before the resumption of system operation?

- Is humidity control designed to maintain the space relative humidities between 30 and 60 percent under all load conditions? (Check the change in space humidity for reduced airflow).

- Does the HVAC design provide adequate water treatment equipment for the prevention of microorganism growth?

6.2.6 IAQ management and design documentation

The management of indoor air quality involves individuals from many disciplines: architects, interior designers, general and mechanical contractors, building product manufacturers, owners, operation and maintenance personnel, and HVAC system designers. The members of the building design team, especially architects and HVAC engineers, have the responsibility for providing a safe and healthy indoor environment for the building's occupants. This is achieved by IAQ-conscious building and system design as well as material specifications.

In this sense, the HVAC engineer needs to design a flexible HVAC system to meet the current and future IAQ requirements of the building. The system thus designed must be properly documented to include design conditions and assumptions, system description, and provision for the changes of occupant density, thermal load, and building usage. Creating the design documentation and making it available to building owners, facility managers, and operators as well as future designers, investigators, and building tenants has the potential of eliminating many IAQ problems and contributing considerably to the total quality management of indoor environment. Table 6.4 is an example of design documentation for a variable air volume (VAV) HVAC system.

6.3 Variable Volume Air Distribution

6.3.1 Traditional versus new design approach

In a typical VAV system design, air distribution is a simple and straightforward process. Usually, a duct layout is made in a certain way, connecting all VAV terminals either by personal preference or by well-established design conventions. An example is shown in Fig. 6.17. Next, trunk and branch ducts are sized and terminals are selected using air

TABLE 6.4   **Example of Design Documentation for a VAV HVAC System**

I. Design Conditions and Assumptions:
- Outdoor design conditions (summer, winter, and intermediate seasons)
  1. Temperatures, dry and wet bulb
  2. Wind direction and velocity
  3. Outdoor air CO_2 concentrations
  4. Potential sources of outdoor air pollutants and their concentrations
- Indoor design conditions (summer, winter, and intermediate seasons)
  1. Temperature
  2. Relative humidity
  3. Description of VAV terminal zones
- Zoning Information
  1. Spaces included in each zone
  2. Occupant densities, activity, and use patterns for each space
  3. People, light, and equipment loads for each space
  4. Miscellaneous loads in each space
  5. Anticipated ventilation effectiveness in each zone for heating, cooling, and intermediate season operating cycles
- Outdoor air ventilation rates in each zone under design and partial load conditions
- Thermal characteristics of building envelopes, including U values and shading coefficients of windows
- Outdoor air infiltration in summer and winter
- IAQ-related codes and standards used in the design
- Fire and safety requirements
- NC or RC noise criteria

II. System Description:
- A diagrammatic representation of the HVAC system showing major components, supply, return, outdoor and exhaust air ducts, local exhaust, VAV terminal zoning, and controls
- A list of major component schedules (capacity, size, and power requirements)
- System and component control strategies in each seasonal and weekly cycle operating mode
- Occupied and unoccupied modes of operation
- Setpoints of control system and their adjustable ranges under various operating modes
- Emergency life safety operating modes of major system components

III. Provision for Future Changes:
- Methods available to increase supply air quantity to each VAV terminal zone, along with their limitations
- Methods available to increase outdoor air quantity to each VAV terminal zone or spaces in each zone, along with their limitations
- Methods available for local contaminant control, along with their limitations
- Design alternatives available to improve indoor air quality through local ventilation supply, filtration, and exhaust, along with their limitations
- Controls documentation
  1. System documentation typically consisting of operators, maintenance, and reference manuals developed by the control manufacturer.
  2. Application documentation typically consisting of system layouts and schematics, checkout, and other materials pertaining to the specific project.

quantities calculated from peak and diversified cooling loads. At this point, the system designer may or may not design a demand-controlled ventilation system (see Sec. 6.2.3). If not, airflow calculations may be performed for ventilation and humidity control purposes. Again, an example for such calculations is shown in Fig. 6.17b. If the resulting airflow is too low, then some design modifications are made to ensure proper ventilation and/or humidity control. This conventional design process is shown briefly in Table 6.5, column A. This is a simple and proven method for VAV air distribution design. However, it is not optimized for total environmental quality following the latest ASHRAE, ISO, and other environmental quality recommendations (e.g., ASHRAE 62, 55, ISO 7730, and Guidelines for Ventilation Requirements in Buildings, Report No. 11 by Commission of the European Communities).

A new design approach is summarized in Table 6.5, column B. This approach emphasizes system analyses for environmental quality. These analyses are performed twice: once at an early stage of design to select a system with a reasonable certainty that it will meet all the environmental criteria when the design is finalized, then at the final stage of design to prove the validity of the design from energy as well as quality viewpoints. The new design approach is somewhat tedious and time-consuming. Still, it has been proven time and again in practical design situations that the new approach can be made compatible with an increasing emphasis on the environmental quality of VAV air distribution design. The system designer typically spends 20 to 30 percent of the design budget for air distribution design, coordinating with architects and other professionals for component integration and space allocation. Considering the fact that the initial and operating costs of an air distribution system may in certain cases reach 40 percent of the entire HVAC costs, it should be quite justifiable to spend an extra 10 percent of time on these system analyses, which will certainly minimize indoor environmental problems in the end and ultimately lead to customer or end-user satisfaction.

6.3.2 Design issues of VAV air distribution

Space air distribution and air transportation (Fig. 6.18a and b) form the two interrelated parts of the VAV air distribution system and pose different design problems. For space air distribution, prediction of comfort, ventilation effectiveness, humidity problems, and acoustics—all related to the quality of environment—are the major design issues. For air transportation, the system designer is concerned about the energy efficiency of air transportation, duct pressure fluctuation, and control stability, as well as ventilation air distribution, noise generation, and transmission. Clearly, each of these considerations presents different design issues, and they will be treated separately in the following sections.

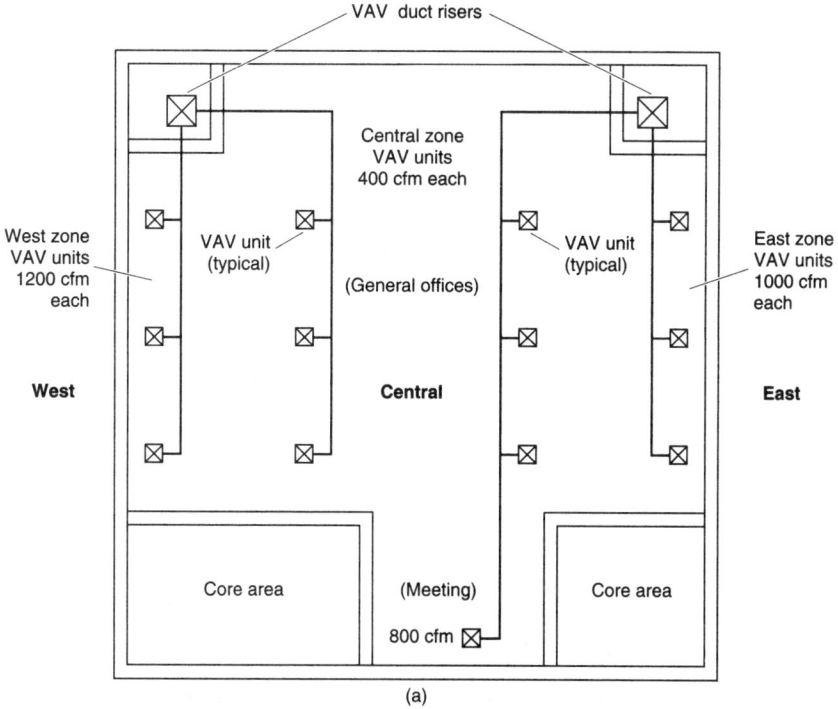

(a)

Time	West zone	Central zone	East zone	Meeting zone	Subtotal
8 AM	1300	2400	*3000	500	7300
4 PM	*3600	*2400	1700	*800	8500

Notes: 1. Peak air supply summation = *sign air quantities
                          = 9800 cfm
        2. Diversified load summation = 8500 cfm
        3. Diversity factor = $8500 \div 98.0 \times 100 = 87\%$
        4. West zone min. air supply = $1.8 \text{ cfm/ft}^2 \times 1300 \div 3600 = 0.65 \text{ cfm/ft}^2$
        5. East zone min. air supply = $1.5 \text{ cfm/ft}^2 \times 1700 \div 3000 = 0.85 \text{ cfm/ft}^2$

(b)

Figure 6.17 VAV air distribution design by conventional strategy. (a) Duct layout and VAV terminal selection. (b) Air supply calculations based on peak and diversified loads (cfm).

VAV space air distribution. The purpose of space air distribution is to create a quality indoor environment by supplying adequately conditioned, cleaned, and ventilated air to the occupied spaces. This principle applies to both constant and variable volume systems, but it requires special attention for VAV air distribution because of its con-

TABLE 6.5   VAV Air Distribution System Design

A. Conventional Strategy

- Select an air distribution system by personal preference or established conventions.
- Peak and diversified load calculations for VAV unit selection and duct sizing.
- Determine the feasibility of demand-controlled ventilation by a CO_2 or gas sensor.
- Usually no analyses performed on thermal comfort, IAQ, energy efficiency or noise. At best, space load analyses for air flow and humidity variations.

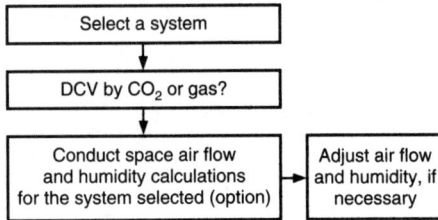

```
        ┌─────────────────────┐
        │    Select a system  │
        └──────────┬──────────┘
                   ↓
        ┌─────────────────────┐
        │   DCV by CO₂ or gas? │
        └──────────┬──────────┘
                   ↓
┌─────────────────────────┐   ┌──────────────┐
│ Conduct space air flow  │   │ Adjust air flow│
│ and humidity calculations│→ │ and humidity, if│
│ for the system selected  │   │  necessary    │
│        (option)          │   │              │
└─────────────────────────┘   └──────────────┘
```

B. Optimized for Total Environmental Quality

- Select an air distribution system and simulate it for comfort, IAQ and noise. If not satisfactory, select another system for simulation. Repeat the process to select an optimal system.
- Determine the feasibility of demand-controlled ventilation by a CO_2 or gas sensor.
- Perform comfort, IAQ, control, energy, and noise analyses for the system selected.
- If the vaults do not meet design criteria, reselect a system and repeat the above process for distribution system optimization.

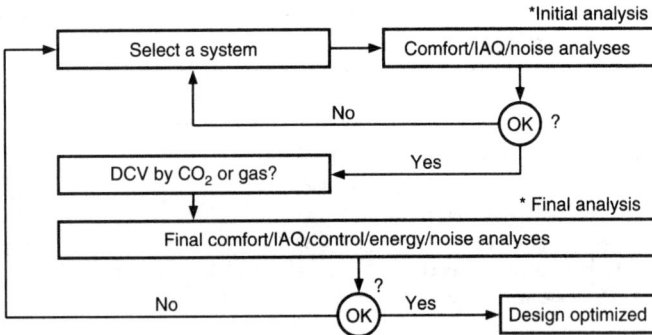

```
                                           *Initial analysis
  ┌─→┌─────────────────┐      ┌──────────────────────────┐
  │  │  Select a system│ ───→ │  Comfort/IAQ/noise analyses│
  │  └─────────────────┘      └─────────────┬────────────┘
  │          ↑                              ↓
  │          │              No        ┌────────┐
  │          │          ┌─────────────│  OK  │ ?
  │          │          ↓             └────────┘
  │  ┌─────────────────┐      Yes
  │  │ DCV by CO₂ or gas?│ ←──────────────
  │  └────────┬────────┘
  │           ↓                    * Final analysis
  │  ┌────────────────────────────────────────────┐
  │  │ Final comfort/IAQ/control/energy/noise analyses│
  │  └──────────────┬─────────────────────────────┘
  │                 ↓ ?
  │      No    ┌────────┐  Yes   ┌──────────────────┐
  └────────────│  OK  │ ─────→  │ Design optimized │
               └────────┘        └──────────────────┘
```

tinuous modulation of supply air. The following subjects are discussed with emphasis on their relationships with supply air volume variation.

Air diffusion performance index (ADPI). In a temperature- and humidity-controlled indoor environment, space air distribution performs two functions: (1) maintaining thermal comfort for the occupants and (2) air quality within the breathing zone (Fig. 6.19). The type and selection of air distribution outlets affect ADPI values and consequently thermal

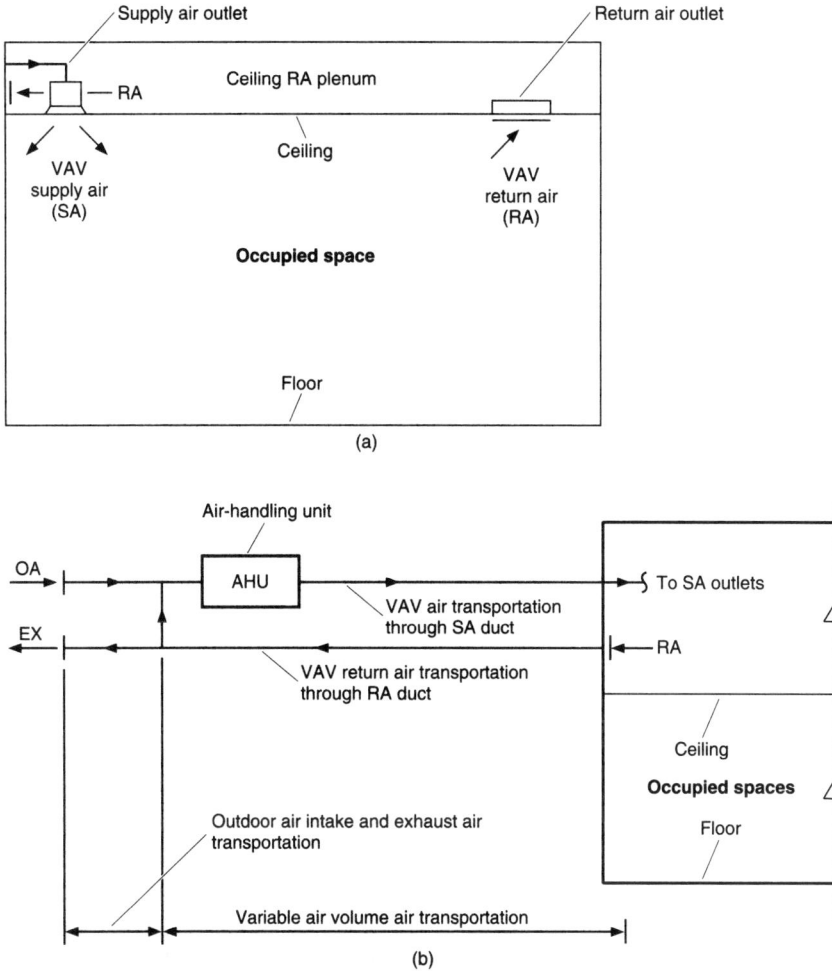

Figure 6.18 VAV air distribution. (*a*) Space air distribution. (*b*) Air transportation.

comfort, and the location as well as method of air introduction and removal from the space affect the air quality in the breathing zone.

ADPI is a performance index which relates space temperature and velocity variations to occupants' thermal comfort. It is expressed as a percentage of thermally comfortable people. Outlet selection and location are considered satisfactory if the resulting ADPI is 80 percent or higher (Fig. 6.19*a*). Figure 6.20 shows the relationships between ADPI and T_{50}/L for four types of air outlets, where

$$T_{50} = \text{air throw with 50 fpm terminal velocity}$$
$$L = \text{space characteristic length in feet}$$

1. A large percentage of people are comfortable where the effective draft tempera-ture (θ) is between -3 and $+2°F$ and the air velocity is less than 70 fpm (Ref. 3, p. 32.8).

$$\theta = (t_x - t_c) - a\,(V_x - b)$$

where t_x = local temperature
     t_c = room average temperature, $°F$
     V_x = local velocity, fpm
     a = 0.07
     b = 30

2. ADPI is defined as the percentage of location where measurements are taken that meet the effective draft temperature and air velocity specified in item 1 above.

3. The goal of VAV space air distribution is to achieve a high ADPI value (80% or bet-ter) for thermal comfort.

(a)

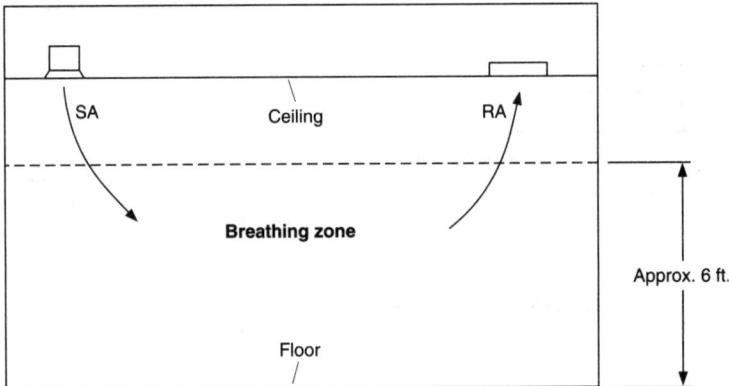

(b)

Figure 6.19 Two functions of space air distribution. (*a*) Maintaining thermal comfort for occupants. (*b*) Air quality within the breathing zone is affected by supply air quality (cleanliness, percent of ventilation air), and ventilation effectiveness.

Table 6.6 shows an example of outlet selection and the resulting ADPI values corresponding to various airflow rates for VAV air distribution.

Figure 6.20 indicates the importance of selecting an optimal throw ratio (T_{50}/L) for the best ADPI value or thermal comfort. For example, round diffusers produce the best ADPI value at 0.8 throw ratio. Once an air outlet is selected for the optimal throw ratio, its ADPI values increase at reduced air quantities for VAV operation (Table 6.6). Still, certain air outlets are better than others for avoiding the undesirable "dumping" phenomenon at reduced airflows, and should be used for VAV applications. Ceiling slot-type diffusers are considered to have the best air distribution characteristics for large air volume reduction. A

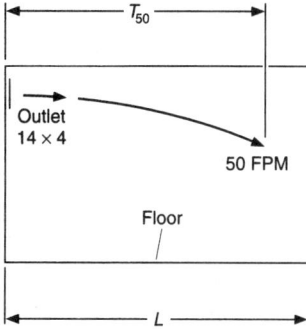

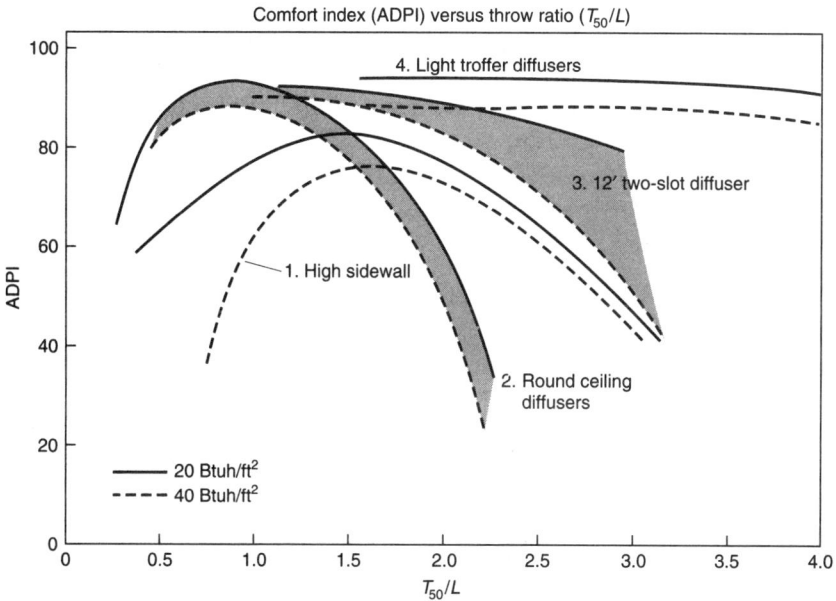

Figure 6.20 ADPI–space characteristics–outlet throw relationships. (*Courtesy of Titus Division of Tomkins Industries Inc.*)

more detailed discussion of this subject is covered in Ref. 5, Six Thermal Comfort Parameters.

Ventilation effectiveness. Merely paying attention to the selection of air outlets to produce higher ADPI values may not improve the quality of indoor air within the breathing zone of occupants. The clean, uncontaminated, conditioned air may not reach the breathing zone because of the ventilation effectiveness of the ventilation system employed. Figure 6.21 shows changes in ventilation effectiveness for four types of

TABLE 6.6   Outlet, Air Supply, NC, and ADPI Relationships for VAV Space Air Distribution

14″ × 4″ outlet cfm	fpm	NC	T_{50}/L	ADPI	Load Btuh/ft^2	Remarks
300	1000	26	2.1	75	40	
210	700	16	1.7	80	28	
150	500	—	1.5	85	20	
90	300	—	1.1	86	12	ADPI estimation (see note below)
0	0	—		100	0	

Note: ADPI is estimated from Miller, P. L. and Nash, R. T., "A Further Analysis of Room Air Distribution Performance," *1972 ASHRAE Transactions.*
SOURCE: Titus Division of Tomkins Industries, Inc. Reprinted with permission.

ventilation methods and temperature differences between supply air and air in the breathing zone. Ventilation effectiveness (Ev) is defined as the ratio of the contaminant concentration in the return or exhaust air (C_e) and in the breathing zone (C_i), or Ev = C_e/C_i.

If the ventilation method is not effective, the contamination concentration in the breathing zone may rise to an unacceptable level, even though a sufficient amount of ventilation air is circulated through the space and the contamination level of air leaving the space is within an acceptable limit. This phenomenon is caused when some supply air is drawn into return or exhaust openings without mixing with room air in the breathing zone. Many factors cause this short-circuiting: examples are location and type of supply, return and exhaust air outlets, windows, doors, partitions, local heating and air conditioning units, and other items that may interfere with the thorough mixing of supply air with space air.

Because of its complexity, there are no authoritative data currently available to estimate accurately the ventilation effectiveness of a ventilation method in a given design situation. However, a range of 0.5 to 1.0 was suggested in Ref. 6 for a typical VAV ceiling supply and return arrangement. Ventilation effectiveness affects the amount of ventilation air required and the selection of air filters if air filtration is also used to supplement ventilation air for IAQ control.

Air movement. The air movement and comfort relationship has been a controversial design issue for air-conditioning systems in general and for VAV systems in particular. The ASHRAE Comfort Chart sets no air movement limits for human comfort. However, other sources indicate some specific air movement/human comfort relationships. Comfort lines by P. O. Fanger (Ref. 2) show a definite relationship between air movement and air temperature (Fig. 6.22). Another source states that low air velocity affects the ability to maintain uniformity of a comfort-

Method of ventilation	$t_s - t_i$	Ventilation effectiveness for cooling operation	

(a)
$C_e < °C$ (32°F)   0.9 – 1.0

(b)
$C_e < -5°C$ (23°F)   0.9   See note 1.

(c)
0.65   See note 2.

(d)
$< 0°C$ (32°F)   1.2 –1.4   See note 1.

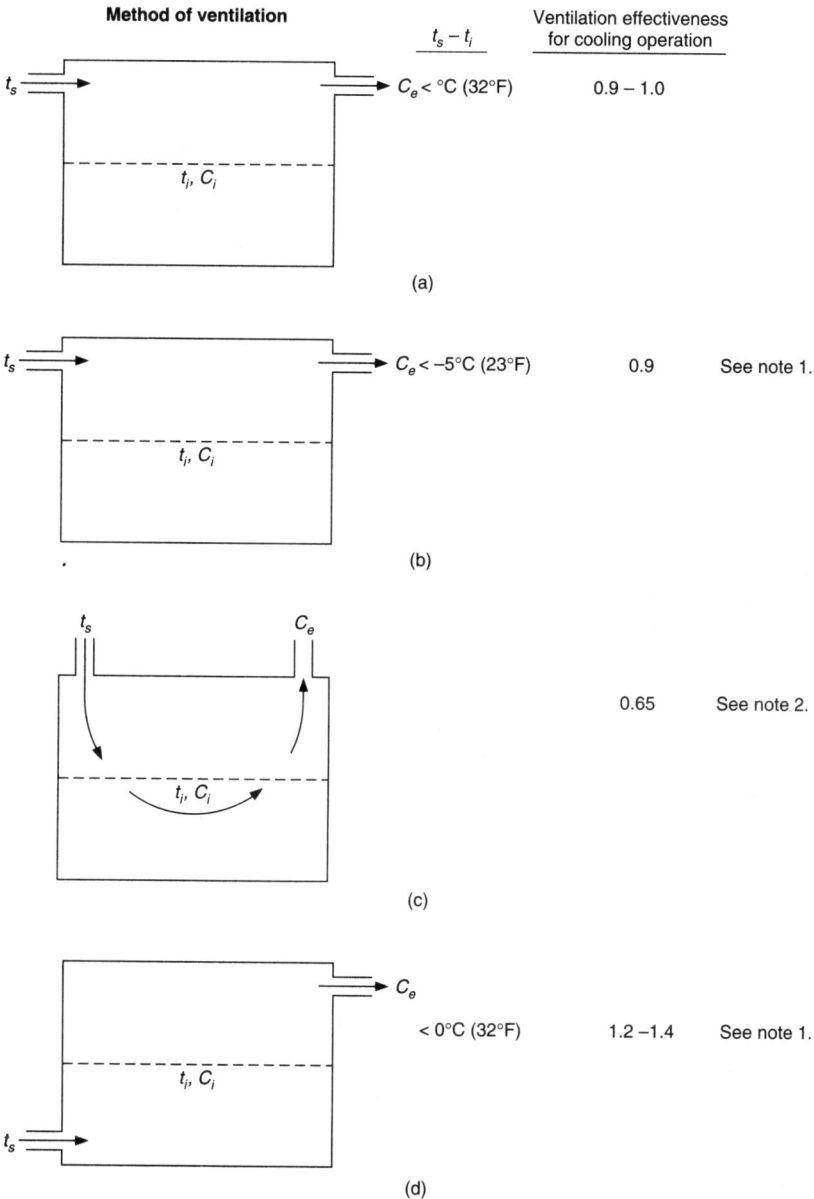

NOTES: 1. Report No. 11, Guidelines for Ventilation Requirements in Buildings, Commission of the European Communities.
2. From VAV/Bypass Filtration System Controls VOCs, Particulates by M. Meckler, HPAC, March 1994.

Figure 6.21   Ventilation effectiveness for different ventilation methods. (*a*) Mixing ventilation. (*b*) Mixing ventilation. (*c*) Mixing ventilation. (*d*) Displacement ventilation.

able temperature throughout the occupied zone. It also states that minimum room air circulation is particularly important in the case of VAV systems. Air movement and thermal comfort (Ref. 4 by M. E. Fountain and G. A. Arens), on the other hand, concludes that it is not clear today what air velocity levels are appropriate for the range of temperatures found indoors.

ISO 7730 standard limits the mean air velocity to 30 fpm during the winter heating conditions and to 50 fpm during the summer cooling period. ASHRAE 55-1992 standard defines allowable airspeeds as a function of ambient air temperature and turbulence intensity (Fig. 6.23). For example, at 80°F air temperature and 0 percent turbulence intensity, the corresponding allowable airspeed can be as high as 130 fpm. On the other hand, both the ISO and ASHRAE standards do not specify a minimum airspeed that is necessary for thermal comfort.

Despite this wide diversity in opinions and recommendations, the air velocity limits definitely affect the various design strategies that are possible for VAV air distribution design.

Here are some guidelines for the system designer: The reduction in supply air quantity will not constitute a comfort problem for VAV systems as long as the space temperature and humidity are maintained within certain ranges. Such temperature and humidity conditions are listed in Table 3 of the ASHRAE 55-1992 Comfort Standard (Table 6.7 of this chapter). Air motion is an important design consideration. However, it is only one of the factors affecting thermal comfort. Airspeeds can be higher than 100 fpm or as low as near 0 fpm depending on other comfort factors (such as space air temperatures, mean radiant temperatures, and humidity) that are associated with it.

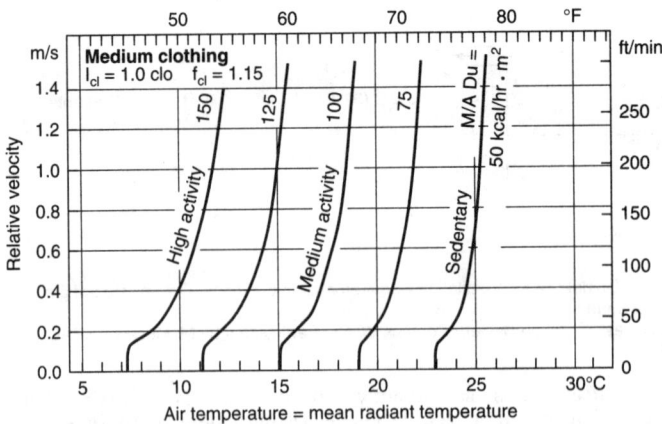

Figure 6.22 Comfort lines by P. O. Fanger (Ref. 2. Reprinted with permission)

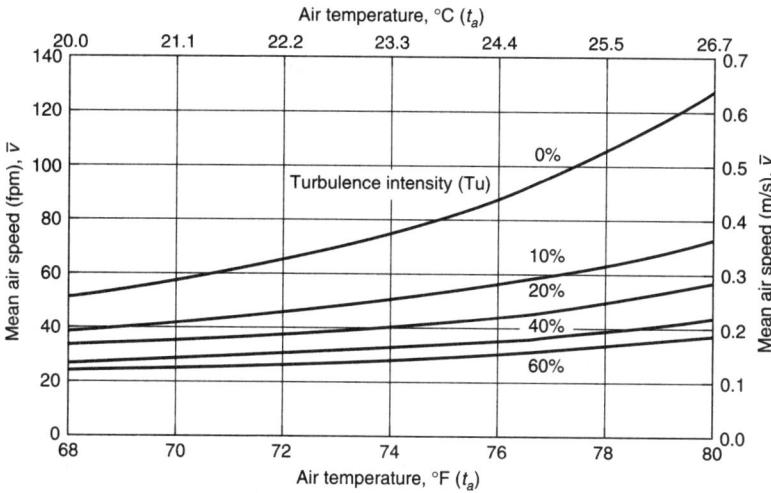

Figure 6.23 Air speed, temperature, and turbulence intensity. Allowable mean air speed ($\bar{v}$) as a function of air temperature (t_a) and turbulence intensity (Tu). The turbulence intensity may vary between 30% and 60% in conventionally ventilated spaces. In rooms with displacement ventilation or without ventilation, the turbulence intensity may be lower. The diagram is based on a 15% acceptable level and the sensation at the head/feet level where people are most sensitive. Higher air speeds may be acceptable if the affected occupants have control of local air speed (Ref. 1. Reprinted with permission)

Higher airspeeds are actually desirable for certain applications. Kitchen air-conditioning is a good example. In this case, the kitchen work area temperature is usually maintained at 88°F, and the desirable airspeed can be as high as 200 fpm at this temperature (Ref. 7: Thermal comfort requirements under hot and humid conditions, by Tanebe, S., Kimura, K., 1987).

TABLE 6.7   Optimum Temperatures for Thermal Comfort

(Excerpt from Table 3, ASHRAE 55-1992 Standard)

Season	Description of clothing	I_d (clo)	Optimum operating temperature	Operative temperature range (10% dissatisfaction criterion)
Winter	Heavy slacks, long sleeve shirt and sweater	0.9	71°F	68–75°F
Summer	Light slacks and short sleeve shirt	0.5	76°F	73–79°F

Notes:
1. I_d garment insulation value.
2. Operating temperature. See ASHRAE 55-1992 Standard, p. 4, definition of operating temperature.

Humidity problems. Humidity becomes a health and comfort problem for VAV space air distribution when the space relative humidity exceeds the optimum ranges (40 to 60 percent for health and 30 to 60 percent for comfort). Temperature control must be combined with humidity control to provide a comfortable indoor environment. This is done for VAV space air distribution by adding moisture in the cold, dry weather in winter. Once a base relative humidity of 20 to 30 percent is maintained by humidification, minor variations in relative humidity due to changes in the supply air quantity will not impact the comfort level of the occupied spaces in winter, even for high-density-occupancy areas.

High-humidity conditions in summer are controlled by the normal dehumidification process of the air-conditioning system employed. This design approach is usually effective as long as the air in relatively high-density-occupancy areas is dehumidified by a separate air-conditioning system to offset a rise in relative humidity when the space-sensible load decreases, but the latent load remains high. Figure 6.24 shows the resulting space relative humidity for a VAV system operating under a high latent and low sensible load condition in summer. The system designer must be aware of this possibility when high-occupancy areas are combined with low-occupancy areas and air-conditioned by a common VAV air-handling unit. A simple psychometric analysis, as shown in Fig. 6.24, usually reveals the load and system conditions under which high humidity occurs.

VAV air transportation. Few air-conditioning books and references discuss the subject of variable air volume air transportation. Even for constant volume air transportation, it is treated either as a duct design problem or considered to be an issue of airflow through ductwork and fittings. In HVAC system design the impact of air transportation on the quality of indoor environment is seldom discussed. This problem has even more serious implications for VAV air transportation systems. Knowing that variable airflow to each outlet affects ventilation air distribution, the system designer needs to know how various air transportation systems affect ventilation air distribution (qualitatively and quantitatively) and the key design parameters involved in order to design a quality VAV air transportation system.

Various air transportation designs for VAV operations. Five duct designs, Designs 1 through 5 for VAV ventilation analysis (Figs. 6.5 through 6.7) still can serve as prototypes for VAV air transportation analysis. These designs share a similar duct layout. However, their objectives are different. In addition to the changes in ventilation air distribution under various flow conditions, the transportation analysis is mainly concerned with energy efficiency and operational flexibility of air trans-

Design situation: Auditorium air conditioning

- Space relative humidity = 56% at design load
-                 "   "   " = 72% min. sensible load
- Supply air flow at min. sensible load = 40% of design flow
- People loads remain stable, but sensible loads fluctuate widely

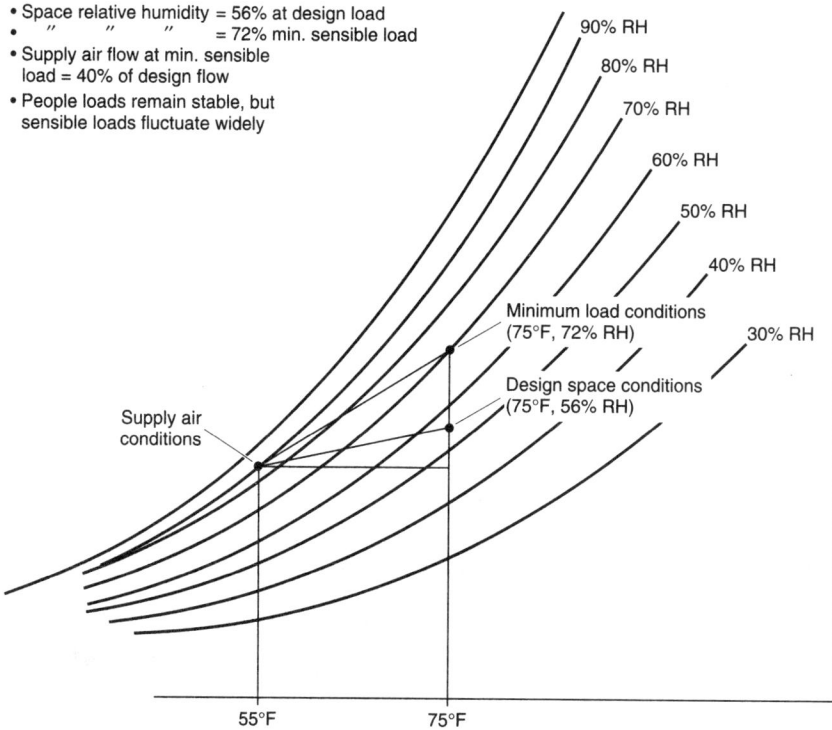

Figure 6.24 Relative humidity changes in an auditorium.

portation, duct pressure fluctuation, noise generation and propagation, and above all, air transportation's relationship with building pressurization. To analyze these additional design parameters, the duct designs are regrouped in two categories, namely,

1. *Centralized air transportation* with conventional or loop duct layout (Fig. 6.25*a*).

2. *Decentralized air transportation* with conventional or loop duct layout (Fig. 6.25*b*). Completely decentralized, in-space air-handling unit (AHU² in Fig. 6.25*b*) is a subset in this category.

A brief outline of the guidelines for centralized and decentralized air transportation systems follows.

1. Guidelines for VAV air transportation design served by a common, multifloor air-handling unit

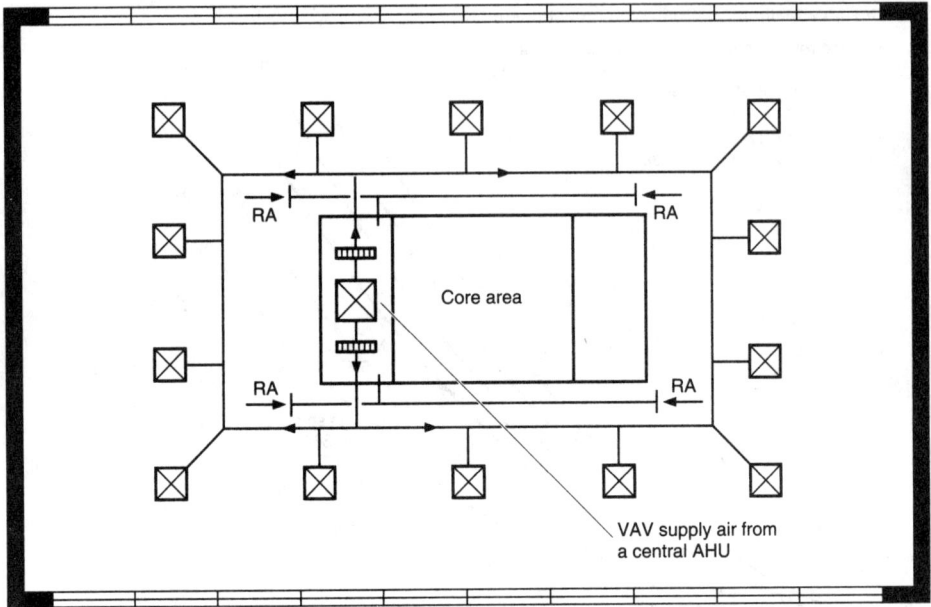

(a)

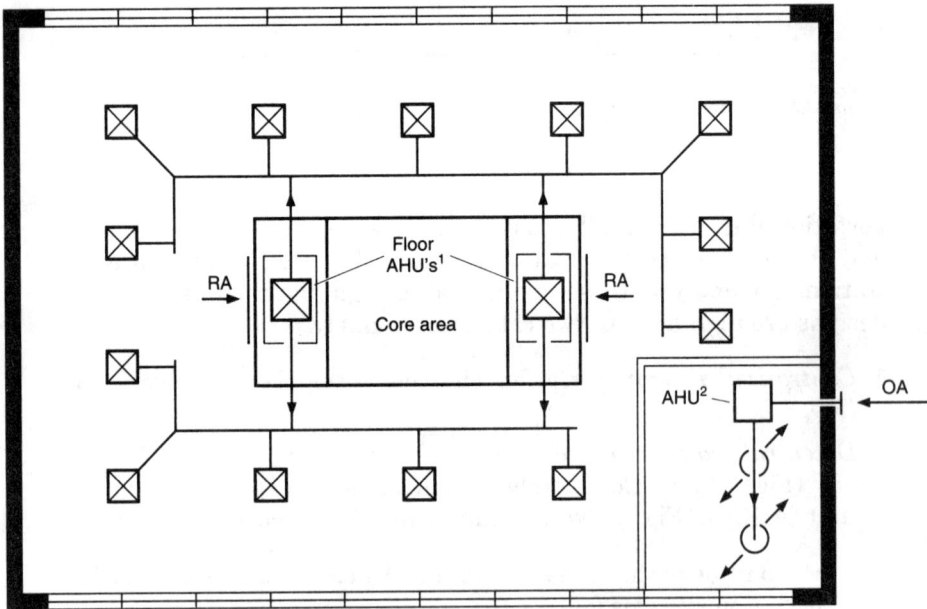

[1] Semi-decentralized floor air-handling units
[2] Completely decentralized room air-handling units (AHU) with or without ductwork

(b)

Figure 6.25 Design choices of centralized versus decentralized air transportation. (*a*) Centralized air transportation. (*b*) Decentralized air transportation.

a. Advantages

(1) The central, multifloor air-handling approach (Fig. 6.26) reduces the number of mechanical floors and AHU rooms. This approach provides a quieter and more reliable system because its major components, fans, coils, and filters are centralized for easy access and maintenance. However, its floor area savings are partially offset by the vertical supply and return air shaft requirements on each floor. Also, branch takeoffs and noise control might be more difficult at the floor next to the air-handling units. Generally, the capacity of each air-handling unit should be limited to 50,000 cfm or less, though more than 100,000-cfm air-handling units have been used to cover more than 30 stories of a high-rise building. It is a good design practice to limit the number of floors covered by a single air-handling unit to 15 to 20 floors, and the mechanical ceiling heights to less than 20 feet.

(2) It is often advantageous to use a loop duct layout to assure duct cost savings and relatively uniform duct pressure under all load conditions. This is especially important when pressure-dependent, thermostat-controlled VAV outlets are used (Fig. 6.27). The noise level of these outlets may increase 5 to 10 dB when the duct pressure increases from 0.2 to 0.35 inches w.g.

b. Application limitations

(1) As long as the duct pressure is properly controlled, there are practically no limits to the number of air terminals that can be connected to this type of air transportation arrangement. However, depending on the actual occupancy and thermal load combination, this type of air transportation may cause a serious deficiency in ventilation air supply and may require the rearrangement of ductwork and rezoning of air-handling units.

2. Guidelines for VAV air transportation design served by floor air-handling units

a. Advantages

(1) A modular, floor-by-floor air transportation approach served by one or more air-handling units on each floor is much easier to design and has the following advantages:

(*a*) Architecturally, it is more flexible in locating air-handling units.

(*b*) Vertical duct spaces can be completely eliminated if the outdoor air intake and exhaust openings can be architecturally arranged on each floor.

(*c*) Extra ceiling height is not required, as in the case of central air-handling units.

(*d*) This arrangement offers easy energy metering for each floor and/or each air-handling unit.

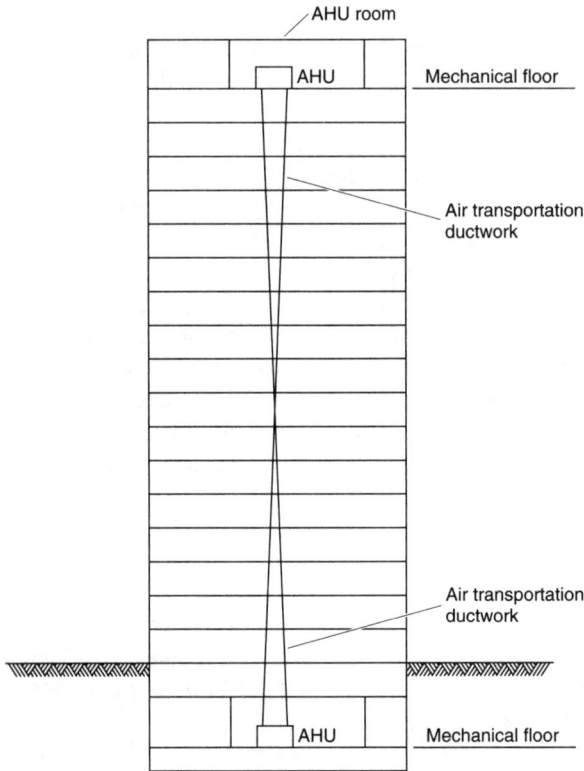

Figure 6.26   Multifloor air-handling system.

   b. Design precautions

     (1) Air-handling rooms are generally located near the occupied areas; construction requires careful coordination with architects and structural engineers for noise and vibration control.

     (2) Because of space and budget limitations, it is usually not practical to provide a standby fan capacity; therefore, when the fan fails, the system loses 100 percent of its capacity. Conversely, a large central system generally has more than one fan and will always retain a certain percentage (usually 70 percent or more) of the total air-handling capacity.

     (3) This type of air transportation arrangement may still have ventilation deficiency problems similar to those of multifloor air transportation designs.

Energy efficiency of VAV air transportation.   Basically there are two design options available for VAV air transportation:

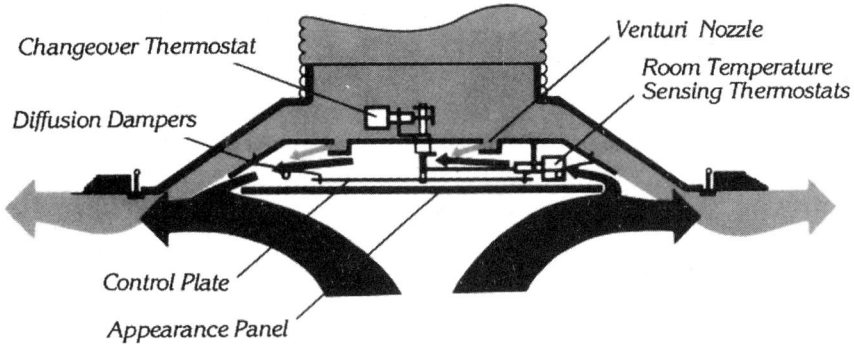

Figure 6.27   Pressure-dependent VAV outlet. *(Courtesy of Acutherm. Reprinted with permission.)*

1. Design a centralized system (Fig. 6.25a) and transport air to all VAV terminals from a central location, but try to minimize transportation losses by duct looping, limiting flow velocity, and use of energy-efficient fans, motors, and duct fittings.

2. Design a decentralized system (Fig. 6.25b). It is not unusual to achieve 50 percent or more energy savings with a decentralized air transportation system when it has substantially less transportation losses (Table 6.8).

Flexibility and other design issues of VAV air transportation.   Although it is conceivable to save substantial energy by decentralizing an air transportation system, the system designer needs to address other, often conflicting, design issues such as flexibility of adding or deleting VAV terminals, delivering sufficient outdoor ventilation air to each terminal under all flow conditions, and still limiting the pressure changes inside the ductwork and noise propagation through air transportation.

6.3.3   Cold air distribution

Cold air distribution offers greater transportation energy and space savings than conventional methods of air distribution. However, cold air means a reduction in airflow, and this airflow reduction may affect occupant comfort and indoor air quality, particularly for VAV operations which further reduce the airflow at low load conditions. The system designer's concern is whether supply outlets currently available are suitable for cold air VAV applications or not.

Reference 5 states that diffusers of current design will (with proper attention to their limitations) perform well with cold air. Figures 6.28

TABLE 6.8   Examples of Energy Consumption Central versus Decentralized System Air Transportation

Item	Central system	Decentralized system
Supply air		
At design conditions	10,000 cfm	1000 cfm
At minimum air flow	4,000 cfm	400 cfm
Design static pressure	6″ w.g.	2½″ w.g.
Air power required[1]		
At design conditions	9.43 hp	0.39 hp
At minimum flow	3.11 hp[2]	0.09 hp[3]
Air power required per 1000 cfm air transportation		
At design conditions	0.943 hp	0.39 hp
At minimum flow	0.311 hp	0.09 hp

Notes:

1. Air power = $\dfrac{\text{static press.(in)} \times \text{airflow (cfm)}}{6360}$ hp

2. Based on a typical central VAV air transportation system with variable fan speed drive and duct static pressure control (Fig. 6.25a).

3. Based on a completely decentralized VAV air transportation system with variable fan-speed drive but supply fan output is regulated by VAV terminal airflow requirements rather than the need to meet a predetermined duct static pressure setpoint (AHU[2] in Fig. 6.25b).

through 6.31 show the performance of three types of diffusers and a high wall grille over a wide range of airflows and temperatures with ADPI lines. The hatched regions in Figs. 6.28 through 6.31 (Ref. 5) represent the area where there is uncertainty about the performance of the distribution system under cold air, and the system operation should not fall within this hatched region. This means that at an approximately 45°F supply air temperature, the flow rate should not be reduced by more than 50 percent for a slot diffuser. Under this turndown ratio, the diffuser performance may become uncertain. When this condition is expected, raising the supply air temperature a few degrees or applying reheat will solve this problem. For example, raising the supplying temperature from 45 to 49°F ($\Delta T = 26°F$) will avoid the hatched area even at a 50 percent flow rate (point a in Fig. 6.28 corresponding to 10 Btu/hr·ft² half-load time). In this example a full load of 20 Btu/hr·ft² is assumed.

6.4   Acoustics for VAV Systems

The component selection for VAV systems is often influenced by acoustical considerations. It is not unusual for a VAV component to change its sound level more than 10 dB over a practical range of airflow. As a sound level increase of 5 dB or more is clearly noticeable by a human ear, the system designer essentially has two design choices:

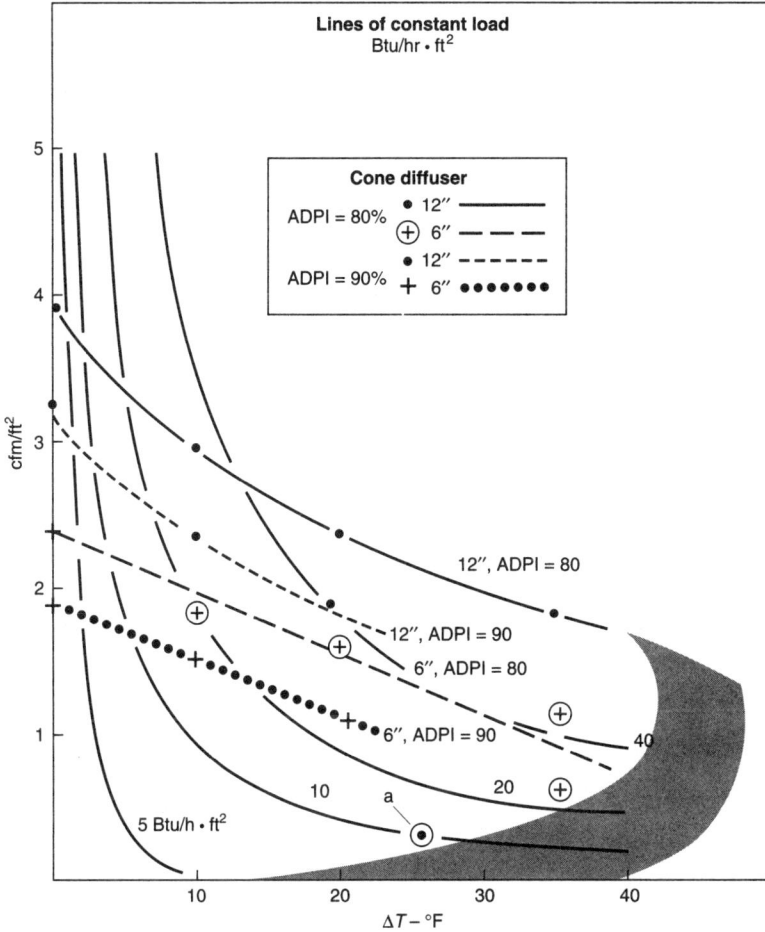

Figure 6.28 Circular cone diffuser performance. (ΔT = room air − supply air.)
Source: Ref. 5. Reprinted with permission.

1. Limit flow changes through VAV components. A smaller flow change produces a smaller sound change, which is less noticeable.

2. Select the sound level at the design flow considerably below (5 dB or more) the background sound level requirement to make the sound level changes hardly noticeable.

The suggested solutions result in derating of components or using a number of smaller units, and may not be economically justifiable, but must always be considered for VAV system design.

For example, in Fig. 6.32, contour A is the space sound pressure spectrum for a VAV box at design airflow and contour B is the space sound pressure spectrum for the box at minimum airflow. The contours repre-

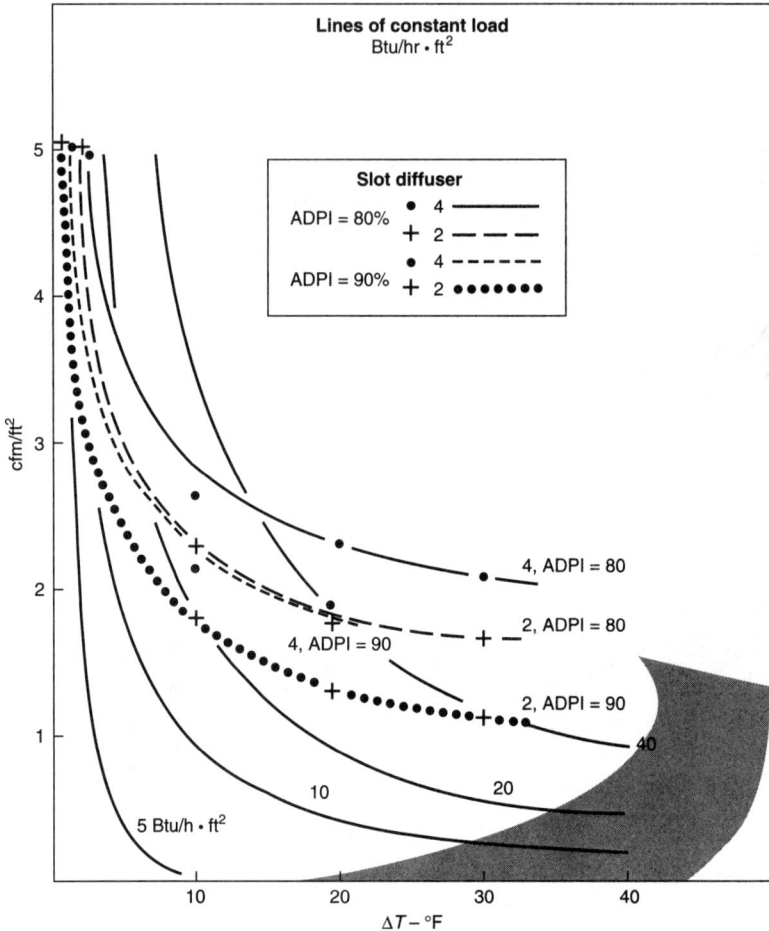

Figure 6.29   Slot diffuser performance. (ΔT = room air − supply air.) *Source: Ref. 5. Reprinted with permission.*

sent the sounds heard by the space occupants. The box should be used only in spaces where the background sound level is RC(NC) 45 or higher. Otherwise the occupants may become annoyed by the sounds because they are clearly noticeable by the occupants. In a practical design situation, noise criteria are usually given by the design objectives, and the procedure explained in the preceding paragraph should be followed.

Except for the aforementioned extra consideration, all other design concepts, methods, and techniques are still applicable to VAV system acoustic design. In the following sections, fundamentals of HVAC and building acoustics are covered first; then an example of acoustic design for a VAV system is briefly described to show how a quality indoor environment can be achieved with an integrated design approach, carefully

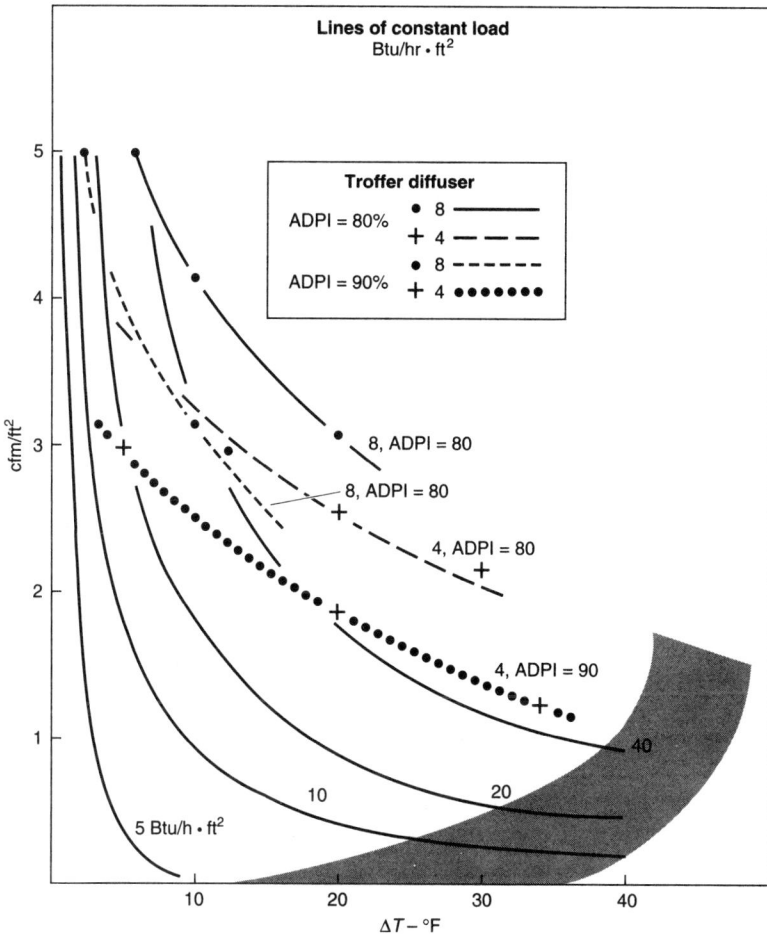

Figure 6.30 Light troffer diffuser performance. (ΔT = room air − supply air.) *Source: Ref. 5. Reprinted with permission.*

mixing VAV acoustic design with proper selection of building construction and materials.

6.4.1 Source-path-receiver concept

As shown in Table 6.9, the source-path-receiver concept describes the relationship of three key elements in HVAC acoustics: (1) noise sources, (2) transmission paths, and (3) the receiver who hears the sounds that travel through various paths to reach the occupied space and form the space sound spectrum. Figure 6.33 shows only two sources: a fan and a VAV terminal unit. However, there are usually more than two noise sources that coexist in parallel and series paths.

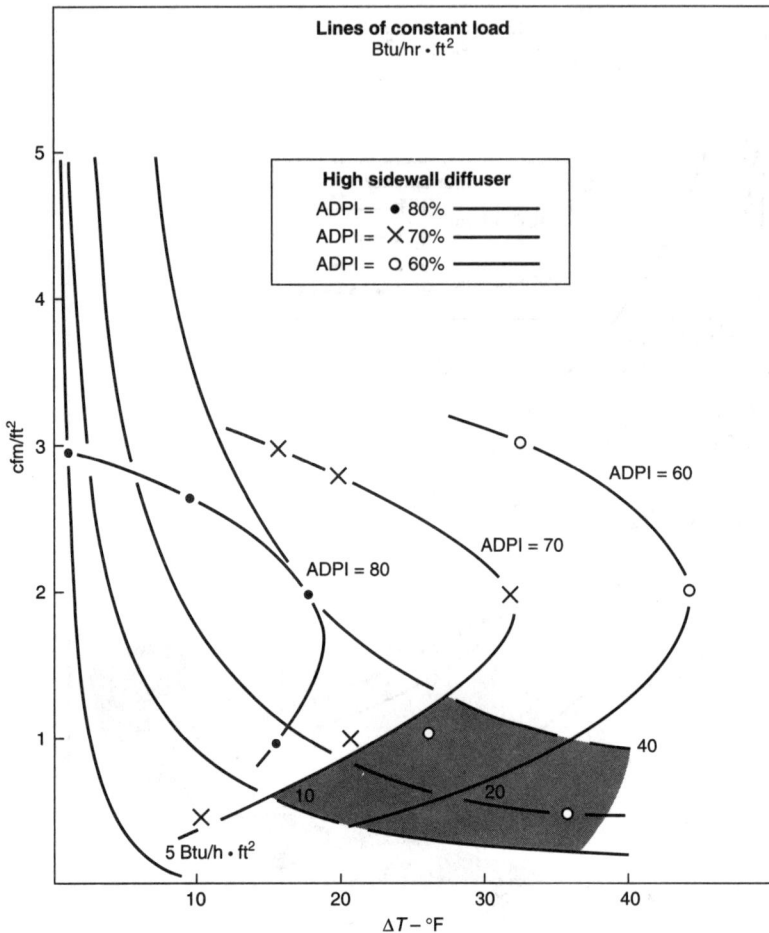

Figure 6.31 High sidewall diffuser performance. (ΔT = room air − supply air.) *Source: Ref. 5. Reprinted with permission.*

Some of these sounds are quite dominant and must be considered, while others may be minor and can be ignored in a practical design.

For example, in the acoustical model shown in Fig. 6.34, sounds transmitted through paths 1 and 3, and 5 may be included in a safety factor of 3 to 5 dB and can be omitted from the actual model to be used for acoustical analyses. Thus, the choice of sound sources and transmission paths (what to include and what not to include) becomes an extremely important design consideration and literally determines the quality of acoustic design.

6.4.2 Partial and total acoustical models

An acoustic model is a schematic representation of all significant sound sources and their transmission paths in a given HVAC/building design.

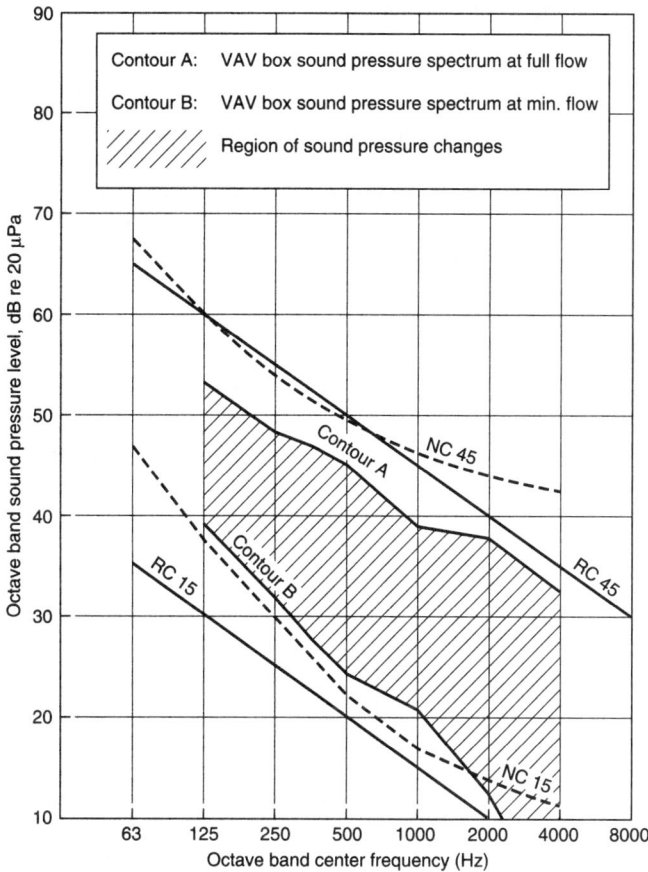

Figure 6.32 Sound-level changes for a VAV box.

As each design is unique, the acoustic model for a given design is also unique and has its own specific sound sources and transition paths. But generally speaking, there are two types of models: partial and total models. The acoustical models shown in Figs. 6.34 and 6.35 are mainly concerned with only part of the HVAC/building design and, therefore, belong to the partial model category. On the other hand, Fig. 6.36 considers the entire HVAC/building design. Thus, it is a total acoustical model.

In describing an acoustical model, one must first decide which noise sources, and therefore which transmission paths, to include in the design. Often, insignificant sounds can be omitted to simplify the design process and time required. However, oversimplification may hurt the accuracy and therefore the quality of the acoustic design.

In general, it is always preferable to take a total model approach by including all significant sounds and transmission paths such as the one shown in Fig. 6.36.

TABLE 6.9 **The Source-Path-Receiver Concept**

Source (Lw)	Path (dB losses)	Receiver (= Lp)
SOUND SOURCE(S) produce SOUND POWER	The SOUND PATH has elements that attenuate the SOUND POWER transmitted to the receiver's space	The SOUND POWER that arrives at the RECEIVER'S SPACE produces a resulting SOUND PRESSURE LEVEL at the RECEIVER'S EAR which we hear

6.4.3 Noise criteria

Three noise-rating systems are commonly used by the HVAC industry: A-weighted sound level (dBA), noise criteria (NC), and room criteria (RC). All use a single-number rating to represent an octave-band spectrum of background, equipment, or component noise.

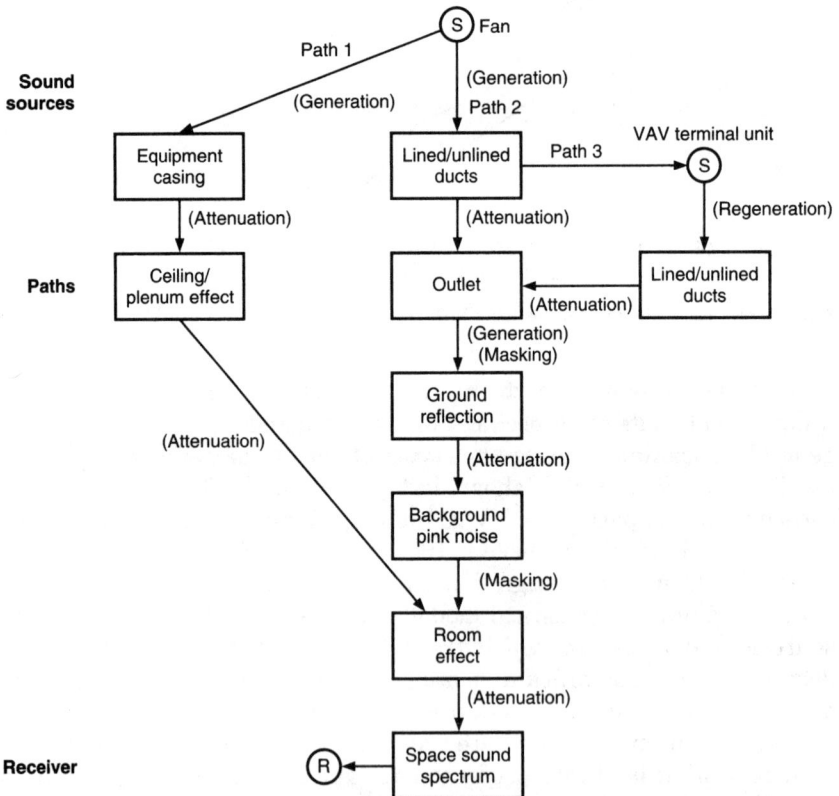

Figure 6.33 Noise generation (regeneration) attenuation, and masking.

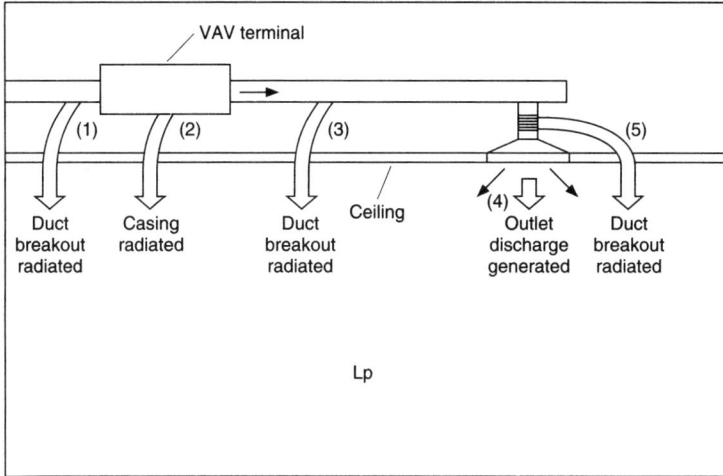

Figure 6.34 Fan-powered VAV terminal or induction VAV terminal unit acoustic model.

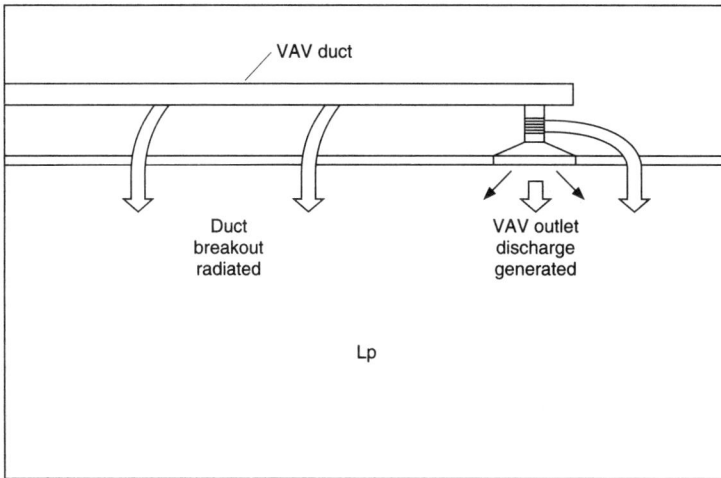

Figure 6.35 VAV diffuser acoustic model.

A-weighted sound level (dBA). The A-weighted sound level is generally used to rate noise in a community environment and to evaluate noise for hearing conservation. This rating method is seldom used for HVAC acoustic design which requires analyses of sound levels at each octave band. However, this rating method is useful in comparing relative loudness of a given HVAC system operating under different conditions but in the same acoustical environment. A good example is a VAV system installed in a building with a considerable variation in supply air

Primary noise sources: S1, S2, S3, S4

Secondary noise sources: S_{e1}, S_{e2}, S_{e3}, S_{e4}

S1   Fume hood
      exhaust fan

S_{e1}

Silencer

S3

S_{e2}

S2   Return air
      VAV unit

Ductwork

Ceiling suspended
fan coil unit

S_{e3} (Elbow, typ.)

S_{e4}

S4

Supply air
VAV unit

Acoustic elements
* Elbows, ducts, silencer

(a)

(b)

Path 1: Fan S1 ———► Ductwork ———► Hood ———► Receiver

Path 2: Fan coil unit S2 ———► Receiver

Exhaust fan

S1

Exhaust air duct

S3

RA VAV
unit

Fume hood

Fan coil unit

S2

S4

SA VAV unit

R

Path 3: RA VAV unit ———► Ductwork ———► Receiver

Path 4: SA VAV unit ———► Ductwork ———► Receiver

(c)

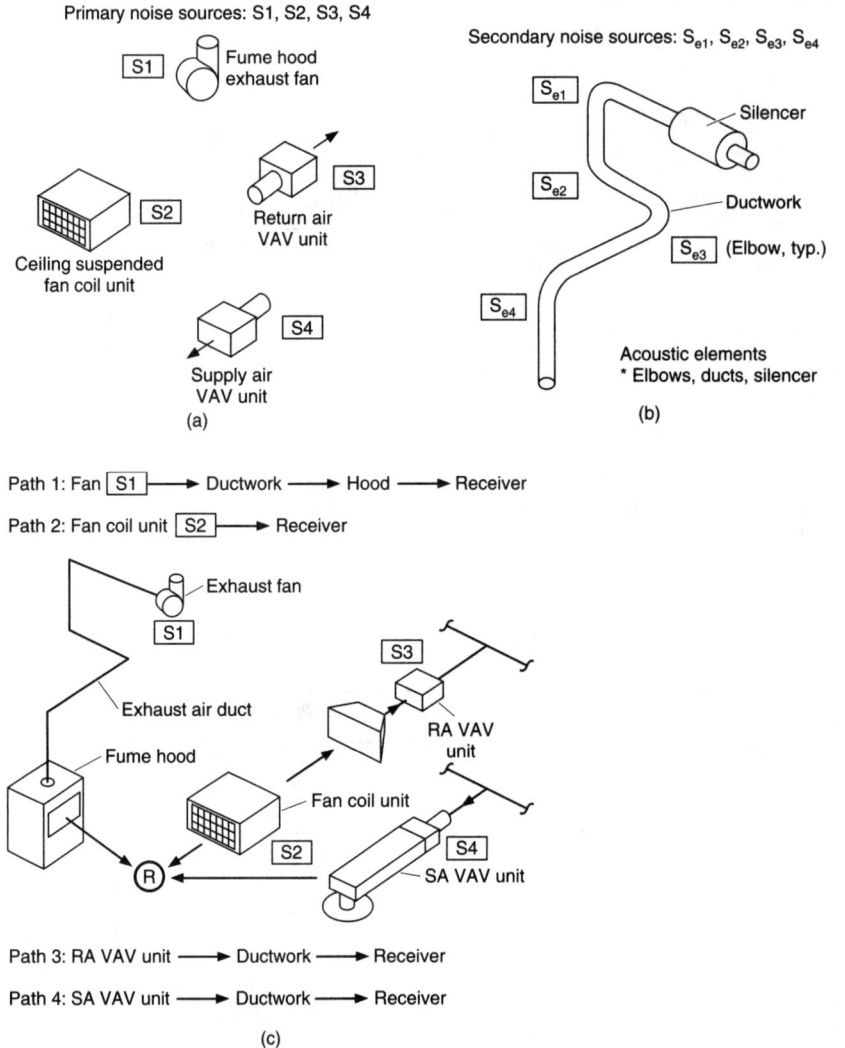

Figure 6.36 Noise sources, transmission paths, and receiver in a laboratory. (a) Primary noise sources. (b) Secondary noise sources. (c) Transmission paths.

quantity. In this situation, dBA ratings can be taken at different air deliveries to determine if raising or lowering the air quantity has a significant impact on the noise level of the space.

Noise criteria (NC).   The NC noise rating is one of the most commonly used methods to rate the noise level of an indoor space. This rating method is also widely adapted by most terminal and outlet manufacturers. However, Ref. 8 cautions the system designer concerning the following limitations of the NC rating method:

1. Listeners generally agree that noise with the spectrum shape of an NC curve sounds too rumbly and too hissy.

2. The NC curves do not extend to 16- and 31-Hz octave bands, the regions where some of the most troublesome HVAC noise problems occur.

3. By reporting a single number based on a tangency to a rating curve at one octave band without assigning a subjective quality to the sound spectrum, one could mistakenly rate two widely different sound spectra as equally acceptable. (See Fig. 6.37.)

Room criteria (RC). The RC rating system consists of a numerical value and a letter suffix. The numerical value is calculated from the average of the 500, 1000, and 2000 octave band sound levels. There are four letter suffixes: R, H, N, and T. Definitions of these letters are best explained by examples shown in Fig. B-1 of Ref. 8, p. 141. The same reference also provides excellent explanations, recommendations, and applications of the RC rating system. The readers are encouraged to read it for detailed treatment of this subject.

6.4.4   Acoustical characteristics of HVAC equipment, system components, and building elements

Definition of an acoustic element.   An acoustic element is a sound source, a sound attenuator, or a sound receiver. Sometimes, it is a sound attenuator as well as a sound generator. HVAC equipment, duct fittings and devices, and building elements such as walls, windows, floors, and ceilings are all acoustic elements. Some acoustic elements are pure sound generators, but many of them are mainly sound attenuators that also generate some noise. HVAC equipment is the former; silencers, duct liners, lined ducts, elbows, and plenums are the latter.

Sound characteristics of HVAC equipment: Each acoustic element has its unique sound-generation/attenuation characteristics, which can be represented by sound power spectra. Figure 6.38 shows sound-generation characteristics of typical HVAC equipment. Figure 6.39 shows sound attenuation and regeneration characteristics of certain acoustic elements. It is important for the HVAC designer to understand these characteristics, because they greatly influence the quality of the space sound spectrum—the very reason and objective of acoustic design.

For example, a fan with a concave upward-shaped sound power spectrum with high sound levels at first, second, and third bands tends to induce vibration or cause rumbly noise in lightweight wall and ceiling constructions. On the other hand, a fan with relatively flat characteristics but with its peak at midbands may produce hissing at the speech bands when combined with a silencer and lined ductwork or a flexible

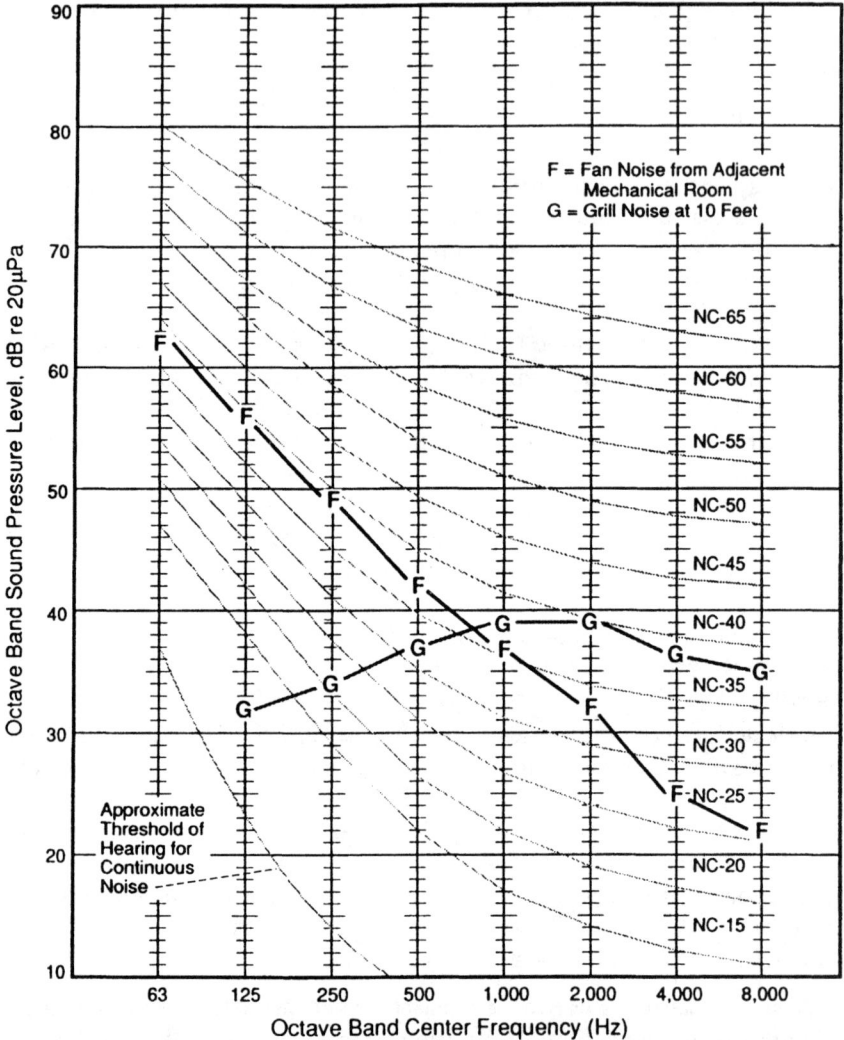

Figure 6.37   Two different HVAC noise spectra with the same NC rating. (*Source: Ref. 8. Reprinted with permission.*)

duct. Thus, care must be exercised in selecting the combination of sound-generating and attenuating acoustic elements.

Sound attenuation and regeneration by system components.   Air distribution components such as ducts, fittings, dampers, terminal units, and outlets all produce self-generated or regenerated noise. Some components generate more noise than others and need to be analyzed. Figure 6.39*b* shows typical regeneration noise levels for duct elbows. Noise generated by duct system components are relatively low levels in magnitude

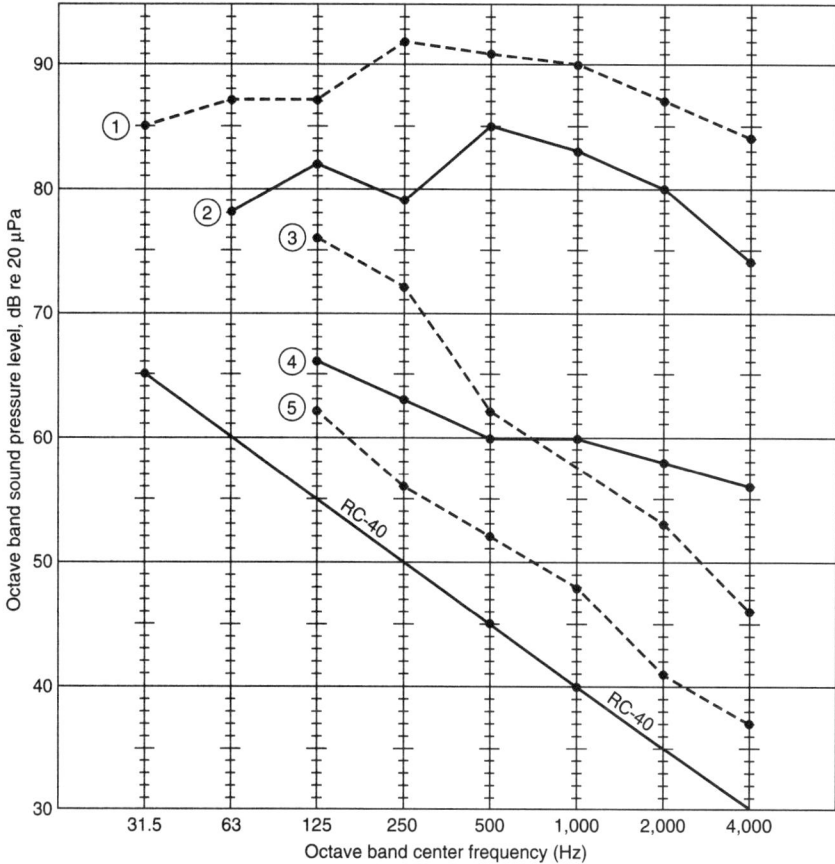

(1) Typical sound pressure spectra near 25-hp fans (Ref. 8)

(2) Typical sound pressure spectra near 300 ton chillers (Ref. 8)

(3) Typical fan room sound pressure spectra (Ref. 11)

(4) Typical fan powered terminal room sound pressure spectra (Ref. 4, Chap. 9)

(5) Typical VAV terminal room sound pressure spectra (Ref. 4, Chap. 9)

Figure 6.38 Typical HVAC noise.

and require less attention in noise control unless they are located close to occupied spaces, and the airflow velocity is excessively high. Elbows and dampers may become a critical noise source if they are located near the conditioned space where extra quietness (usually NC/RC 25 or less) is required. In this case, they should be installed a considerable distance from the space and the duct velocity should be kept low (typically 800 fpm or less). Fans are generally considered to be the major source of noise for VAV systems. However, in practical design situa-

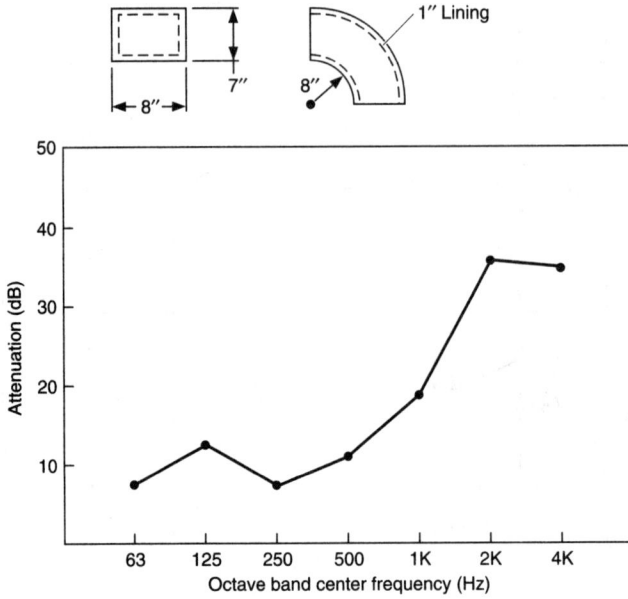

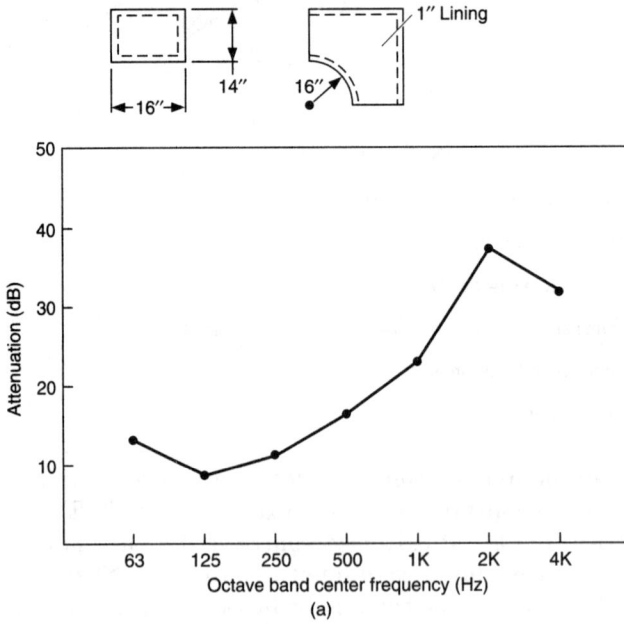

Figure 6.39 Sound attenuation and regeneration of elbows. (a) Attenuation.

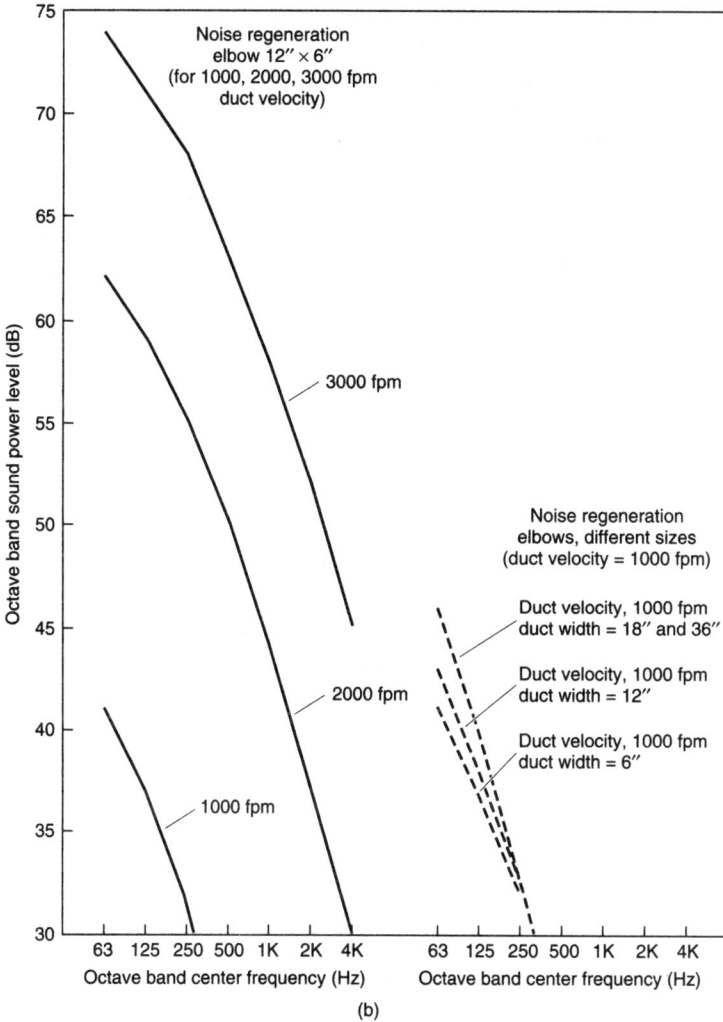

Figure 6.39 (*Continued*) (*b*) Regeneration.

tions, fans are not necessarily the most critical noise elements and the relative importance of fans versus other components must be carefully analyzed case by case.

6.4.5 Building acoustics

Building acoustics plays an important role in HVAC acoustic design. It affects the intensity and quality of sound generated by HVAC equipment and transmitted through ductwork and other propagation paths.

In this book, *building acoustics* refers to building construction methods and materials which directly or indirectly modify and reduce the characteristics and levels of noise that reach the occupied space. Typical examples are wall, ceiling, and floor constructions and materials. For a broader understanding of building acoustics, the reader is encouraged to read Refs. 8, 9, and 10.

In building acoustics, the VAV system designer is mainly concerned with noise, or unwanted sounds, in the occupied space, and how these sounds are influenced by architectural space acoustics, as well as the effectiveness of building construction methods and materials that isolate or reduce equipment noise generated inside and outside the occupied spaces.

Sound magnitude. Sound magnitude is a subjective, ear-oriented reaction not linearly related to the sound intensity, and often expressed by loudness levels such as phons and sones. The equal-loudness contour in Fig. 6.40 shows why certain sounds are perceived as being louder than others, despite the sound-pressure levels that would indicate the contrary. In practical acoustic design, this sound magnitude must be translated into a measurable physical quantity. Sound power and sound pressures are used for this purpose.

Sound reduction. Sound reduction in buildings is achieved by the following methods:

1. Reduction of noise generation at the source by proper selection and installation of equipment or, if necessary, provision of a sound-isolating enclosure over the equipment

2. Reduction of noise transmission from point to point along each transmission path by proper selection of construction materials and methods

3. Reduction of noise in the occupied spaces through acoustic treatment of the space

Building elements and HVAC acoustics. The major function of building elements in HVAC acoustics is absorption and isolation of HVAC equipment and component noise (Fig. 6.41). Generally, noise isolation (Fig. 6.41c) is far more effective than noise attenuation by surface absorption (Fig. 6.41b). Noise reduction by architectural barriers (walls, partitions, and floors) is also very effective in HVAC noise control (Fig. 6.42).

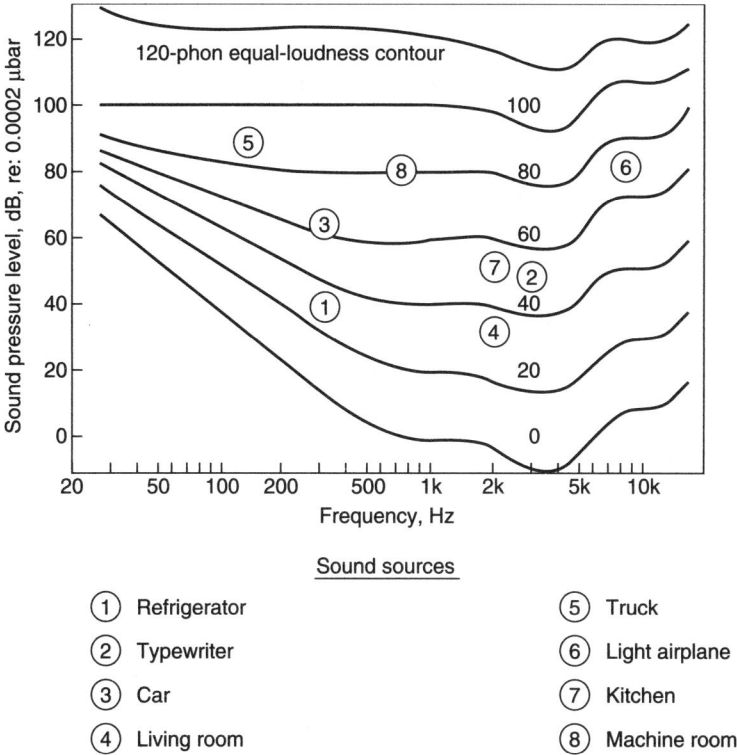

Sound sources

(1) Refrigerator   (5) Truck

(2) Typewriter   (6) Light airplane

(3) Car   (7) Kitchen

(4) Living room   (8) Machine room

Figure 6.40 Various sound sources and their equal-loudness contours.

As demonstrated in the preceding examples, the traditional way of separating HVAC acoustics from building acoustics may not provide an economical and high-quality acoustical environment. In certain design situations, some architectural/structural subsystems or elements may become the weakest path or paths of noise transmission. Increasing the attenuation effectiveness of these elements may be far more important than acoustically treating other HVAC-dominated transmission paths. In Fig. 6.43, the noise transmitted through the AHU room wall is the dominating source of noise transmission. Case A in Table 6.10 shows the space sound spectra resulting from each of three transmission paths 1, 2, and 3 in Fig. 6.43 and what the occupant hears, the combined space noise rating of NC-41 or RC-42(N). In this case, the supply and return air paths contribute one octave band each (first and second bands), but the wall path contributes five octave bands in formulating the final space sound pressure spectrum, or the noise level perceived by the occupant. By installing a heavier ceiling and tightening the wall construction, the space sound spectrum can be reduced to

Values shown all at 500 Hz

89 dB	84 dB	40 dB
88 dB at 500 Hz		
Untreated room surface (a)	Treated room surface (b)	Isolated noise source (c)

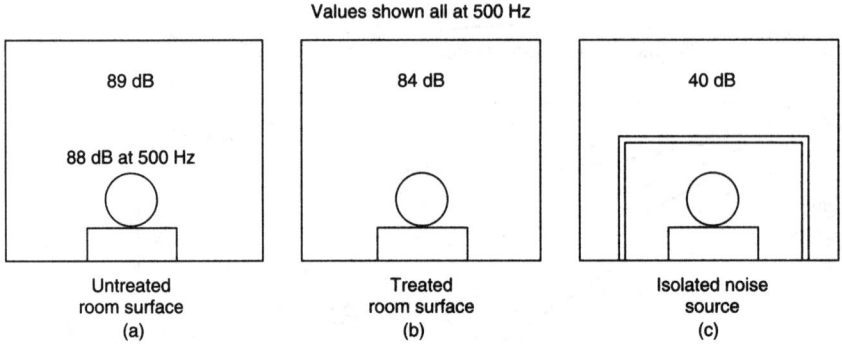

Figure 6.41 Examples of noise control by absorption and isolation. (*a*) All bare surfaces. (*b*) Ceiling treated. (*c*) Bare room.

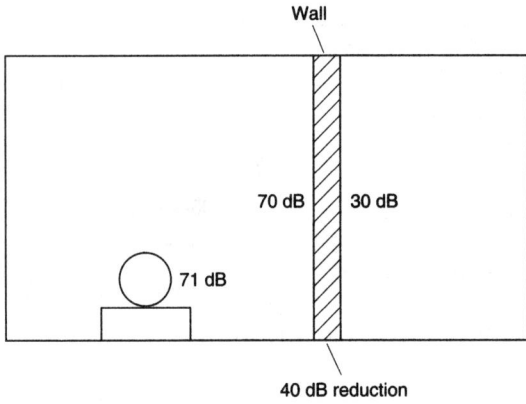

Wall

70 dB   30 dB

71 dB

40 dB reduction

Figure 6.42 Noise reduction through a building element.

the level shown in case B. As a result, the corresponding space noise rating is now reduced to NC-35 or RC-35(N) which is within the ASHRAE-recommended ranges for normal office applications.

Masking. When two noise sources coexist and are perceived simultaneously, each sound is less distinguishable because of the presence of the other. This phenomenon is called *masking*. Open office environments often employ background sound masking by introducing *white noise* to mask unwanted sounds or noise between the offices. Frequently, fans and air outlets can be utilized to provide the masking effect. However, for VAV applications, the noise levels of system components fluctuate continuously, and the system designer needs to select VAV components carefully so that the noise levels of the components will always stay below the background sound level. In this case the background sound provides a masking effect.

6.4.6   Reading and applying
manufacturers' sound data

Data presentation.   Manufacturers of HVAC equipment present sound data in many different ways. The following describe some representative methods of data presentation.

AHU and fan sound data

1. *Computerized format for air-handling units (AHU).*   Figure 6.44 shows an AHU sound data sheet. It includes detailed information on discharge, inlet, and radiated sound power spectra and other pertinent data for a given set of fan cfm and static pressure.

2. *Limited computer format for fans.*   Figure 6.45 shows data in this category. It usually shows discharge sound power spectra only.

3. *Sound data to be calculated from charts and tables.*   Figure 6.46 is an example in this category. In this case, the data do not include the information on radiated sound data.

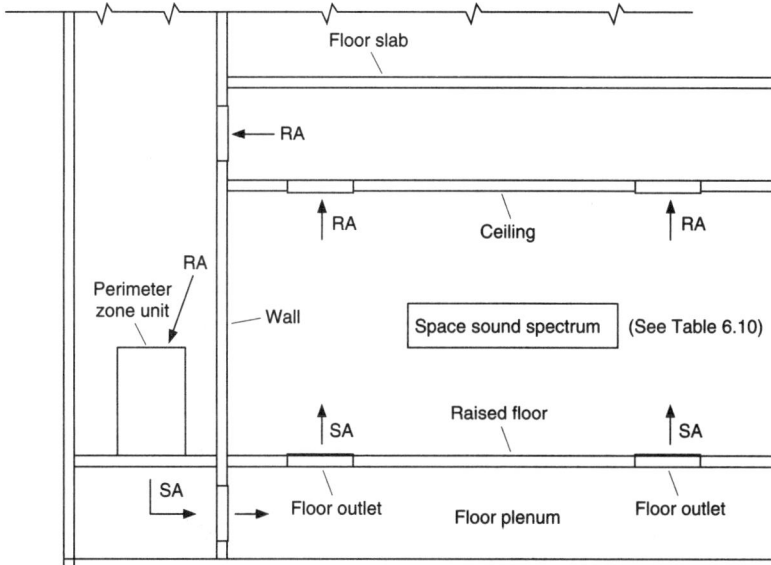

Transmission paths

① AHU → Ceiling plenum → SA outlet → Room

② AHU → Floor plenum → RA outlet → Room

③ AHU → Wall → Room

Figure 6.43   Multiple-path noise transmission.

TABLE 6.10 Multipath Noise Transmission Through HVAC and Architectural Elements

A. Average wall construction, acoustical ceilings

| | Space Sound Spectrum Octave Band | | | | | | | Description | | | | |
Path No.	63	125	250	500	1K	2K	4K	Source	–	Transmission Path	–	Receiver
1 (SA path)	48	52	44	– less than 10 –				Perimeter AHU Fan	→	Floor SA Plenum → SA Outlet	→	Space occupant
2 (RA path)	60	50	43	42	36	33	24	Perimeter AHU Fan	→	Ceiling RA Plenum → RA Outlet	→	Space occupant
3 (Wall path)	46	44	46	43	38	35	29	Perimeter AHU Fan	→	AHU Room Wall	→	Space occupant
Combined	61	55	50	46	41	38	32					

Space Noise Ratings: NC-41 or RC-42(N)

B. Tight wall construction, gypsum board ceilings

| | Space Sound Spectrum Octave Band | | | | | | | Explanation |
Path No.	63	125	250	500	1K	2K	4K	
1 (SA path)	48	52	44	– less than 10 –				No change in construction
2 (RA path)	54	29	16	15	12	11	–	A heavier ceiling has reduced noise transmission
3 (Wall path)	46	41	37	33	29	25	20	A tighter construction has reduced noise leakage
Combined	56	52	45	33	29	25	20	

Space Noise Ratings: NC-35 or RC-35(N)

4. *Tabulated sound pressure spectra.* A typical example in this category is the sound data presentation for cooling tower noise. Figure 6.47 is such an example.

VAV box and fan-powered unit sound data. Figure 6.48 is an example of sound data in this category.

Data application. Obviously the format of presentation and data actually presented vary from manufacturer to manufacturer according to the types of equipment they supply. The system designer needs to apply such data selectively and, if necessary, obtain additional information required for VAV system design. The following serve as specific guidelines for sound data application:

1. For air-handling-unit noise analysis, it is often more important to know the noise propagated through the unit casing or the unit return opening (Fig. 6.44). Compare three noise transmission paths: supply, return, and through the casing. If the noise transmission through the casing or the return air opening is suspected, obtain necessary sound data and conduct a noise analysis to determine attenuation requirements and provide spaces for silencers and other noise control devices.

2. Frequently, architectural solutions are more effective than mechanical solutions. For example, use of acoustically better quality wall and ceiling constructions and materials may be more cost-effective than a quieter air-handling unit.

3. VAV boxes and fan-powered units are frequently installed very close to or even directly above the occupied spaces. Avoid this situation if the occupied space requires a design sound rating of less than RC (or NC) 35. If this is not feasible, use quieter VAV boxes and fan-powered units. Improve the acoustical quality of ceilings and surface finishes. Consult with the architect to see if the design sound rating can be redefined. When these measures are combined with careful selection of terminal units (VAV boxes and fan-powered terminals), acceptable sound levels for most applications (RC 35-40) can be achieved even with the terminal units installed directly above the ceiling or below the floor of the occupied spaces.

4. Different types of fans produce different sound-spectrum shapes (Fig. 6.49). Also, the same fan produces different sound spectrum shapes if the fan is allowed to operate under different conditions. Estimate the fan operating conditions for the design and minimum airflow rates. Obtain sound data from two or three fan manufacturers. Compare their performance data for sound levels and spectrum shapes. Select the best fan for your application.

39L Performance Summary Report	
Tag Name: UVA2–LAB (100%)	07–01–93
Carrier ACAPS Program v1.40	Page 1 of 1

Unit Size	6
Fan Model	LA
Inlet Guide Vanes	NO
Standard Airflow rate (cfm)	2000
Site Altitude (ft)	0
Altitude Airflow Rate (cfm)	2000
External Static (in. w.g.)	0.25
Other Static Pressure	
Cooling Coil (in. w.g.)	1.00
Heating Coil (in. w.g.)	0.00
Accessory (in. w.g.)	0.00
Total Static Pressure (in. w.g.)	1.25
Fan Power (bhp)	.7
Fan RPM	917
Acoustic Data	
Wheel Diameter (in.)	12.63
Number of Blades	43
Blade Passsage Freq.	657.49

Sound Power		Discharge	Inlet	Casing
	31.5 Hz	87	84	84
	63 Hz	82	74	76
	125 Hz	83	65	71
	250 Hz	79	60	62
	500 Hz	81	61	61
	1000 Hz	73	56	57
	2000 Hz	69	58	56
	4000 Hz	66	53	53
	8000 Hz	61	45	51

Note: Casing-Radiated sound power levels assumes gasketed casings
Legends: LA=LA

Figure 6.44 Air-handling unit sound data sheet (I). (*Courtesy of Carrier Corporation. Reprinted with permission.*)

6.5 Building Pressurization and VAV Systems

Building pressurization is a system problem involving considerations for objectives of pressurization and space usage as well as building construction, porosity, floor separation, wind, stack, and internal partition effects. To properly control building pressures, these factors must be assessed individually and their combined effects evaluated through adequate mathematical models and simulation techniques.

Series: S21–H   Size: 17

C.F.M.: 16500   DENSITY: 0.075   ALTITUDE: 0
S.P. : 4.00   MATERIAL: Mild Steel   TEMPERATURE: 70

	UNIT #1	UNIT #2	UNIT #3
Fan Type	BAF-DW	BAF-DW	BAF-DW
Fan Size/Model	270	245	222
Wheel Diameter	27	24.5	22.25
Class	I	I	II
Outlet Velocity	2188	2657	3223
Fan R.P.M.	1338	1610	1952
Max R.P.M. for class	1499	1693	2372
BHP @ STD Density	13.66	16.01	17.96
BHP @ OP Density	13.66	16.01	17.96
Static Efficiency	76.04%	64.88%	57.82%

PLOT CURVE POINTS

SEL. UNIT #1
CFM	26346	23621	20895	18170	15444	12719	9993	7268	4542	1817
SP	0.23	1.51	2.63	3.56	4.23	4.63	4.83	4.87	4.82	4.73
BHP	12.26	13.73	14.27	14.06	13.33	12.27	11.04	9.61	8.09	6.48

SEL. UNIT #2
CFM	23940	21463	18987	18510	14034	11567	9081	6604	4128	1651
SP	0.26	1.70	2.93	4.00	4.88	5.62	6.00	6.08	6.01	5.86
BHP	14.61	15.95	16.37	16.01	15.13	14.00	12.65	11.01	9.11	7.04

SEL. UNIT #3
CFM	21735	19487	17238	14990	12741	10493	8244	5996	3747	1499
SP	0.32	2.07	3.55	4.85	5.92	6.81	7.27	7.37	7.28	7.10
BHP	16.07	17.55	18.01	17.61	18.64	15.40	13.92	12.12 .	10.02	7.75

SOUND POWER LEVELS

OCTAVE BANDS
	1	2	3	4	5	6	7	8	LWA
UNIT #1	97	95	91	86	83	79	76	71	89
UNIT #2	97	96	93	89	84	81	78	76	92
UNIT #3	98	97	96	92	87	84	80	77	94

Figure 6.45 Air-handling unit data sheet (II). (*Courtesy of Miller Picking Corporation and Twin City Fan and Blower Company. Reprinted with permission.*)

6.5.1 Objectives of building pressurization

The entire building or part of the building is pressurized for several different reasons. For example, pressurization may be required for IAQ reasons, local contamination control, unidirectional airflow, smoke, or combinations of these requirements (See Fig. 6.50).

FAN CODE: MX40

(performance chart — Fan Static Pressure vs Volume Flow; AIR DENSITY 0.075 lb/ft³, MOMENT OF INERTIA 2.4 lbft²)

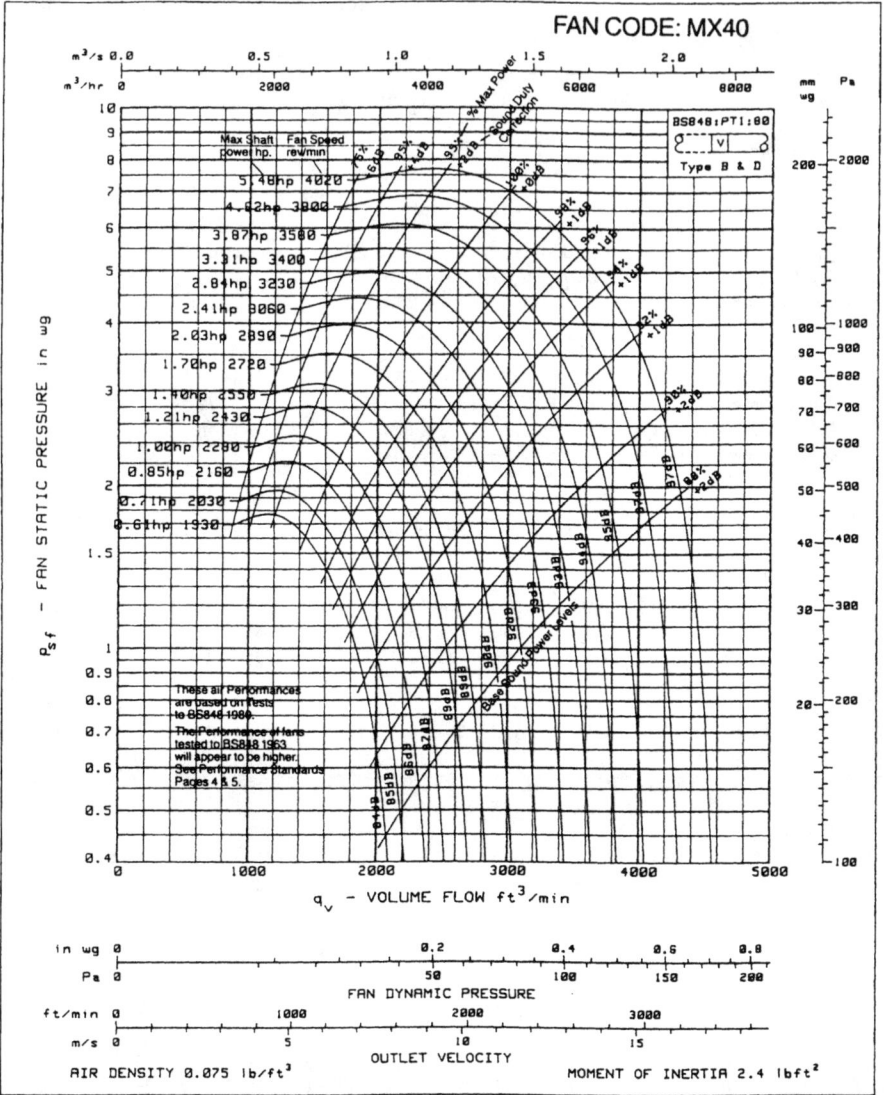

For Type A & C Connections and Outlet Diffuser Performance see Table I and II page 20

Speed Range rev/min	Inlet Type	INLET SOUND POWER SPECTRA Octave Band Mid-Frequency Hz								Outlet Type	OUTLET SOUND POWER SPECTRA Octave Band Mid-Frequency Hz							
		63	125	250	500	1 K	2 K	4 K	8 K		63	125	250	500	1 K	2 K	4 K	8 K
1880–2540	Ducted	− 5	− 6	− 6	− 9	−13	−14	−19	−25	Ducted	− 5	− 3	− 3	− 6	−10	−13	−19	−25
	Free	−14	−15	− 9	− 6	−12	−13	−18	−24	Free	−12	− 5	− 5	− 5	− 7	−11	−17	−23
2680–3130	Ducted	− 7	− 8	− 6	− 6	−11	−13	−17	−23	Ducted	− 7	− 6	− 3	− 2	− 8	−12	−17	−23
	Free	−16	−17	− 9	− 3	−10	−12	−16	−22	Free	−14	− 8	− 5	− 1	− 5	−10	−15	−21
3320–3790	Ducted	− 9	− 9	− 7	− 5	− 9	−13	−15	−21	Ducted	− 9	− 8	− 3	− 1	− 6	−10	−15	−21
	Free	−18	−18	−10	− 2	− 8	−12	−14	−20	Free	−16	−10	− 5	0	− 3	− 8	−13	−19

Inlet Sound Levels (dB re 1pW)
Outlet Sound Power Levels are approximately equal to Inlet Sound Power Levels + 2dB

Figure 6.46 Tables and chart for sound-power-level calculation. (*Courtesy of Woods Company. Reprinted with permission.*)

SOUND RATING DATA SHEET
BALTIMORE AIRCOIL COMPANY, INC.

Date _____

Model: (sample only)

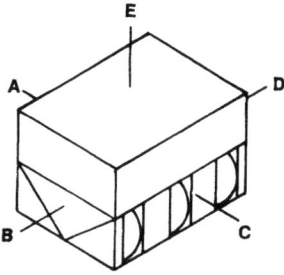

CALCULATED SOUND POWER LEVEL (PWL)

Octave Band	Center Frequency	dB re 10⁻¹² Watt
1	63 Hz	95
2	125	95
3	250	90
4	500	91
5	1000	85
6	2000	82
7	4000	78
8	8000	72

A

OCTAVE BAND	DISTANCE 5 ft	50 ft
1	77	64
2	76	63
3	73	58
4	75	60
5	73	53
6	68	50
7	63	45
8	58	39

B

OCTAVE BAND	DISTANCE 5 ft	50 ft
1	73	63
2	73	62
3	68	58
4	71	59
5	67	53
6	64	49
7	60	46
8	55	40

E

OCTAVE BAND	DISTANCE 5 ft	50 ft
1	79	61
2	77	61
3	75	59
4	74	56
5	69	52
6	67	49
7	61	46
8	53	39

D

OCTAVE BAND	DISTANCE 5 ft	50 ft
1	73	63
2	73	62
3	68	58
4	71	59
5	67	53
6	64	49
7	60	46
8	55	40

C

OCTAVE BAND	DISTANCE 5 ft	50 ft
1	77	64
2	76	63
3	73	58
4	75	60
5	73	53
6	68	50
7	63	45
8	58	39

Octave band sound pressure levels (SPL) in dB RE 0.0002 microbar for 5 ft and 50 ft distances from unit sides (A, B, C, D) and top (E).

Figure 6.47 Cooling tower sound data. (*Courtesy of Baltimore Aircoil Co. Inc. Reprinted with permission.*)

Models: PESV, AESV, DESV ■ Sound Data ■ NC Values

Inlet Size	CFM	Discharge ΔP_s 0.5"	1.0"	2.0"	3.0"	Radiated ΔP_s 0.5"	1.0"	2.0"	3.0"
4	75	-	-	-	-	-	-	-	21
	125	-	-	20	23	-	-	25	28
	175	-	20	25	28	-	23	29	33
	250	21	26	30	33	22	28	34	38
5	125	-	-	-	-	-	-	-	-
	175	-	-	-	23	-	-	-	-
	250	-	-	25	28	-	-	23	26
	300	-	21	27	31	-	22	27	30
	350	-	24	30	33	21	26	31	34
6	175	-	-	-	-	-	-	20	22
	225	-	-	-	22	-	-	23	26
	300	-	-	23	26	-	23	27	30
	350	-	20	26	28	20	25	29	32
	400	-	23	28	31	22	27	31	34
	450	-	24	29	32	24	29	33	36
	500	21	26	31	34	26	30	35	37
7	250	-	-	-	-	-	-	20	23
	300	-	-	-	21	-	-	22	26
	350	-	-	20	24	-	-	24	27
	400	-	-	23	26	-	21	26	29
	500	-	20	27	30	-	23	28	31
	600	-	23	30	34	20	25	30	34
	650	-	25	31	35	21	26	31	34
8	350	-	-	-	-	-	-	-	21
	400	-	-	-	-	-	-	20	23
	450	-	-	-	20	-	-	22	24
	500	-	-	-	22	-	20	24	26
	600	-	-	21	24	-	23	26	29
	700	-	-	23	27	21	25	29	31
	800	-	20	26	29	23	27	31	33
9	450	-	-	-	22	-	-	22	26
	500	-	-	-	23	-	-	23	27
	600	-	-	21	25	-	22	25	28
	700	-	-	22	26	22	25	28	29
	800	-	-	23	27	24	27	30	31
	900	-	-	24	28	26	29	32	33
	1000	-	20	25	29	27	30	33	35

Inlet Size	CFM	Discharge ΔP_s 0.5"	1.0"	2.0"	3.0"	Radiated ΔP_s 0.5"	1.0"	2.0"	3.0"
10	550	-	-	-	-	-	20	24	28
	600	-	-	-	-	-	21	24	29
	700	-	-	-	-	22	24	26	29
	800	-	-	-	20	25	27	29	30
	1000	-	-	21	25	29	31	33	34
	1200	-	-	25	29	32	34	36	37
	1400	-	22	28	32	35	37	39	40
12	800	-	-	-	20	-	-	25	29
	900	-	-	-	21	-	20	26	30
	1000	-	-	-	22	-	22	27	30
	1200	-	-	-	23	21	25	29	32
	1500	-	21	25	25	25	29	33	36
	1800	-	-	22	26	28	32	37	39
	2100	-	-	23	28	31	35	39	42
14	1000	-	-	-	-	-	-	24	29
	1200	-	-	-	-	22	24	27	30
	1500	-	-	-	20	27	30	32	34
	1800	-	-	-	22	32	35	37	39
	2100	-	-	-	23	36	39	41	42
	2400	-	-	21	25	40	42	44	46
	3000	-	-	23	27	45	48	50	52
16	1400	-	-	-	-	-	-	23	27
	1600	-	-	-	-	-	20	25	28
	2000	-	-	-	-	-	23	28	31
	2400	-	-	-	21	20	26	31	34
	2800	-	-	-	22	23	28	34	37
	3200	-	-	21	24	25	30	36	39
	4000	-	-	25	28	28	34	39	42
24 X 16	3000	21	25	29	32	24	29	33	36
	3500	23	27	31	34	27	32	36	39
	4000	25	29	33	36	30	35	39	42
	5000	28	32	36	39	34	39	44	47
	6000	31	35	39	42	38	43	48	51
	7000	33	37	41	44	42	46	51	54
	8000	35	39	43	46	44	49	54	57

➤ ΔP_s is the difference in static pressure from inlet to discharge.
➤ Dash (-) in space denotes NC value less than 20.
➤ All Sound Data are based upon tests conducted in accordance with ARI 880-94 in the Laboratory at TITUS, Richardson Texas.

Octave Band Sound Attenuation Factors:

Radiated Sound	Octave Band 2	3	4	5	6	7	
Environmental Effect	3	2	1	1	1	1	Per ARI 885-90
Ceiling Effect	9	10	12	14	15	15	Mineral Fiber Tile, 5/8"-35#/ Cu. Ft.
Room Effect	9	10	11	12	13	14	3000 Cu. Ft. Space, 10 Ft. from Source
Total dB Reduction	21	22	24	27	29	30	

Discharge Sound	Octave Band 2	3	4	5	6	7	
Environmental Effect	3	2	1	1	1	1	Per ARI 885-90
Duct Lining	1	3	8	21	20	12	5 Ft., 1" Fiberglass Duct Lining
End Reflection	11	6	2	0	0	0	8" Termination to Diffuser
5 Ft, 8" Flex Duct	6	10	17	19	19	12	Vinyl Core Flex
Room Effect	9	10	11	12	13	14	3000 Cu. Ft. Space, 10 Ft. from Source
Total dB Reduction	30	31	39	53	53	39	

Additional dB reduction per octave band in sound resulting from 300 cfm flow division:

Inlet Size	(dB)
7,8	3
9	5
10	7
12	8
14	10
16	11
24 X 16	14

Figure 6.48   VAV box sound data. *(Courtesy of Titus Division of Tomkins Industries, Inc.)*

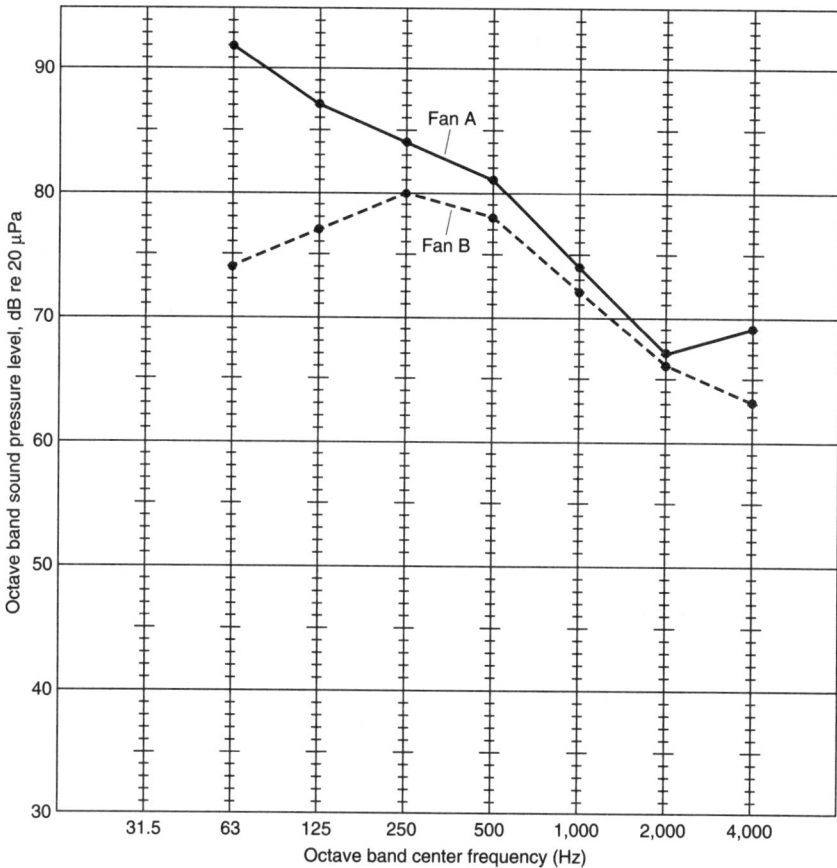

Figure 6.49 Sound pressure spectra for different fans for the same application.

6.5.2 Space usage

Space usage determines the type of pressurization required. General pressurization shown in Fig. 6.50a usually maintains a building at a positive pressure to prevent outdoor contaminants from entering the building. Sometimes, a space or several spaces of a building are maintained at a negative pressure to eliminate exfiltration to adjacent spaces. For certain patient rooms in hospitals, the strict, one-direction airflow is maintained by the staged pressurization to prevent the reversal of airflow. In Fig. 6.50b, the rooms or spaces in a flow path are maintained at different pressures. For example, in the flow path, patient room a, anteroom c, and corridor b are maintained at +0.2, +0.17, and +0.14 inches w.g. In Fig. 6.50c, the space usage requires positive smoke removal from the space in case of a fire. In this situation,

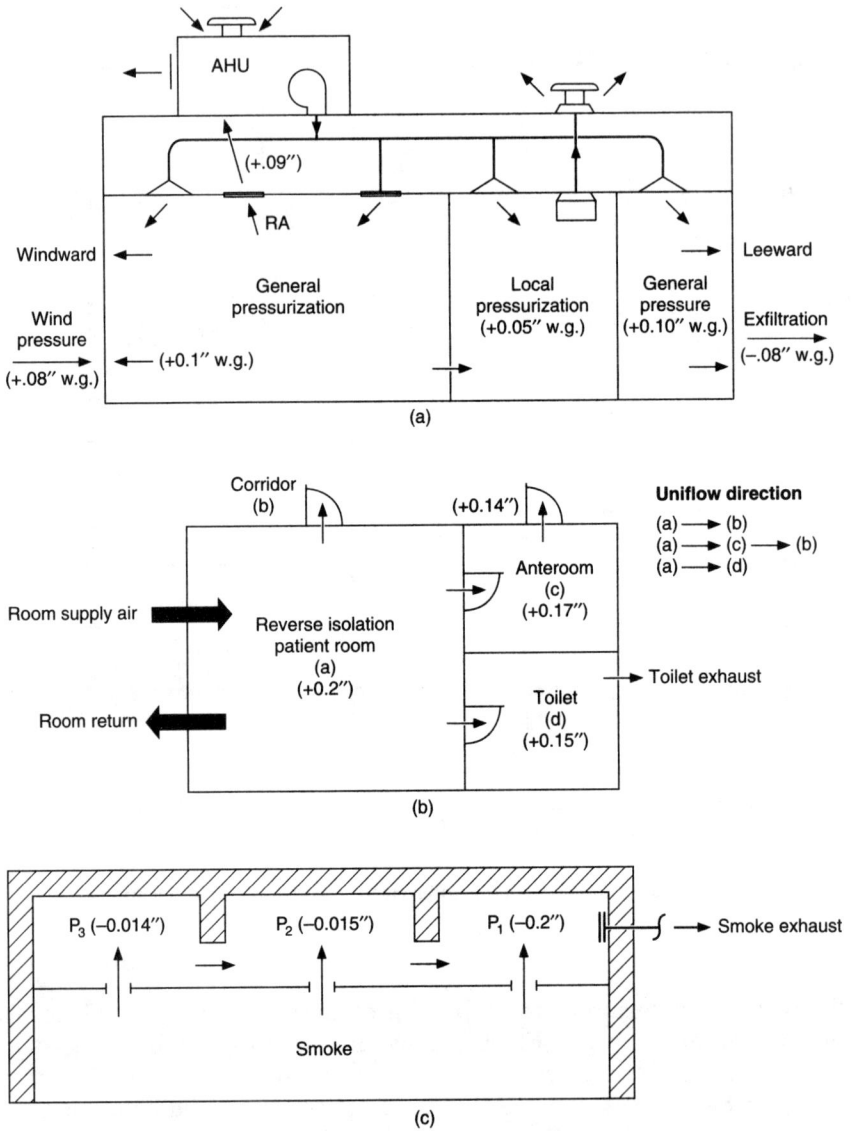

Figure 6.50 Types of building pressurization. (*a*) General and local pressurization. (*b*) Reverse isolation room unidirectional airflow pressurization. (*c*) Cold smoke control pressurization. (Negative ceiling chamber pressure)

the ceiling chamber must be maintained at a definite negative pressure to assure the removal of smoke at a certain predetermined rate.

6.5.3   Building and environmental factors affecting pressurization

Once the pressurization requirements based on space usage and pressurization objectives are determined, the system designer needs to evaluate the effects of building and environmental factors on pressurization. This is usually accomplished by mathematical models, each representing a specific pressurization situation. Many models have been proposed and applied in pressurization calculations. Two such models are shown in Fig. 6.51. Table 6.11 lists typical leakage flow rates in cfm for different types of leakage paths in buildings. By using these models, pressurization requirements are calculated. Then suitable mechanical systems, usually the building's HVAC systems, are employed to achieve the levels of pressurization required.

6.5.4   Impacts of VAV systems on building pressurization

Natural forces, mainly wind and stack effects, generate certain pressures inside the building. The magnitudes and variations of such pressures are determined by the three-way interaction of natural forces, building characteristics, and mechanical systems provided. In this three-way interaction, the environmental and building factors are generally given to the system designer, and the resulting natural building pressures may substantially deviate from the pressurization required by the design objectives. However, the designer can manipulate the natural pressures by the intervention of mechanical systems. Still, this mechanical intervention may become too expensive without appropriate architectural measures to tighten the structure to reduce air leakage. This is particularly true for VAV systems.

6.5.5   Analysis of system–building–environment interaction

As previously stated, the pressures in buildings are affected by three-way system, building, and environmental interaction. If the building is completely airtight, then the building pressure is simply a function of the air pumped in and out of the building. However, no building is completely airtight and a more realistic situation is shown in Fig. 6.52. In this schematic layout, various mechanical systems play a major role in determining the building pressure. Yet the building pressures are also affected by building construction, wind, and nonmechanical factors,

Door model (pressure and leakage by doors)

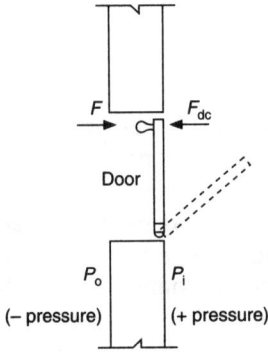

$$Pd = \Delta P = \frac{(F - F_{dc}) \times 2\,(W - D)}{K_d \times W \times A} \text{ ----- (1)}$$

$$Q_{pd} = (Q_{0.3}) \left(\frac{Pd}{0.3}\right)^{0.55} \text{ -------------- (2)}$$

where
$\Delta P = P_i - P_o$
F = Total door opening force (lbs)
F_{dc} = Force to overcome the door closer (lbs)
W = Door width (ft)
D = Distance from the door knob to the knob side of door (ft)
K_d = A constant (5.2)
A = Door area (ft²)
Q_{pd} = Air leakage at pressure difference of ΔP (cfm)
$Q_{0.3}$ = Air leakage at pressure difference of 0.3" w.g. (cfm)
Pd = Design pressure difference (in. w.g.)

F
F_{dc}
Door
P_o P_i
(– pressure)   (+ pressure)

Example

$F = 15$ lbs, $F_{de} = 10$ lbs
$W = 3$, $D = 0.25$
$A = 21$
$Q_{0.3} = 0.5$

$$P_d = \frac{(15 - 10) \times 2\,(3 - 0.25)}{5.2 \times 3 \times 21} = 0.084''$$

$$Q_{pd} = 10 \left(\frac{.084}{0.3}\right)^{0.55} = 5 \text{ cfm}$$

Window model (leakage through window by wind pressure)

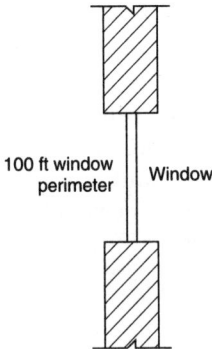

$$P_w = C_w \times k_w \times V \text{ ----------------- (3)}$$

$$Q_{wi} = (Q_{0.3}) \left(\frac{P_w}{0.3}\right)^{0.55} \text{ ------------ (4)}$$

where
P_w = Window pressure (in. w.g.)
C_w = Pressure coefficient, 0.8 for windward walls
    and –0.8 for leeward walls
K_w = A coefficient, 4.82×10^{-4} at 0.075 lb/ft² density
V = Wind velocity in mph
Q_w = Air infiltration at pressure difference of P_w (cfm)
$Q_{0.3}$ = Air infiltration at pressure difference of 0.3" w.g. (cfm)
P_w = Design window pressure difference (in w.g.)

100 ft window perimeter   Window

Example

$V = 15$ mph
$P_w = .0864''$ w.g.

$$Q_w = 0.5 \left(\frac{.0864}{0.3}\right)^{0.55} \times 100$$
$$= 25.2 \text{ cfm}$$

Figure 6.51   Mathematical models for building pressurization analysis.

TABLE 6.11   Leakage Flow at Various Pressure Differences

Leakage paths for directional flow

Source and type	Directional flow for pressure difference (inches w.g.)			
	0.05	0.10	0.20	0.30
Walls—Leakage allowed by specifications of National Association of Architectural Metal Manufacturers	0.02	0.03	0.05	0.06
Walls—Measured values of 8 multistory office buildings with sealed window and spandrel panels of precast concrete or steel.				
"Tight" construction	0.03	0.05	0.08	0.10
"Average" construction	0.09	0.15	0.24	0.30
"Loose" construction	0.19	0.29	0.46	0.59
Wall—Porous brick (8.5 in.) and lime mortar				
Plain	0.08	0.15	0.27	0.40
Wall—Frame construction with bevel painted siding, sheating, building paper, wood lath and three coat gypsum plaster.	0.002	0.002	0.004	0.005
Wall—8-in. concrete block—good workmanship	0.05	0.10	0.20	0.30
Walls—Elevator shaft				
Cast in place concrete	0.18	0.22	0.33	0.39
Concrete block	0.65	1.00	1.40	1.70
Walls—Stair shaft				
Cast in place concrete (parged)				
High value	0.003	0.004	0.006	0.008
Low value	<0.001	<0.001	<0.001	<0.001
Windows—Leakage allowed by specs.				
ANSI A 134.1 aluminum				
Maximum—sliding	0.28	0.41	0.60	0.75
ANSI A200.1 wood—all types	0.19	0.27	0.40	0.50
Door—Stairwell, 3.0 × 7.0 ft				
0.10 in. crack (door frame)	4.85	7.1	10.4	13.0
Door—Swinging				
⅛-in. crack	10.0	15.0	20.0	25.0
¼-in. crack	15.0	26.0	38.0	46.0
Floor to wall joint (direct method with pressure balance)	0.63	1.1	1.7	2.2

SOURCE: *Heating/Piping/Air Conditioning*, March 1982. Reprinted with permission.

and their interactions with mechanical systems can become considerably complicated. Systematic analyses to measure these interactions are necessary in many practical design situations. More detailed treatment of the subject is covered in Chap. 10, Sec. 10.6 Building Pressurization Analysis.

References

1. ASHRAE 55-1992 Standard, Definition of Thermal Comfort, p. 3.
2. Fanger, P. O., *Thermal Comfort,* Denmark Technical Press, Copenhagen, Denmark, 1970.
3. ASHRAE Handbook 1985, Fundamentals, Chap. 8.
4. Fountain, M. E., Arens, E. A., "Air movement and thermal comfort," *ASHRAE Journal,* August 1993.

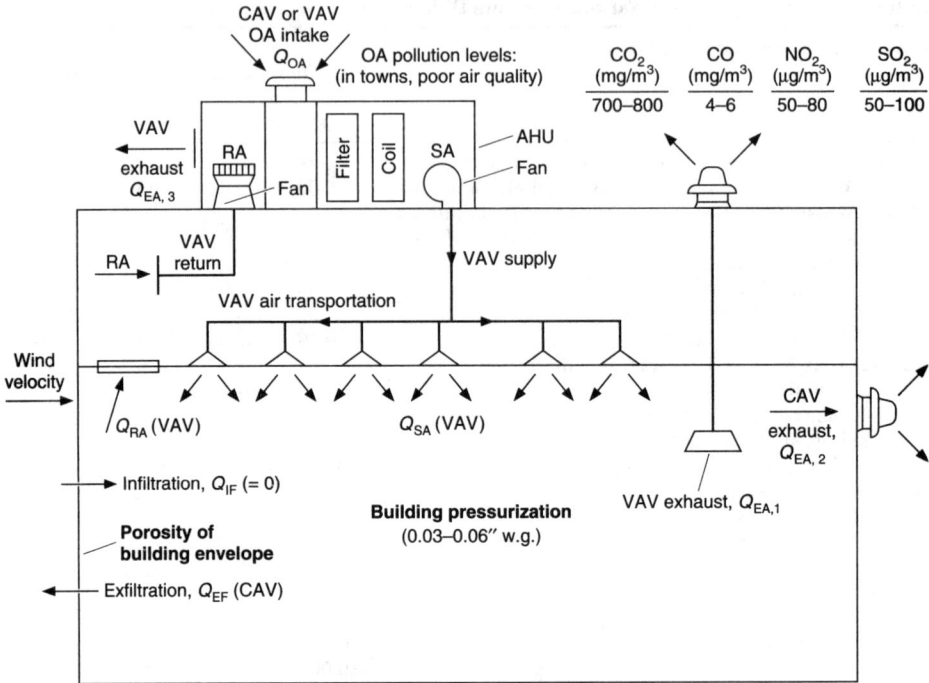

Flow balance and building pressurization

$$Q_{SA} - Q_{RA} = Q_{EF} + Q_{EA,1} + Q_{EA,2} = Q_{OA} - Q_{EA,3}$$

• To pressurize building, exfiltration Q_{EF} must be large enough to maintain a reasonable building pressure.

Figure 6.52 Mechanical systems and building pressurization.

5. Copyright © 1992. Electric Power Research Institute. EPRI TR-101480. *Determination of the Operational Characteristics of Cold Air Diffusers, Figs. 6.4 through 6.7.*
6. Liu, R. T., Raber, R. R., Yu, H. S., "Filter selection on an engineering basis," *Heating/Piping/Air Conditioning,* May 1991, p. 41.
7. Tanabe, S., Kimura, K., "Thermal comfort requirements under hot and humid conditions," *Proceedings of First ASHRAE Far East Conference on Air Conditioning in Hot Climates, Singapore,* ASHRAE, Atlanta, Georgia, 1987.
8. Schaffer, M. E., *A Practical Guide to Noise and Vibration Control for HVAC Systems,* ASHRAE, Atlanta, 1991.
9. Stein, B., Reynolds, J. S., *Mechanical and Electrical Equipment for Buildings,* 8th ed., John Wiley & Sons, Inc., New York, 1992, chap. 26.
10. Grimm, N. R., Rosaler, R. C., *Handbook of HVAC Design,* McGraw-Hill Publishing Co., New York, 1990, chap. 49.
11. *Practical Noise Control Guide,* Japan Institute of Architects, Japan, 1975, p. 109.
12. International Standard ISO 7730-1984(E).

Design Considerations

7.1 Architectural versus Mechanical Design Considerations

The architect and the system designer share a common goal of designing quality buildings with functioning systems. Yet they may have considerably different viewpoints and concerns for achieving the same objective. The architect is more interested in economical space utilization, aesthetic appearance, pleasant indoor environment, and overall economy of the project. The system designer is concerned about more urgent mechanical design considerations such as code and standard compliance, system reliability, maintainability and constructability, and above all, space and technical coordination.

These somewhat incompatible architectural and mechanical considerations should be merged into an integrated design at an early stage—preferably during the schematic and preliminary design phases—and applied throughout the entire design process. These integrated considerations should include (1) architectural space planning for mechanical floors, rooms, vertical shafts, and horizontal distribution spaces, (2) comfort, IAQ and acoustics, (3) life safety and other code/standard compliance, (4) proper and efficient functioning of mechanical systems, and (5) system and component selections and their impacts on building initial and operating costs. The interrelationship among architectural, mechanical, and integrated design considerations is shown diagrammatically in Fig. 7.1, item A, Predesign Thought Process. This thought-merging process should take place before the start of item B, Integrated Building/System Design Process.

The proper formation and implementation of integrated design considerations affects the quality and performance of mechanical systems in buildings. Often, system design is based on standardized, prescribed

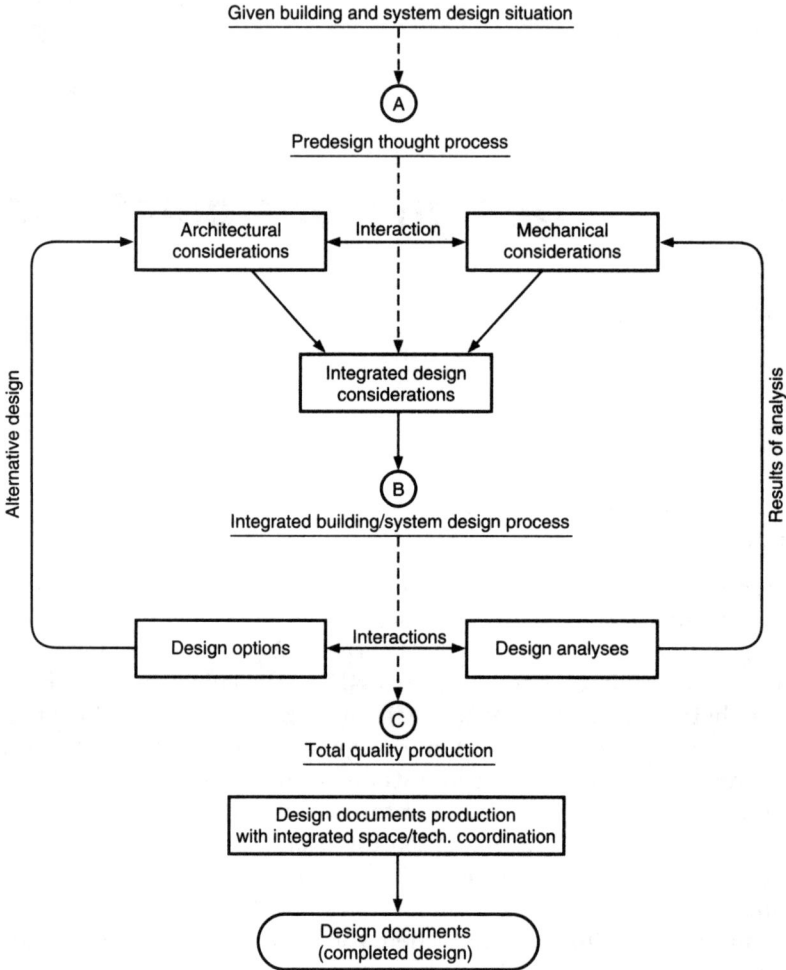

Figure 7.1　Integrated and total quality production.

design considerations irrespective of the specific design situation given to the system designer. This approach frequently results in defective design and even operational problems. Figure 7.1 shows diagrammatically the importance of integrated design considerations. It is a fluid process considerably influenced by each specific design situation, design options selected, and tools available for building/system analyses. This interactive relationship between the predesign thought and actual design process is illustrated in items A and B of Fig. 7.1, and further explained in Fig. 7.2, which shows details of the formation and application of integrated design considerations.

7.2   Integrated Design Considerations

7.2.1   Architectural space planning
for mechanical systems

The architect's main concerns in mechanical space planning are (1) location and (2) size of the space. However, for truly integrated design, several other considerations need to be added: (3) relative location of mechanical rooms and shaft spaces, (4) selection of building materials and construction methods for noise and vibration control, and (5) access to mechanical spaces for system operation and maintenance. Some of these items are more important than others, and their relative importance—and therefore the amount of resources to be allocated—will be determined by each specific design situation and system options to be considered. For example, for a rooftop VAV unit design option in a small office building, item 4 may be the most important consideration, while other items are less critical for system design.

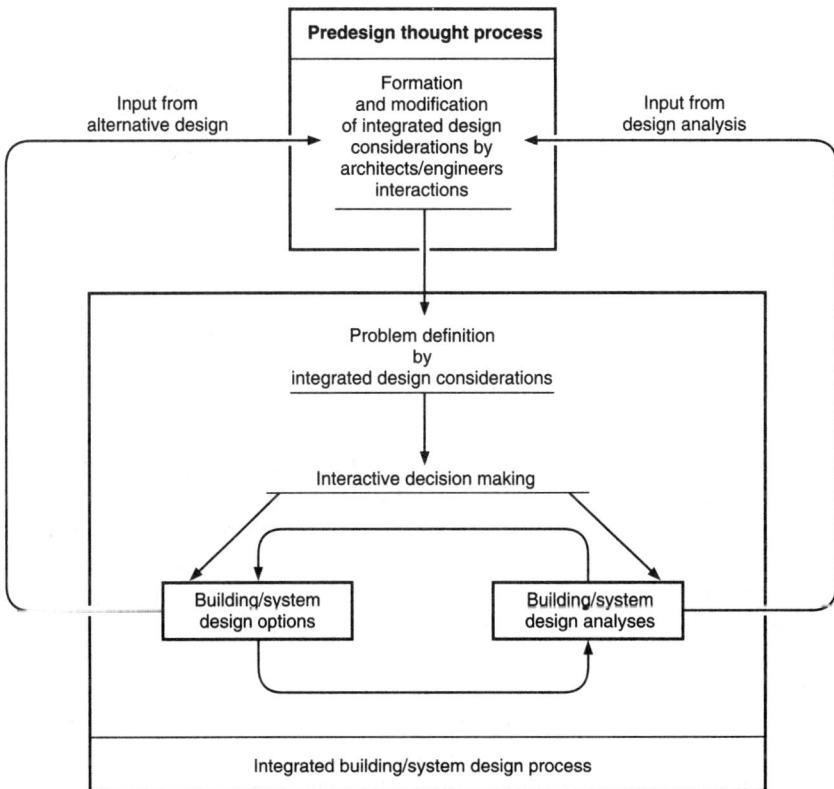

Figure 7.2   Integrated thought process for quality design.

7.2.2 Comfort, IAQ, and acoustics considerations

These considerations are often intimately related to system design and deserve special attention, especially for VAV system design. These design parameters are usually treated simplistically and mechanically using prescriptive design guidelines or standards. For example, the IAQ parameter is treated simply as a ventilation problem where a certain cfm/person figure is assigned to each room and a total ventilation air quantity in cfm is calculated by adding cfm assigned to each space. This simplistic approach totally ignores the fact that IAQ is affected by many other parameters such as (1) variation in ventilation air supply, (2) ventilation effectiveness, (3) outdoor air and room contamination levels, (4) local exhaust, (5) filtration efficiency, and others. Again, the relative importance of these parameters should be evaluated before committing design resources. Often, a preliminary design analysis for each design option can reveal the relative importance of various design parameters or even detect a need to consider a special parameter (for example, use of an acoustically improved ceiling or wall). The design considerations in this category require the highest degree of interaction between building/system design and analysis. It is therefore important for the system designer to understand how to select proper design considerations to achieve system design for total environmental quality.

7.2.3 Compliance with codes and standards

Protecting the public's health and safety is the primary purpose of codes and standards. Compliance with codes and standards is common to all design situations and should be addressed early in the schematic design phase. Codes and standards can influence each phase of the design process from schematic design to document preparation for final design. Without early identification of the proper codes and standards, drawings and specifications are frequently reworked at a later stage of the design process, which can result in undesirable cost and time addition.

Compliance with codes and standards is generally considered a standard design consideration common to all design situations, and can be summarized in a table or two. Table 7.1 is a sample of code/standard tabulation for VAV system design. In this case, compliance is simply a matter of following these codes and standards. However, the VAV system designer is cautioned to evaluate the impacts of air volume variation on the compliance requirements. For example, merely supplying

TABLE 7.1   Codes and Standards for VAV System Design

- The BOCA National Building Code—1990
- The BOCA Mechanical Code—1990
- National Fire Protection Association (NFPA)
- American Society for Testing and Materials (ASTM)
- American National Standards Institutes (ANSI)
- Sheet Metal and Air Conditioning Contractors' National Association (SMACNA)
- American Society of Mechanical Engineers (ASME)
- Air Conditioning and Refrigeration Institute (ARI)
  Acoustics Application/ARI Standard 885-90
- Underwriters Laboratory (UL)
- All applicable local codes
- Ventilation for Acceptable Indoor Air Quality—ASHRAE Standard 62-1989
- Thermal Comfort/ASHRAE Standard 55-1992
- Room Air Distribution/ASHRAE Standard 113-90
- ISO International Standard 7730

the total amount of outdoor air required by the standard does not guarantee the amount specified will reach conditioned spaces. For certain VAV systems, the amount of outdoor air actually supplied to each conditioned space varies continuously, often at a level far below the standard specified minimum quantity.

Code and standard compliance for VAV systems is not a straightforward process and must be analyzed individually for each design situation/option combination, using appropriate models for simulation. This review process must be carried out before the end of each design phase, and inadequacies of the design discovered must be corrected before the final design documents are submitted to the owner or its agency. When the analyses and simulations to prove the compliance with codes and standards are included in the final design documents, such documentation will greatly enhance the credibility of the design team.

7.2.4   Proper and efficient system functioning

Certain design considerations are reserved for the proper and efficient functioning of air systems in general and VAV systems in particular. Good examples are (1) building pressurization, (2) unidirectional flow control, (3) control stability, (4) efficient air transportation, (5) space air distribution, (6) duct static pressure variation, and (7) system reliability and redundancy. These design considerations are discussed in the following chapters and sections:

1. *Building pressurization:* Part 2, Chap. 6, Sec. 6.5

2. *Unidirectional flow control:* Part 2, Chap. 6, Sec. 6.5

3. *Control stability:* Part 2, Chap. 11, Sec. 11.9

4. *Efficient air transportation:* Part 2, Chap. 6, Sec. 6.3

5. *Space air distribution:* Part 2, Chap. 6, Sec. 6.3

6. *Duct static pressure variation:* Part 2, Chap. 11, Sec. 11.8

7. *System reliability and redundancy:* Part 2, Chap. 11, Sec. 11.10

7.2.5 System and component selection

Certain design considerations are specifically associated with the system and components selected for the design. This is because each system has its own unique layout and operating characteristics which result in specific space and technical requirements. It is important for the system designer to develop a personal database of special design considerations, each associated with a specific VAV system. Figure 7.3

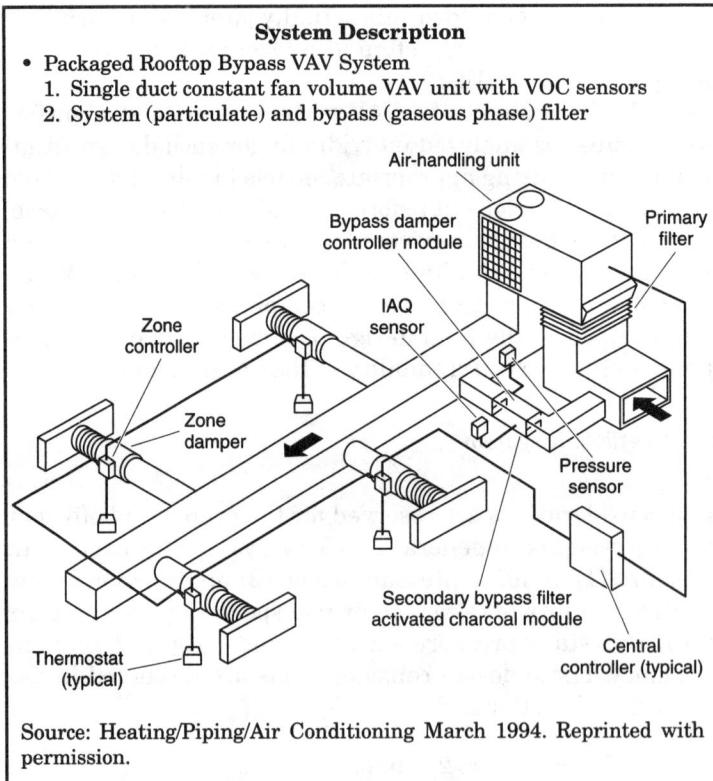

System Description
- Packaged Rooftop Bypass VAV System
  1. Single duct constant fan volume VAV unit with VOC sensors
  2. System (particulate) and bypass (gaseous phase) filter

Source: Heating/Piping/Air Conditioning March 1994. Reprinted with permission.

Figure 7.3 Selection of specific design considerations.

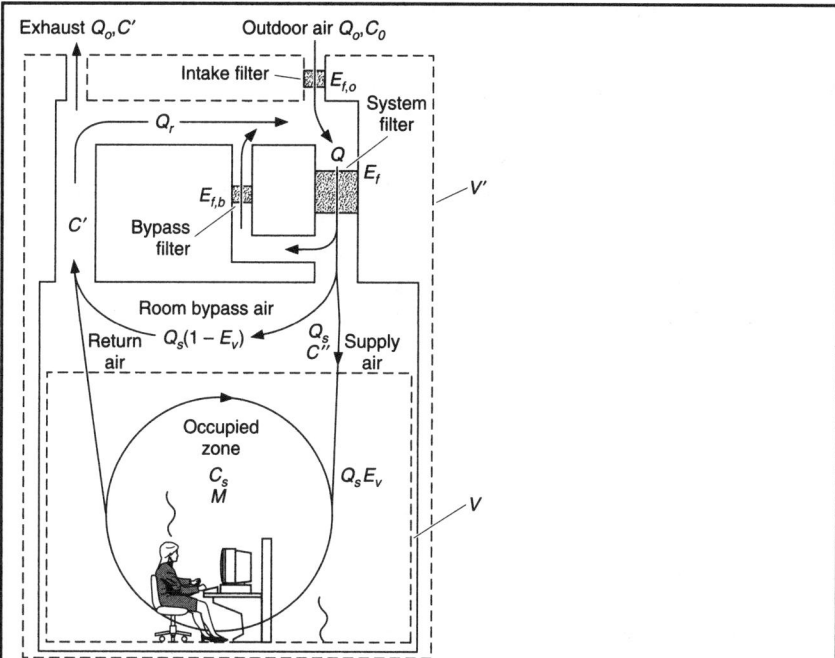

Constant volume HVAC system with a VAV by-pass loop.

Design Situation	Design Consideration
• Office Applications	1. Analysis of local air contaminant concentration 2. Location of VOC sensors 3. Air cleaning strategy, system and bypass filter efficiencies 4. Ventilation effectiveness 5. Local exhaust/contaminant control 6. Building pressurization 7. Demand-controlled ventilation
• Laboratory Applications	Items 1 through 6 above plus 7. Unidirectional flow control

Figure 7.3 *(Continued)*

is an example of design considerations associated with a specific VAV system. A personal knowledge base of this nature can be gradually compiled based on each individual's own design experiences and other information sources. The databases of these special design considerations can assist the system designer in identifying important design issues to be considered at each design phase.

7.3 Design Guidelines versus Design Considerations

Whenever an air system design is attempted, either design guidelines or design considerations are used to guide the system designer. Design guidelines are usually a set of design policy statements which have authority over design activities. Guidelines can be a few lines, key-point statements, or an extensive several-hundred-page statement in the form of a design manual or book. For a conventional design approach, design guidelines are universally applied to all design situations (Fig. 7.4). Sometimes design guidelines are selectively used. In this case, the judgment is the designer's own and usually based on personal experience and knowledge of similar design situations. Generally, no analysis or simulation is performed to justify the omission of certain guidelines.

Design considerations are, on the other hand, much wider in scope. Personal experience, rules of thumb, design guidelines and standards, integration techniques, computer simulations, and analyses are used as the basis for decision making to initiate a design. Design considerations differ from design guidelines in one important aspect. They are not applied to any design situations indiscriminately. As illustrated in Fig. 7.2, applicable design considerations are selected by going through the predesign thought process, which screens various design considerations using personal experience, individual and shared knowledge bases, and analytical tools. As a result, the integrated design approach for decision making is a better method in attaining high-quality building and system design. In theory, the conventional design approach (Fig. 7.4) can also achieve the same results. However, it would be more time-consuming and resource intensive, if it is to attain the same level of design quality.

7.4 Experience-, Knowledge-, and Simulation-Based Design Considerations

As explained in Sec. 7.3, design considerations are broader in scope, more selective, and more design-situation-oriented than conventional, prescriptive design guidelines. They are generally derived from three sources: experience, knowledge, and manual or computer simulation. Each source has its own uniqueness in formulating design considerations and produces different impacts on the quality of building and system design.

7.4.1 Experience-based design considerations

Experience is considered a valuable asset to design professionals, and the dos and don'ts of design experience serve as effective design con-

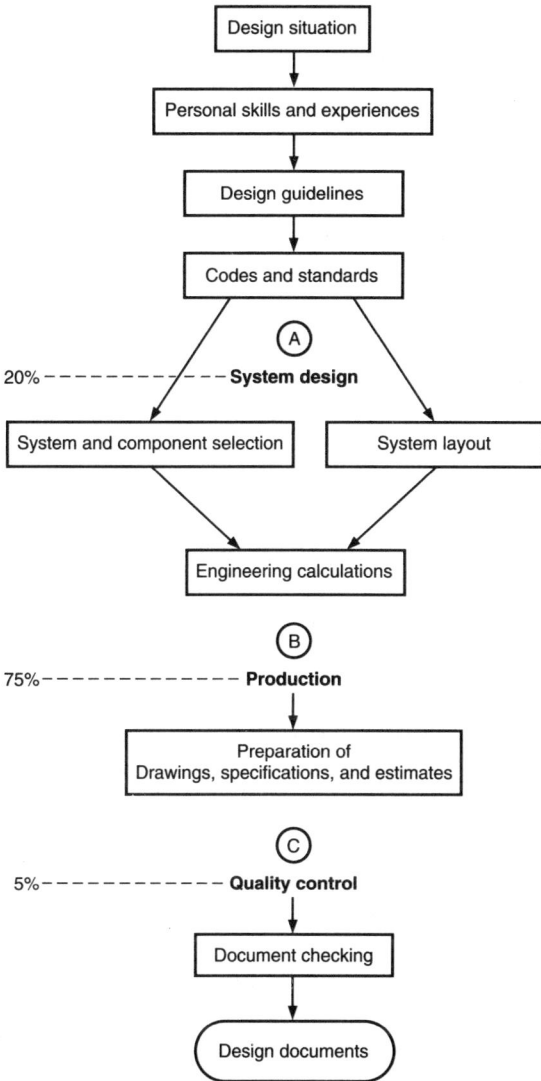

Figure 7.4 Conventional process for system design and production.

siderations to avoid certain design pitfalls. Still, experience is not always applicable to a particular situation. On the contrary, design and resulting field experiences are often extrapolated and universalized to apply indiscriminately to all design situations. Figure 7.5 shows two cases of misapplication for experience-based design considerations.

Nevertheless, experience-based design considerations are quite useful for certain design situations, especially for those situations where

Design Situation 1
- *Experience:* Low total supply air to the conditioned spaces is perceived to be a main source of occupant discomfort.
- A specific *design consideration* in this case is to prevent the total space air circulation from dropping below an experience guided value, say 0.4 cfm/ft^2 for certain applications.
- Recent studies and also current comfort standards indicate that the minimum airflow requirement for comfort is influenced by many factors, and there should be no minimum airflow, provided that the other comfort and IAQ criteria are satisfied.

Design Situation 2
- *Experience:* Lining of the low-pressure ductwork and flexible ducts is considered a good design practice for effective noise attenuations.

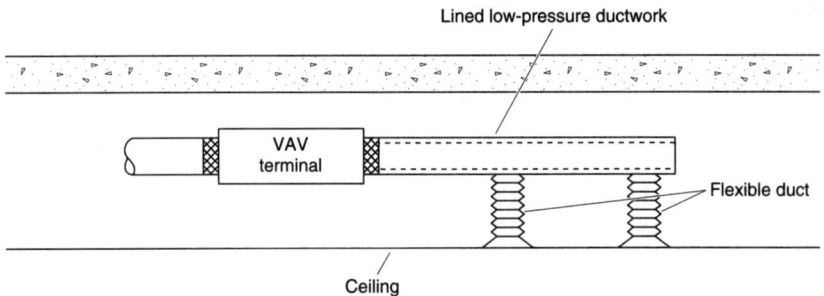

- Lined flexible ducts show an abrupt reduction in insertion loss from 2000 to 4000 Hz octave bands, which means a sudden noise increase perceived as an objectionable hissing sound.

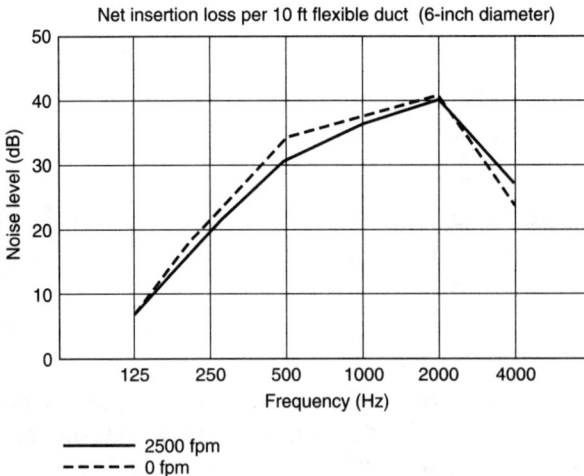

- Minimize the length of flexible ducts.

Figure 7.5   Misapplication of experience-based design considerations.

building and system designers' major concerns are space allocation and coordination among various disciplines. Table 7.2 illustrates a case in point. For this particular design situation, designers' past experiences are the decisive factors in developing a quality layout. However, in some other design situations where design decisions are more readily impacted by physical laws, knowledge-based design considerations should be rigorously applied to solve design problems.

TABLE 7.2   Application of Experience-Based Design Considerations

Design Situation 3
* Architect needs to conceptualize a floor plan showing utility core space allocation including mechanical/electrical rooms (if any).
* System designer provides drawings for duct distribution and architectural requirements for total environmental quality.

Design Considerations
* Architectural: Least interference with architectural functions, min. floor and ceiling space requirements. Provides a pleasant indoor environment. Meet IAQ requirements.
* Mechanical: Environmental quality considerations (comfort, IAQ, noise). Symmetry of mechanical room/shaft layout desirable. Energy efficiency, constructibility, serviceability of the layout (These considerations are influenced by the past experiences.)

Architect / System Designer Interactions
* Design interactions between architect and system designer to develop a best core plan jointly (usually a few trial designs are required).

Best Design Solutions
* Experience-based design considerations are most suitable for design situations where the major concern is space allocation and coordination and also where the situations are less sensitive to environmental quality factors such as thermal comfort, IAQ and noise problems.

7.4.2   Knowledge-based design considerations

Engineering knowledge on HVAC design in general and to a certain extent on VAV system applications is widely scattered in books, journals, technical publications, manufacturers' engineering data, and of course personal experiences. Yet often this knowledge base is not systematically organized for specific design situations. Before the commencement of actual design, the system designer should list applicable knowledge-based design considerations, keeping in mind that the adoption of proper design considerations for the given design situation can greatly enhance the quality of the design.

Table 7.3 summarizes how knowledge-based design considerations are applied to a specific design situation, and how the architect and the system designer interact with each other in this design situation, and finally, the best design solutions that can be evaluated by knowledge-based design considerations.

TABLE 7.3   Application of Knowledge-Based Design Considerations

Design Situation 4
- Architect needs to know the feasibility of installing multiple, ceiling-mounted VAV air-handling units in a large office area.
- System designer provides drawings for HVAC design and architectural requirements for total environmental quality.

Design Considerations
- Architectural: Least interference with structural members. Maximum ceiling height for a given floor-to-floor height. Life safety considerations (fire and smoke zonings, smoke control, etc.) Building envelope thermal efficiency.
- Mechanical: Shaft and mechanical room constructions. Sufficient space for proper system functioning and noise prevention. System energy efficiency and ventilation effectiveness. Ceiling construction for acoustical considerations.

Architect / System Designer Interactions
- Knowledge Base: Duct design, construction, and balancing know-how. Engineering knowledge on architectural acoustics, noise control, ventilation fundamentals, energy calculations, duct air pressure, and flow controls.
- Design Interactions: Knowledge bases are applied to evaluate the mechanical soundness of the design proposed. If deficiencies are found, propose necessary improvements to architect.

Best Design Solutions Evaluated by Knowledge-Based Considerations
- Acoustical performance of building materials and system components, such as duct fittings, terminals, and outlets.
- Ventilation design.
- Building pressurization.
- Air movement inside the building.
- HVAC system selection.
- Energy conservation and optimization.
- Thermal storage design (both ice and chilled/hot water).

7.4.3 Simulation-based design considerations

Simulation is a simple concept. Consider a case of how a system design is optimized. First, the designer describes the system mathematically or constructs a mathematical model for a specific simulation purpose (for example, a model for noise, IAQ, or comfort analysis). This model can be a simple manual model or a complex computer model. Which model is used for simulation is immaterial. The key issue is how realistic a simulation should be for the purpose contemplated, and the answer to this question will automatically decide which engineering tool—manual or computerized—should be used for simulation. For example, the outdoor temperature can be simulated by actual hour-by-hour weather data, or just by a set of summer and winter design conditions. One is simple and the other is complicated, but both are unmistakably simulations.

The question of how sophisticated a simulation should be depends on many factors. Most of these factors are situation dependent and must be evaluated individually for each specific design situation. Mathematical models of varying sophistication are readily available for the system designer to construct a simulation for most applications. The key issue here is the ability to select a proper model, which is generally a function of familiarity with mathematical model construction and application as well as experience in applying simulation techniques in practical design situations.

Figure 7.6 illustrates an example of applying a computer simulation technique in an acoustical design situation.

Design Situation 5

- After going through knowledge- and experience-based design considerations, and having made necessary modifications and adjustments, the proposed layout is essentially a sound design. However, for certain aspects of design, it may be advisable to go through simulation-based technical evaluation. Typical examples are multiple-path noise transmission and controllability of the system proposed. The following example identifies the worst noise path by a computer-simulated design technique.

System Description: 1. Central ducted supply air system
                         2. Ceiling plenum and shaft return air system

Major Noise Paths: 1. Supply fan–duct–diffuser–space: ①–②–③–④
                         2. Space–ceiling plenum–return air shaft–RA fan: ⑥–⑦
                         3. Supply duct–RA shaft wall–space: ⑤–⑦–⑥

Worst Noise Path: Computer multiple-noise path analyses identify the space–ceiling plenum–return air shaft–RA fan as the worst noise path when the space is the nearest to the RA fan.

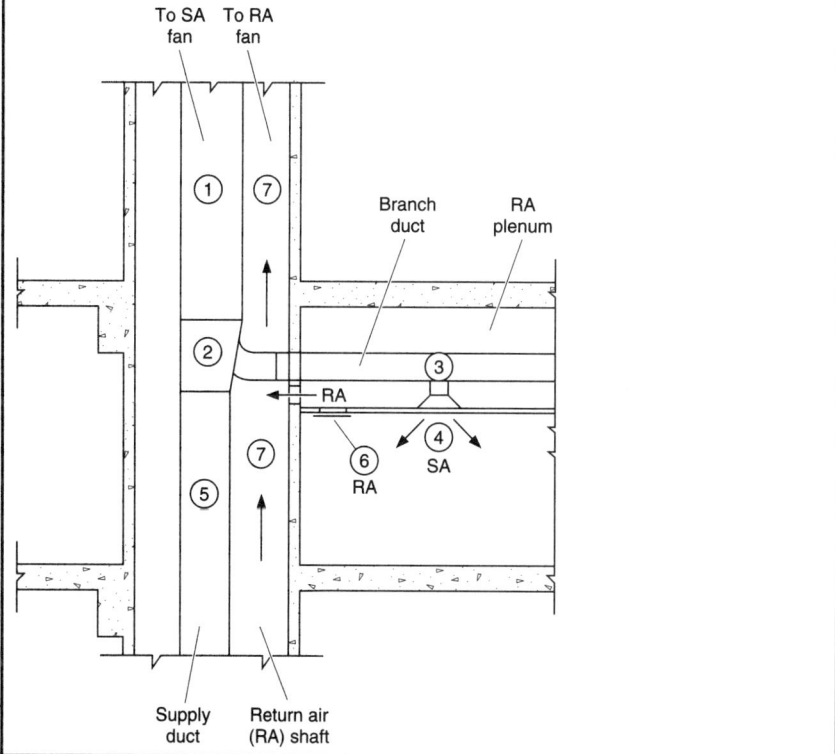

Figure 7.6 Acoustical design for an acoustical design situation.

The greatest advantage of using computer-aided simulation is its versatility in simulating a variety of situations or its ability to demonstrate the combined effect of multiple design parameters on the outcome of simulation and therefore the quality of the design. In other words, simulation will allow the user to gain "simulated experiences" similar to the ones actually encountered in the field, or to confront situations that rarely happen but *could* happen as a result of certain special parameters or conditions built into the design.

Manual simulations are not as powerful as computerized simulations. Still, for most design situations, simple manual simulations are sufficient to produce a desired result for the intended purpose. Whether manual or computerized, simulation-based design considerations can supplement the deficiency of experience- and knowledge-based design considerations, and enable the system designer to adopt the most appropriate design considerations for total quality design.

8

System Selection

8.1 Factors Affecting System Selection

Variable air volume (VAV) systems were initially developed for cooling applications and designed to maintain a constant air temperature while varying the supply air volume. Although several heating options and VAV terminals were added later, the problems associated with varying air quantity, such as deficiency in heating, ventilation, and fluctuating noise and comfort levels, have remained essentially the same and become the major issues in VAV system selection. These, plus some other facts affecting the selection of VAV systems, are covered in the following sections.

8.1.1 Applications

VAV systems are ideal for commercial and institutional buildings where good temperature control and high energy efficiency are the major considerations for system selection. The main drawbacks are poor heating, ventilation, and humidity control under certain weather, load, and occupancy conditions. These drawbacks are caused by the fact that any VAV system must vary air quantity for temperature control, and the system designer needs to balance the advantages (ease of temperature control and high energy efficiency) against the disadvantages (poor heating, ventilation, and humidity control in selecting and applying VAV systems).

8.1.2 Heating alternatives

VAV systems were originally developed for cooling-only applications, and heating needs were met by several heating options in the form of independent perimeter systems. These systems are either heating only

Perimeter heating by hot water, electric, or steam radiators (See Fig. 8.3)

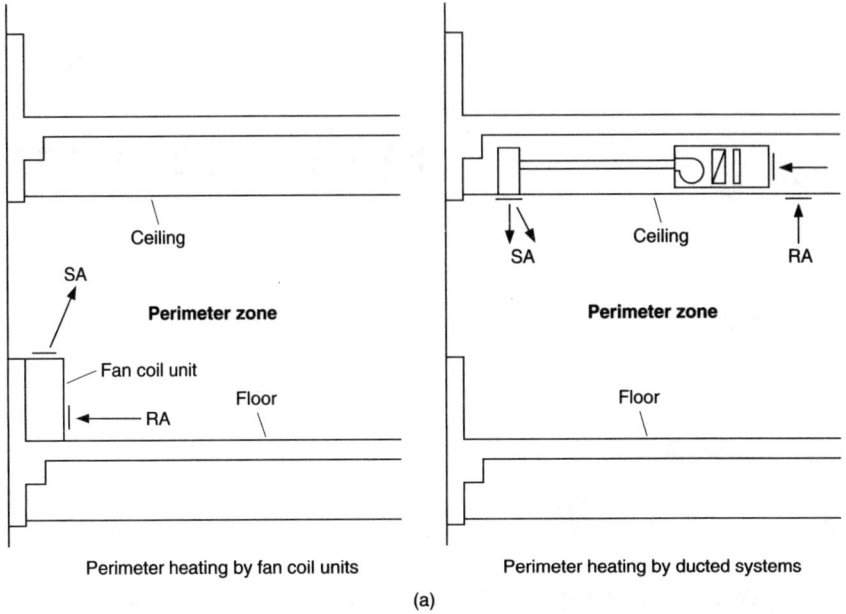

Perimeter heating by fan coil units   Perimeter heating by ducted systems

(a)

Perimeter heating by CAV fan-powered terminals

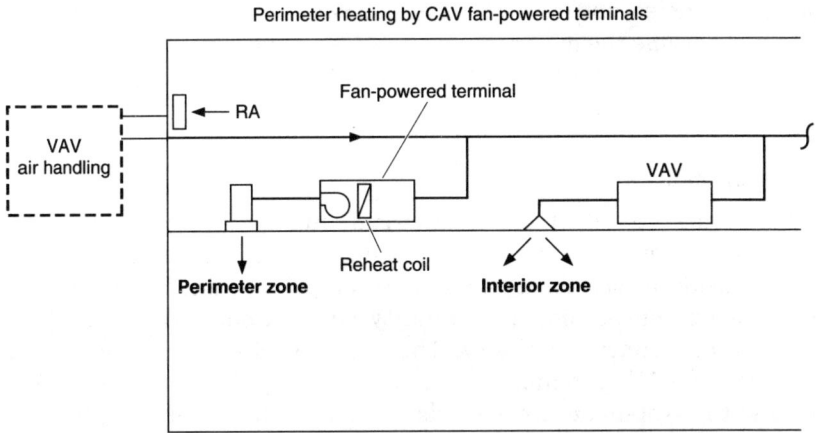

Supplemental heating by reheat coils in VAV or fan-powered terminals (See Fig. 8.2)

(b)

Figure 8.1   Heating options for VAV systems. (*a*) Independent perimeter heating options; (*b*) Supplemental heating option.

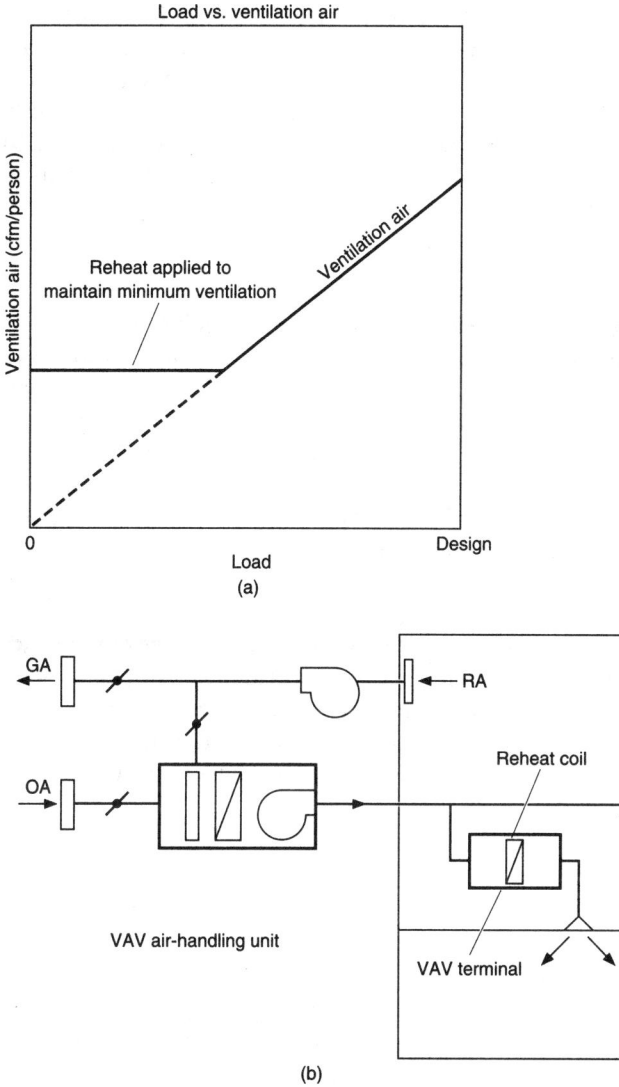

Figure 8.2 Supplemental heating to satisfy ventilation requirements.

or a combination of heating and cooling systems (Fig. 8.1a). Later, supplemental heating options were added to VAV systems to provide overhead heating and to eliminate the use of an independent perimeter heating system (Figs. 8.1b and 8.2).

When selecting a heating alternative, the system designer is cautioned to pay close attention to the possibility of excessive reduction in ventilation in the perimeter zone of the building, particularly in cold climates. Office zones are typically designed for 100 ft^2 per person occupancy, or 0.2 cfm per ft² floor area ventilation air supply. Unless the heating system is carefully selected and controlled, the ventilation air supply may be reduced to less than 0.05 cfm per ft² or 5 cfm per person (100 ft²/person assumed) in the perimeter zone, which is not acceptable from an IAQ viewpoint. Often it is necessary to add an artificial load to increase the air quantity to satisfy the ventilation requirement. This is illustrated in Fig. 8.2a.

8.1.3 Ventilation alternatives

As mentioned in preceding sections, the inherent disadvantage of VAV systems is the continued fluctuation in ventilation air rates supplied to various spaces. In addition, the change in ventilation rates is purely a function of space load variations and is independent of space occupancy variations. Consequently, under certain load and occupancy conditions, the ventilation rate per person may drop far below the 15 cfm per person specified by the ASHRAE IAQ procedure. However, certain types of VAV systems perform better than others in dealing with this problem.

Simple VAV system with perimeter radiation which is under zone or per building exposure control. This system (Fig. 8.3) works fine as long as load and occupancy fluctuation are relatively minor and extremely high occupancy density areas are not mixed with low-occupancy areas and connected to a common VAV system. The system designer should con-

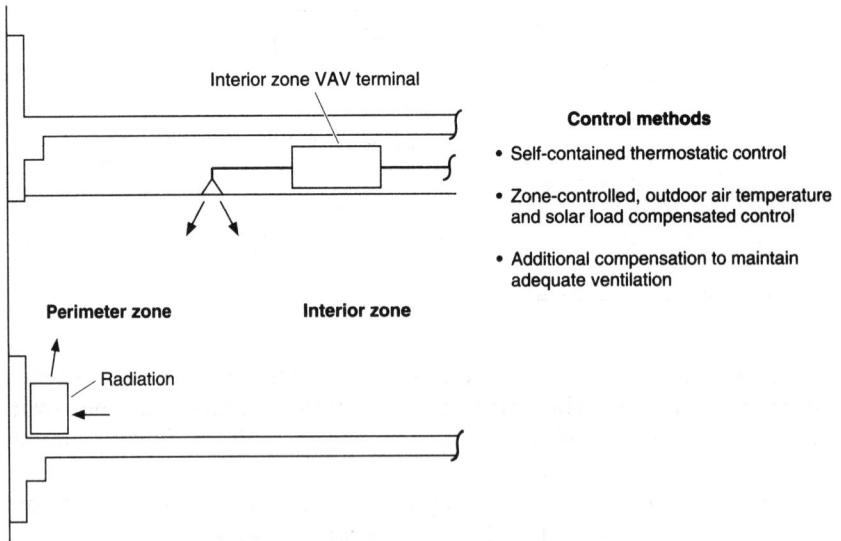

Figure 8.3 Simple VAV system with perimeter radiation.

duct a ventilation analysis both at full and partial load conditions for ventilation rate variation.

VAV reheat system with an independent perimeter system. This system provides an improvement in ventilation quality over the preceding system by adding reheat capability to each VAV terminal. Reheat supplies an artificial cooling load to each zone, thus maintaining a certain level of ventilation. This is illustrated in Fig. 8.4.

VAV reheat with constant zone air circulation. This system maintains a constant air circulation in each VAV zone by an individual fan-powered terminal. However, the ventilation rate is not affected by this arrangement. On the contrary, when fan-powered terminals are used for perimeter heating, the primary air supply is often completely shut off, thus reducing the ventilation rate to zero (Fig. 8.5). In this case, ventilation is totally dependent on interior zone VAV terminals and may create a serious IAQ problem.

Ventilation options combined with conventional VAV system. Essentially there are three levels of ventilation options to improve the ventilation air supply to each VAV zone:

Level 1: Transfer the air from the adjacent areas whenever a high rate of ventilation is required, and exhaust the air introduced or return it to the air-handling unit (Design 3 in Chap. 6, Fig. 6.6). The main advantage of this option is economy. It is relatively inexpensive and

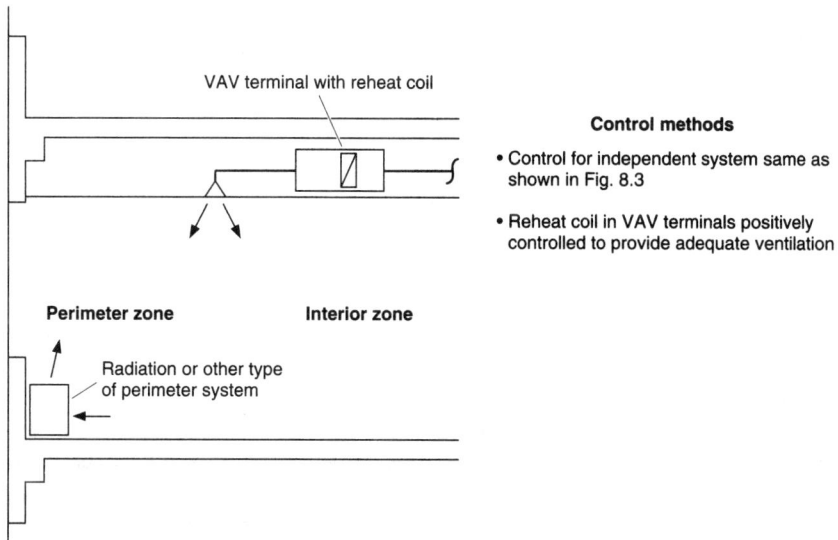

Figure 8.4 VAV reheat system with an independent perimeter system.

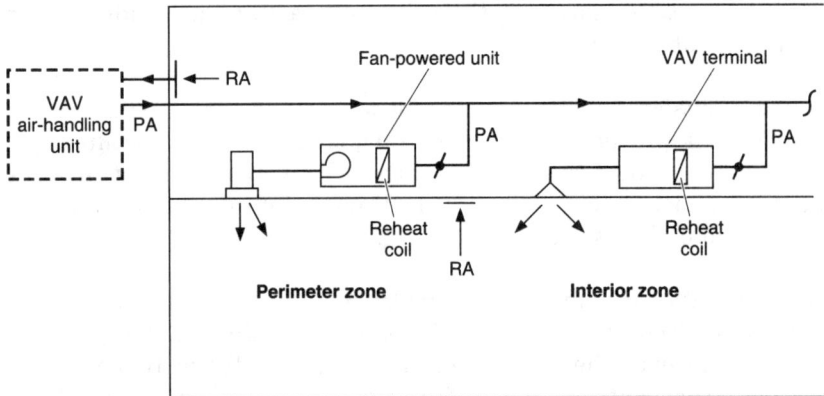

Heating cycle

- Primary air supply (PA) to perimeter zone fan-powered unit is completely shut off. Fan-powered unit circulates heated air without ventilation.

- Interior zone VAV terminals continue to supply primary air, reheated if necessary to maintain suitable ventilation.

- In this scheme, perimeter zones receive ventilation air from interior zones and may not be adequately ventilated.

Figure 8.5   VAV reheat with constant zone air circulation.

energy efficient to run. The drawback is that the air introduced may be contaminated to a certain extent.

Level 2: Provide a separate VAV unit for high-occupancy-density areas (conferences, classrooms, auditoriums, restaurants, etc.). (See Design 1 in Chap. 6, Fig. 6.5.) This is a relatively expensive option. However, it solves a potential ventilation problem, though it is essentially a design compromise because it still combines zones with various load and occupancy areas and supplies ventilation air from a common, centralized air-handling system.

Level 3: Provide ventilation air directly to each VAV zone from an independent outdoor air-handling system. This is the most effective way of meeting the IAQ-based ventilation requirements. Design 2 in Fig. 6.5 shows an example of this ventilation approach for a major office building.

Selection of a ventilation alternative is a subset to VAV system selection. First, the system designer selects a VAV system most suitable for the given design situation, considering factors such as heating alternatives, system size, environmental quality, system serviceability, and budget constraint. Then the designer determines the feasibility of upgrading ventilation efficiency by selecting one of the three ventilation options discussed above. A simple ventilation analysis usually reveals the need to improve ventilation.

8.1.4 System size

Small systems with single or multiple VAV units per floor. System size has a considerable impact on the selection of VAV systems in this case. If the building is a single-story or low-rise structure with 10,000 ft^2 or less area per floor, a packaged VAV system with or without reheat would probably be the best choice for the building. The packaged units could be a conventional constant temperature, variable flow type or a variable temperature, variable flow type. This system can be easily combined with various VAV terminals, including fan-powered terminals. Also, the system is quite compatible with all the independent perimeter systems. The packaged units are typically in the 5- to 20-ton-capacity range. Floor area, space usage, and unit size are some of the key factors in determining number of units per floor. The packaged VAV units can be either chilled water or direct expansion (DX) type (Fig. 8.6). A chilled water VAV system consists of air-handling units served by a central chilled water plant. A VAV/DX system, on the other hand, is a self-contained system with a compressor, DX coil, control and water- or air-cooled condenser (Fig. 8.7). Which type to choose, chilled water or direct expansion, depends upon each specific design situation.

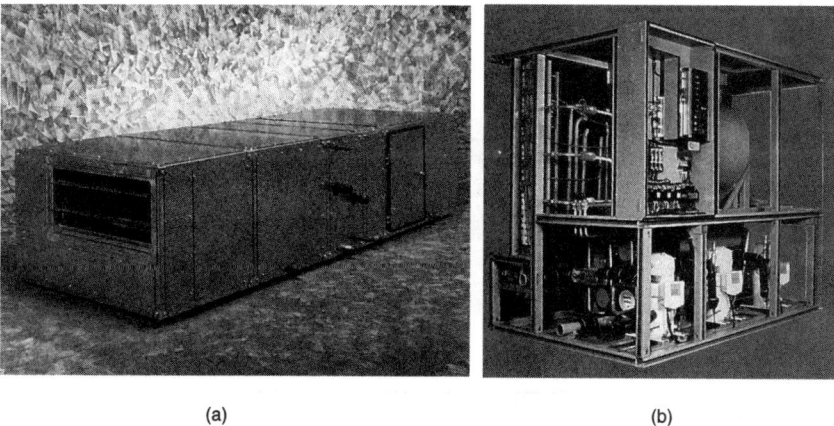

(a)   (b)

Figure 8.6 Packaged VAV units. (*a*) Chilled water VAV air-handling unit; (*b*) DX/VAV packaged unit. (*Courtesy of Trane Co. Reprinted with permission.*)

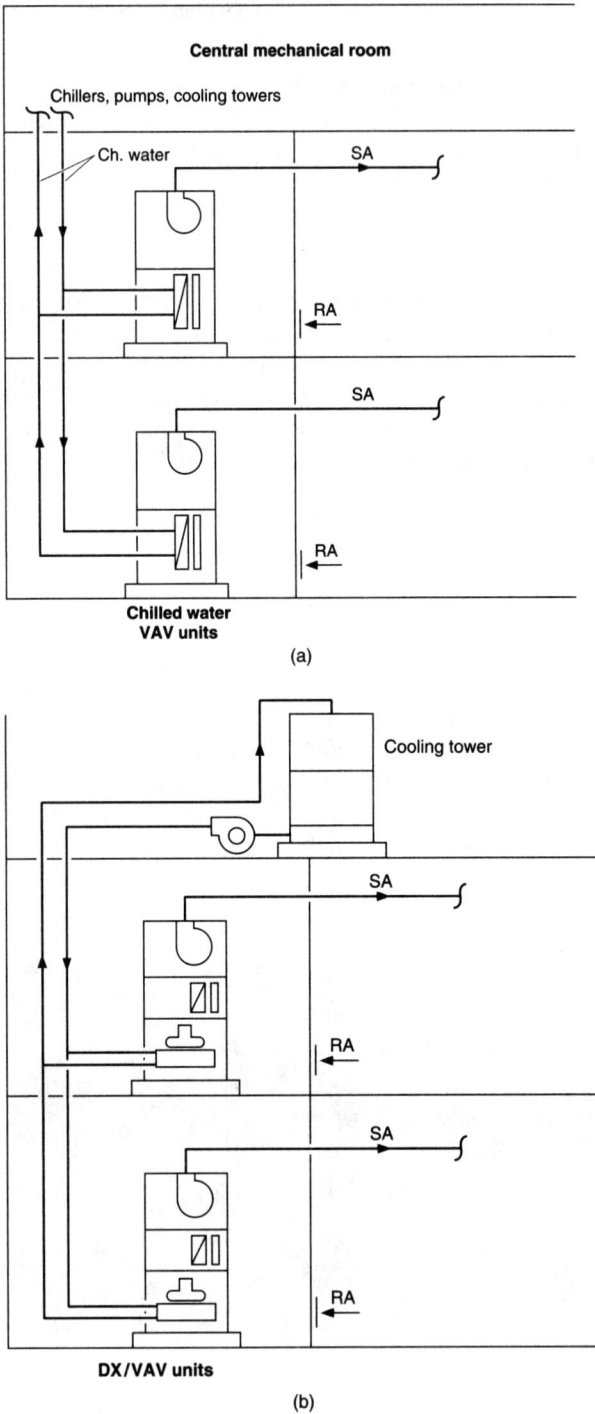

Central mechanical room

Chillers, pumps, cooling towers

Ch. water

SA

RA

SA

RA

**Chilled water
VAV units**

(a)

Cooling tower

SA

RA

SA

RA

DX/VAV units

(b)

Figure 8.7 Packaged VAV air-conditioning systems. (*a*) Chilled water VAV air-conditioning system; (*b*) DX/VAV air-conditioning system.

However, for small systems, it is generally advantageous to use DX units because of lower initial and operating costs.

Advantages.   There are many advantages in using decentralized packaged VAV units:

1. *Floor space required.*   Generally, decentralized VAV systems served by multiple or floor-by-floor packaged units are considered to occupy more floor space than centralized VAV systems. However, depending upon design parameters such as floor area covered by each unit, operating schedule, building load, occupancy diversity factor, and method and ducting of outdoor intake and exhaust, the decentralized packaged system may actually occupy less area than the multifloor central system.

2. *Operating flexibility.*   The multiple-packaged-unit system offers more operating flexibility than the central system. The units can be turned on and off individually, and this may result in less overall energy consumption than continuous partial load operation of a large central system.

3. *Metering of energy consumption.*   Metering is much easier and more accurate for packaged units. This can be more difficult with the multifloor central system.

4. *Staged building construction and usage.*   The multiple-packaged-unit system easily allows staged construction and tenant occupancy. Also, renovation work can be staged to permit construction and occupancy on a floor or per unit basis.

5. *Fire zoning.*   The packaged VAV system has fewer penetrations through concrete floors. This generally means fewer fire dampers are required and more effective fire zoning.

Drawbacks.   There are also certain drawbacks to decentralized packaged VAV units:

1. *Equipment noise.*   The fact that packaged VAV units are necessarily installed close to the conditioned spaces poses a potential noise problem. Noise control requires close coordination among the system designer, architect, structural engineer, and sometimes the acoustical consultant. The design team should work together in the areas of equipment selection, mechanical room sizing, and construction for effective noise control. Details of these subjects are covered in Chaps. 5, 6, and 10.

2. *Standby capacity.*   Packaged units are not equipped with standby fans or compressors. Any malfunction of these vital components will result in unit shutdown, while large central units are generally equipped with multiple fans or compressors which should provide

reasonable standby capacity for most of the normal operating hours.

3. *Maintenance.* Multiple units means multiple equipment to maintain—including fans, compressors, and filters. This may add substantial expenses to the overall operating costs.

4. *Fan capacity modulation.* Generally, large central systems can perform fan capacity modulation more efficiently and economically. Particularly if a bypass type for capacity modulation is used for small packaged units, there are practically no energy savings because the capacity and static pressure of the supply fan essentially remain unchanged.

Medium-to-large systems with large packaged VAV units or central built-up air-handling units. In general, the drawbacks of decentralized packaged systems tend to be the advantages for large central systems. In other words, large systems are less noisy, easier and less expensive to maintain, and more efficient to run. Also, they usually have much

TABLE 8.1   System Comparison: Central versus Packaged Systems for Owning and Operating Costs

Building Description

* Gross area: 550,000 ft^2
* Number of floors: 42
* Type of structure: steel
* Design conditions: summer 79°F db, 63°F wb
  winter 38°F

	System Description	
Item	*Installed*	*Alternate proposal*
Cooling system	900 tons, chilled water	1000 tons, DX packages 30 and 40 tons
Location	33d floor	Each floor
Heating system	Hot water boilers	Hot water boiler
Location	Penthouse (42nd floor)	Penthouse (42nd floor)
Interior zone control	Shutoff VAV	Shutoff VAV
Perimeter zone control	VAV with hot water reheat	VAV with hot water reheat
HVAC cost:		
Core	$4.30/ft^2	$4.32/ft^2
Tenant	$3.10/ft^2 (average)	$3.10/ft^2
Maintenance/repair costs	$0.06/ft^2/yr	$0.09/ft^2/yr
	$0.50/ft^2 present worth	$0.73/ft^2 present worth
HVAC energy costs	$0.26/ft^2/yr	$0.35/ft^2/yr
	$2.10/ft^2 present worth	$2.82/ft^2/yr
Total percent worth	$10.00/ft^2 (12%, 30 yr.)	$10.97/ft^2 (12%, 30 yr.)

SOURCE: Heating/Piping/Air Conditioning, July 1989. Reprinted with permission.

better standby capability and cost less to run and operate. Table 8.1 compares two system alternatives for their owning and operating costs.

Large central systems versus smaller, decentralized VAV systems. In conclusion, if the floor area per floor is very large, multiple units per floor with each unit covering 15,000 to 25,000 ft² are favored. However, if the number of floors is more than 10 stories, central systems are generally less expensive than floor-by-floor packaged systems. For small buildings, packaged systems are more advantageous than central systems (Table 8.2).

8.1.5 Environmental quality

VAV system selection affects thermal comfort, acoustics, indoor air quality, and space air distribution. These factors are the key elements influencing the quality of indoor environment. Figure 8.8 schematically represents the system selection and environmental quality relationship.

TABLE 8.2 Summary of Preferred System: Central versus Floor-by-Floor Packaged Systems

Design consideration	Central	Packaged
Low rise buildings (no. of floors less than 10)	—	Preferred
Medium to high-rise buildings (no. of floors more than 10)	Preferred	—
Economizer ventilation	Preferred	Note 1
Energy-efficient operation	Preferred	—
Standby capability	Preferred	—
Maintenance	Preferred	—
Noise control		
Floor-by-floor metering	—	Preferred
Selective floor operation	—	Preferred
Space requirement	Note 2	Note 2
Small buildings	—	Preferred
High-occupancy-density area in building	Note 3	Preferred
Indoor air quality	Note 4	Note 4

Notes:
1. It requires close building coordination to install an economizer cycle for packaged systems.
2. Inconclusive and cannot be generalized.
3. A separate packaged unit can be installed for high-occupancy area.
4. Can be comparable if systems are adequately enhanced for indoor air quality.

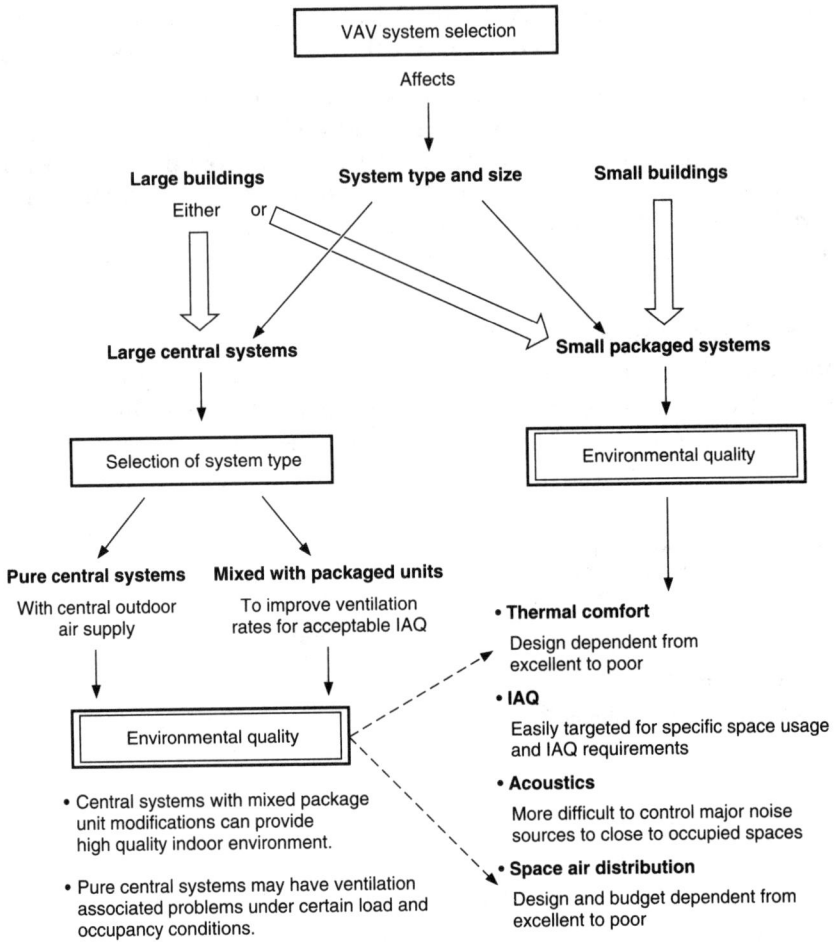

Figure 8.8 System selection and environmental quality.

Building size and usage affect system selection which, in turn, affects system type and size, and these two factors have considerable influence on environmental quality. However, the quality of indoor environment is also affected by the design approach taken by the system designer. By following the procedure described in Fig. 8.8, environmentally quality-conscious system designers can design a VAV system that provides high-quality indoor environment.

8.1.6 Serviceability

Air-handling units and VAV terminals are two major components of VAV systems. Both components require regular maintenance. Thus, service access is a critical design consideration for the system designer.

Decentralized packaged systems are less serviceable because economic pressure to maximize rentable space forces packaged units to be installed in a restricted, hard-to-access space. Selection of a central VAV system usually lessens the service access problem by installing air-handling units in the central mechanical room where an adequate space can be provided for proper equipment maintenance and service.

8.1.7   Budget, staged construction, and other constraints

A system is occasionally selected not for performance or its ability to produce quality environment but by budget, staged construction, and occupancy, as well as other nontechnical considerations. Sometimes it is necessary for the system designer to balance system performance (quality indoor environment and low maintenance and operating costs) with initial costs and other nontechnical considerations in selecting a VAV system. However, such compromise selections are often unjustifiable and lead to underperformance and excessive life-cycle costs of the system selected.

Component Selection

All variable air volume systems vary the supply air volume to meet the needs of cooling and heating in various zones. This variation in air volume is accomplished first at the VAV terminal equipment level; then the overall change in system air volume is matched by the corresponding change in fan air volume. Various techniques have been developed to initiate air volume changes both at the terminal and fan levels. Each technique shows its own unique characteristics when the airflow is modulated from maximum to minimum and each has a different impact on the system performance. As any VAV system consists of fans, ductwork, duct fittings, dampers, and terminal equipment (Fig. 9.1), and each of these system components possesses its distinctive pressure/flow characteristics, their interactions in a VAV system must be carefully studied for the entire range of system operation. This will ensure the compatibility of component selection in a given system and its operating environment.

To this end, each major component's operating characteristics under various control modes must be understood first; then the combined effect of all the components is analyzed for the given system and operating conditions. The following subsections briefly describe the characteristics of major VAV components and how they should be selected for specific applications.

9.1 Fans and Controls

The fan selection for a VAV system requires careful matching of the fan pressure characteristics with the system pressure characteristics for the entire fan operating range. This design issue is best illustrated by an example. In Fig. 9.2, the fan has a rather flat slope at the reduced flow range, and accurate and steady control may be difficult in this

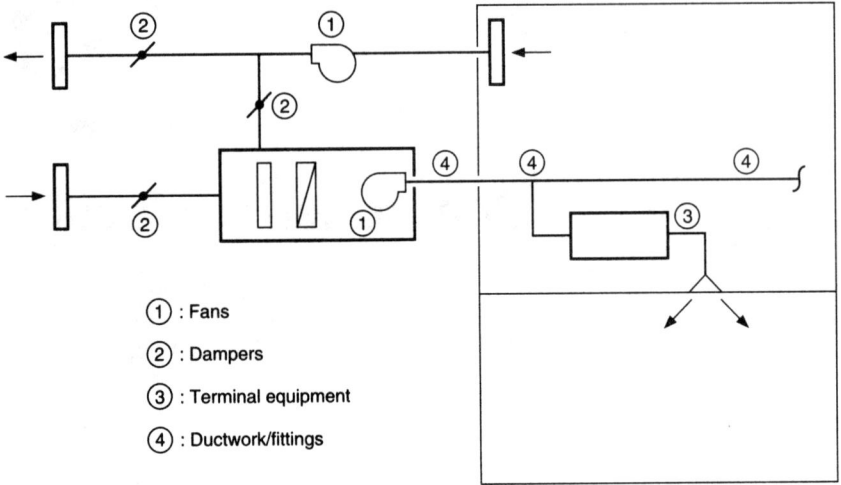

Figure 9.1   Major VAV system components.

Fan control: variable speed by pressure or flow sensor

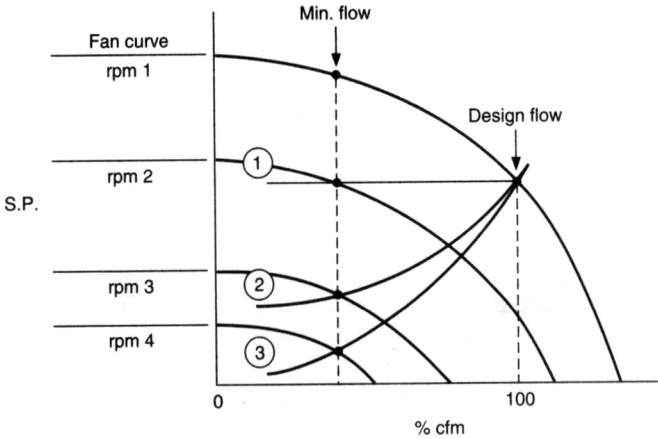

Control schemes

① : Pressure control, sensor at fan discharge

② : Pressure control, sensor at downstream

③ : Flow control, sensors at VAV terminals

Figure 9.2   Fan and system pressure matching.

TABLE 9.1   Fans and Controls for VAV Operation

Fan type	Fan modulation	Control scheme	Sensor location
Centrifugal	Discharge damper	Pressure control	At fan discharge
	Inlet vane	Pressure control or flow control	At a designated location or at terminal equipment
	Speed control	Pressure control or flow control	At a designated location or at terminal equipment
Axial	Variable pitch	Pressure control or flow control	At a designated location or at terminal equipment

range. Thus, if this type of fan is to be selected, control selection and design should be carefully matched to ensure a stable and energy-efficient operation. In the case of Fig. 9.2, control schemes 2 or 3 would be a far better choice than scheme 1.

As shown in Fig. 9.2, fan characteristics, operating range, method of capacity modulation, and location of sensor are all important considerations for satisfactory system operation. It is therefore important to consider all these factors in selecting fans and their controls for VAV systems. The system designer should evaluate each of these factors carefully before making a final choice. Of particular importance is the minimum fan capacity required. If the fan is required to operate at an extremely reduced capacity, the designer needs to make certain the matching control scheme has no control problems. It may be advisable to reselect the VAV system to be used, or reconfigure the system to maintain a minimum flow to avoid unstable operation. This may be accomplished by adding an artificial load by reheating to increase the minimum airflow or by separating the area that has substantial

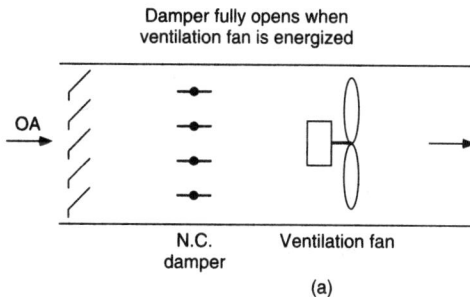

Figure 9.3   Four modes of damper control. (a) Two-position control.

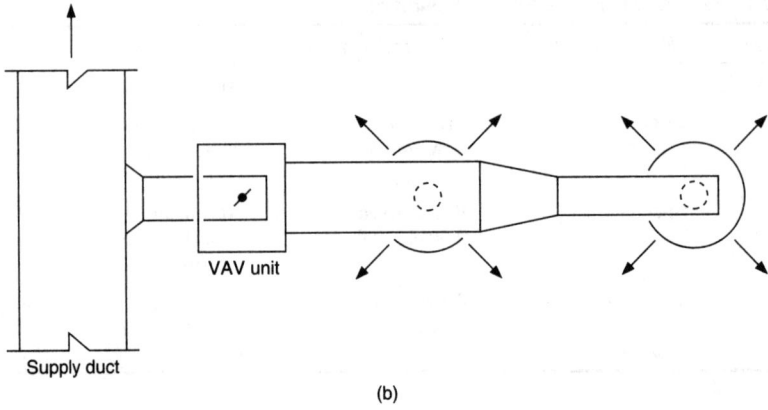

(b)

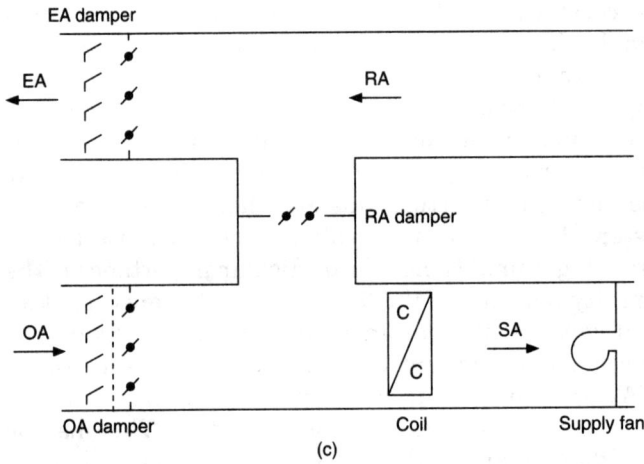

(c)

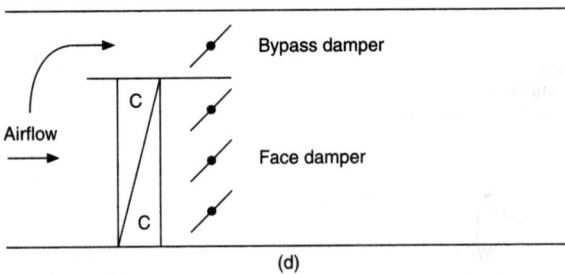

(d)

Figure 9.3 (*Continued*) (*b*) Modulating control. (*c*) Modulating mixed control. (*d*) Modulating bypass control.

loads from other areas with small loads, and providing a separate air-conditioning system for the latter.

Once a fan is selected and its operating characteristics, flow, and pressure requirements are determined, the selection of control scheme, device, and sensor location is a relatively straightforward process. Table 9.1 lists fan types, method of capacity modulation, and recommended sensor location.

9.2 Dampers

The general function of dampers in an air-handling system is to control the flow of air. There are essentially four modes of flow control: two-position, modulating, modulating mixed air control, and modulating bypass control (Fig. 9.3). All modulating controls should produce changes in airflow volume that are proportional to the control signal, that is, the linearity of control should be maintained. As shown in Fig. 9.4, the value of α is defined as a ratio of duct system pressure drop

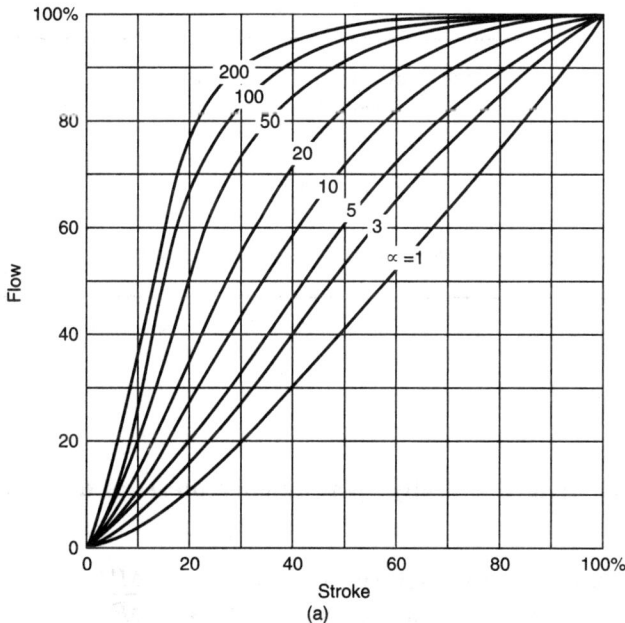

Figure 9.4 Flow, stroke, and α value relationship for modulating dampers. (a) Parallel blade damper installed characteristic curve. *(Courtesy of Honeywell Inc.)*

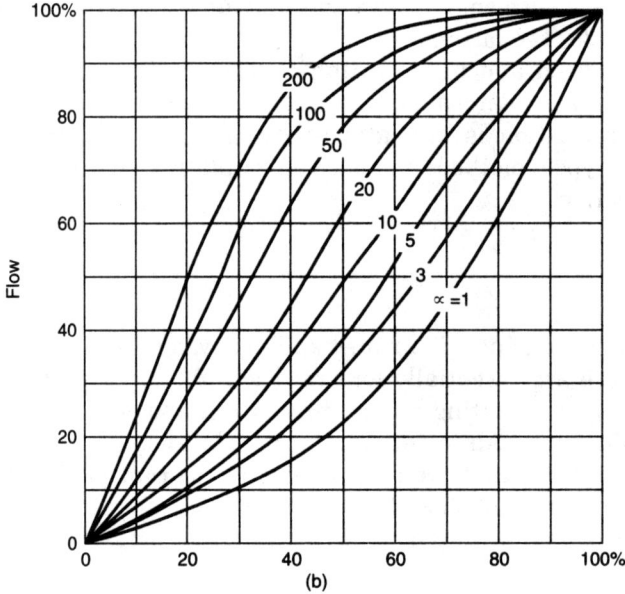

$$Q \text{ (flow)} = \sqrt{\frac{\alpha + 1}{\alpha + K^{1/2}}} \times 100 \qquad \alpha = \frac{\Delta P_2}{\Delta P_1} = \frac{\text{system press. drop exclusive of damper}}{\text{open damper press. drop at max. flow}}$$

K = Inherent flow, in percent of max. inherent flow at a given stroke

$Q = k\sqrt{\Delta P}$ where ΔP = total system pressure drop (Basic flow equation)

Figure 9.4 (*Continued*) (*b*) Opposed blade damper installed characteristic curve. (*Courtesy of Honeywell Inc.*)

exclusive of damper divided by fully open damper pressure drop at maximum flow, or

$$\alpha = \frac{\text{system press, drop exclusive of damper}}{\text{fully open damper press, drop at max. flow}} = \frac{\Delta P_2}{\Delta P_1}$$

Since ΔP_2 has a predetermined value in a given air-handling system, the value of ΔP_1 is important in determining the control characteristics or the linearity of the damper selected. This usually means the damper face area is smaller than the cross-sectional area of the duct in which the damper is installed. This is especially critical for the selection of return or mixing air damper in a modulating mixed flow control. A selection example of this control is illustrated in Fig. 9.5 and Table 9.2. Dampers in this example are selected according to a damper manufacturer's recommendations.

The damper selection procedure for VAV systems is relatively straightforward. First, type of application, damper type, airflow range, and duct size for each damper are determined. Next, dampers are sized by one of several sizing methods available. After the determination of damper sizes, actual pressure drops through each damper are calculated at a few critical design points for each application to see whether the damper selected can function properly or not. This type of analysis is particularly important for VAV economizer cycle operation. Damper pressure and stability analysis is discussed in Chap. 11.

There is no general consensus concerning what type of dampers should be used for a specific application. Table 9.3 lists types of

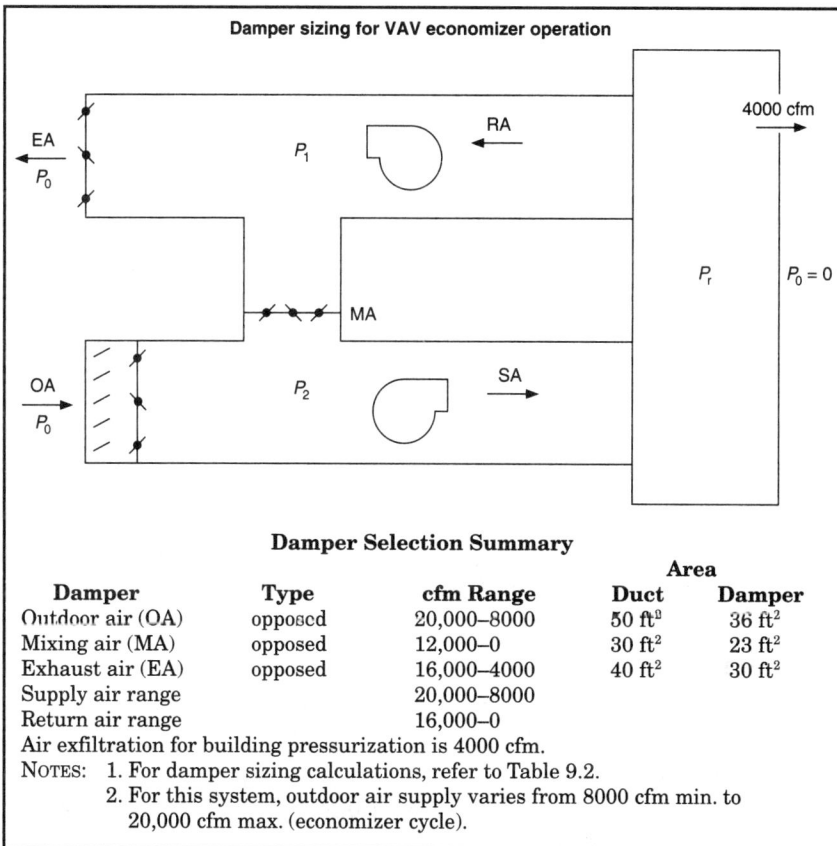

Damper sizing for VAV economizer operation

Damper Selection Summary

Damper	Type	cfm Range	Area Duct	Area Damper
Outdoor air (OA)	opposed	20,000–8000	50 ft^2	36 ft^2
Mixing air (MA)	opposed	12,000–0	30 ft^2	23 ft^2
Exhaust air (EA)	opposed	16,000–4000	40 ft^2	30 ft^2
Supply air range		20,000–8000		
Return air range		16,000–0		

Air exfiltration for building pressurization is 4000 cfm.

NOTES: 1. For damper sizing calculations, refer to Table 9.2.
   2. For this system, outdoor air supply varies from 8000 cfm min. to 20,000 cfm max. (economizer cycle).

Figure 9.5 Damper selection and sizing for economizer operation.

TABLE 9.2   Damper Sizing Calculations for Example in Fig. 9.5

- Duct sizing:

    OA duct = MA duct (plenum) using 400 fpm Force Velocity (FV)

    EA duct is also selected at 400 fpm FV

    OA duct size = $\dfrac{20,000}{400}$ = 50 ft^2, MA duct size = $\dfrac{12,000}{400}$ = 30 ft^2

    EA duct size = $\dfrac{16,000}{400}$ = 40 ft^2

- Damper sizing: Ref. 1 method, p. 414

    OA louver ΔP = EA louver ΔP = 0.2"; Damper $\Delta P_{\text{full open}}$ = 0.02"

    1. OA Damper: Approach velocity = $\dfrac{20,000}{50}$ = 400 fpm

    (opposed blades)

    Correction factor = $\dfrac{10^6}{400^2}$ = 6.25, $\Delta P_{1000 \text{ fpm}}$ = 0.02 × 6.25 = 0.125"

    Free-area ratio = $[1 + (79.7448 \times 0.125)]^{-0.234}$ = 0.571

    Damper area = $\left(\dfrac{50 \times 144 \times 0.571}{0.381}\right)^{0.9217}$ = (10,790.6)$^{0.9217}$ = 5215 in^2 = 36.2 ft^2

    2. MA Damper: Approach velocity = 400 fpm

    (opposed blades) Duct area = $\dfrac{12,000}{400}$ = 30 ft^2

    Damper area = $\left(\dfrac{30 \times 144 \times 0.571}{0.381}\right)^{0.9217}$ = (6474)$^{0.9217}$ = 3256 in^2 = 22.6 ft^2

    3. EA Damper: Approach velocity = 400 fpm

    (opposed blades) Duct area = $\dfrac{16,000}{400}$ = 40 ft^2

    Damper area = $\left(\dfrac{40 \times 144 \times 0.571}{0.381}\right)^{0.9217}$ = (8632)$^{0.9217}$ = 4246 in^2 = 29.5 ft^2

dampers recommended by three sources for typical control applications. However, if α values explained in this section can be estimated, the system designer can easily determine the correct damper type to ensure control linearity from Fig. 9.4a and b.

9.3   Terminal Equipment

VAV terminal equipment performs two functions: air distribution and volume control. These two functions are performed either by a single device or by two separate devices. Figure 9.6 classifies VAV terminal equipment. What type of terminal equipment to use is largely determined by architectural and mechanical design considerations, system size, performance requirements, and budget constraints. The following sections cover the selection considerations as well as a few selection examples for supply outlets (Sec. 9.3.1), self-regulated VAV outlets (Sec. 9.3.2), VAV terminals (Sec. 9.3.3), fan-powered terminals (Sec. 9.3.4), and VAV induction terminals (Sec. 9.3.5).

9.3.1 VAV supply outlets

Generally, VAV supply outlets are selected to produce adequate air distribution for human comfort as well as to meet other environmental quality requirements, such as acceptable IAQ and good acoustics.

General guidelines for supply outlet selections

1. Select outlets for smaller air quantities, ideally 300 to 400 cfm per each outlet. For long continuous linear diffusers, blank off a certain portion of the diffusers. Consult with the diffuser manufacturer for the length and interval of blanked-off diffusers.

2. Select outlets which do not cause dumping at the minimum supply air. This generally means high-entrainment outlets with a relatively small discharge area to achieve higher airflows at the minimum flow.

3. For dual heating/cooling applications, a minimum heating air quantity recommended by the outlet manufacturer must be maintained to ensure proper air and temperature distribution during the coldest winter days.

4. Space air movement and temperature spreads should be within 20 to 50 fpm, and 5°F for the height 6 feet from the floor.

5. For actual outlet selection, follow the procedure described in Fig. 9.7.

TABLE 9.3 Damper Recommendations for Different Applications

| Control application | Damper type recommended | | |
	Source A (Ref. 1)	Source B (Ref. 2)	Source C (Ref. 3)
• Return Air	Parallel	Parallel	Opposed
• Outdoor Air/Exhaust Air		Opposed	Opposed
(with weather louver or bird screen)	Opposed		
(without weather louver or bird screen)	Parallel		
• Coil Face	Opposed		Parallel
• Bypass			Parallel
(with Perforated Baffle)	Opposed		
(without Perforated Baffle)	Parallel		
• Two Position (all applications)	Parallel		

VAV terminal equipment

Air distribution   Air volume control

Section   **Section**

9.3.1 Supply outlet   Return (exhaust)   **9.3.2**   **Integrate with outlet**
                             outlet

• Linear diffusers   • Ceiling or side   **Separate control device**
  Slot diffusers   wall grilles

• Light troffers   • Light troffers   **9.3.3** • VAV terminals with
  single or double   single or double   or without reheat coils

• Square/round   • Square/round   **9.3.4** • Fan-powered terminals
  ceiling diffusers   ceiling diffusers   with or without reheat coils

• Side wall diffusers,   **9.3.5** • VAV induction terminals
  grilles

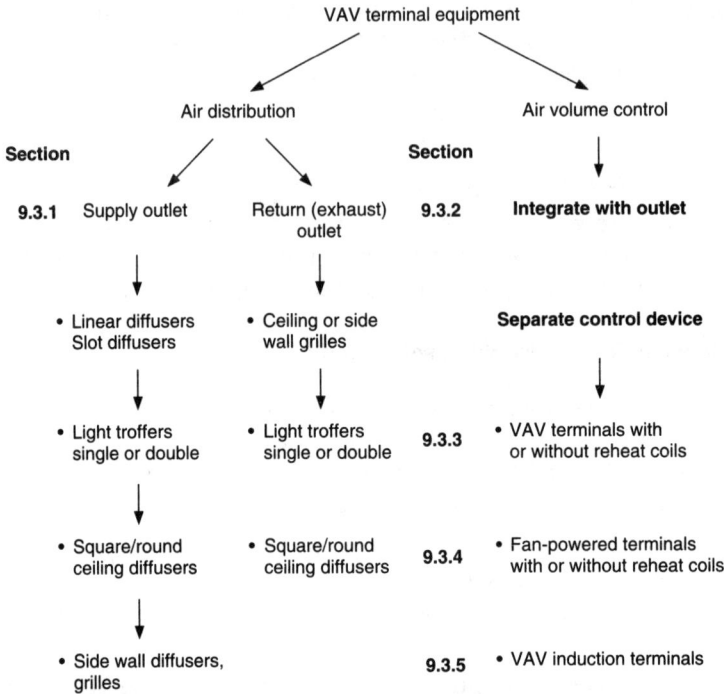

Figure 9.6   Classification of VAV terminal equipment.

9.3.2 Self-regulated VAV outlets

Self-regulated VAV outlets combine the function of air distribution and air volume regulation in a single unit. The selection considerations and procedures for the VAV outlets in this category are essentially similar to conventional VAV outlets described in Sec. 9.3.1. The only difference is that the supply air ductwork must be carefully designed and controlled to maintain a relatively stable low pressure not exceeding 0.2 to 0.3 inches for the entire flow range in order to keep the noise level of outlets below the acceptable level.

The noise level of self-regulated VAV outlets tends to exceed the normally acceptable noise level of NC-35 when the duct pressure reaches 0.25 inches w.g. or higher. Ductwork should be generously sized for low-velocity distribution, ideally without size reduction and, if possible, looped to minimize duct pressure fluctuation as the airflow increases or decreases. For larger systems with extensive duct routing, the installation of pressure-reducing dampers may become necessary. The system designer must carefully estimate the possible duct pressure build-up or the excess pressure available and select proper pressure-reducing dampers accordingly. Figure 9.8 shows several cases of pressure-reducing damper installations.

Selection example

Process	Example
Determine design cfm and ΔT	300 cfm, $\Delta T = 20°F$

Sidewall grille location

Process	Example
Select outlet type and location in the room	Plan 10′ ← 15′ → Elevation 9′ ← 15′ →
Determine L and T_{50}/L from arch. plan and Table 3 of Ref. 4	$L = 15'$ $T_{50}/L = 1.3 - 2.0$
Calculate required throw	$T_{50} = 20$ ft to 30 ft
Select outlet size and model number	Size = $14'' \times 4''$ Aeroblade grille
Determine noise level and ΔP	NC 26, $\Delta P = 0.107''$ w.g. (0° deflection setting)
Check ADPI, if the necessary information is available	ADPI = 75 at 300 cfm = 85 at 150 cfm

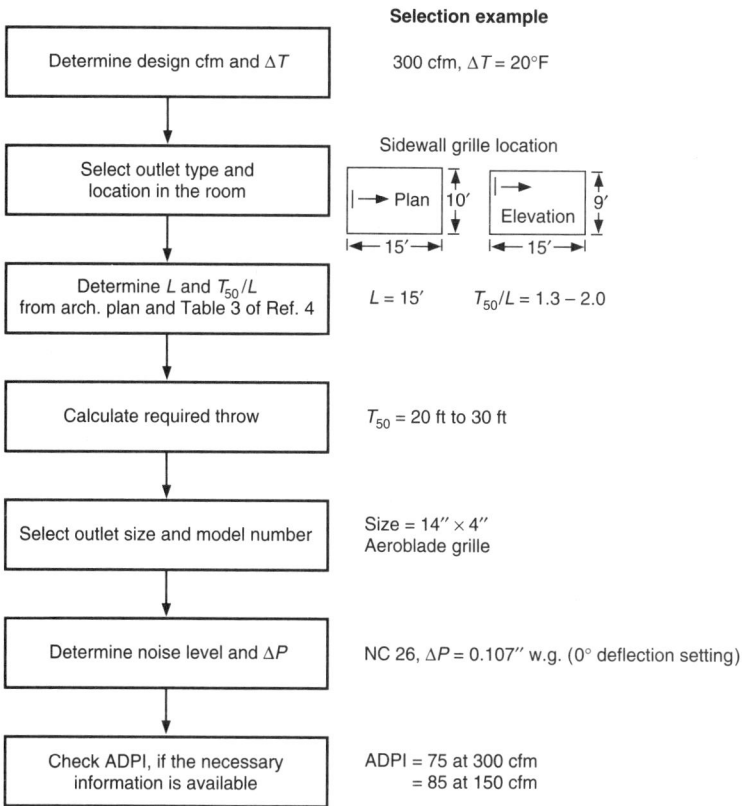

Figure 9.7 VAV outlet selection procedure and example. (*Example data reprinted with permission from Ref. 4, Titus Division of Tomkins Industries, Inc., publication.*)

9.3.3 VAV terminals

VAV terminals are essentially an airflow regulating device, dependent or independent of incoming air pressure and equipped with or without reheat coils (Fig. 9.9). Important selection considerations are pressure drops, noise, and reheating to maintain an adequate ventilation and air circulation.

Pressure drop and noise information are tabulated in the catalog and readily available. However, a few steps of engineering calculations are necessary to determine the amount of reheating required. Figure 9.10 shows an information flow for reheat calculation. Once reheat is determined, the selection process is completed. For noise-sensitive applications, a noise analysis may be required to determine the additional sound attenuation required. Figure 9.11 shows the complete procedure of VAV terminal selection.

Legend

Air-handling unit

Self-regulating outlet

Pressure-reducing damper

SP
Static pressure sensor

Trunk duct

SP

(a)

Trunk duct

SP

(b)

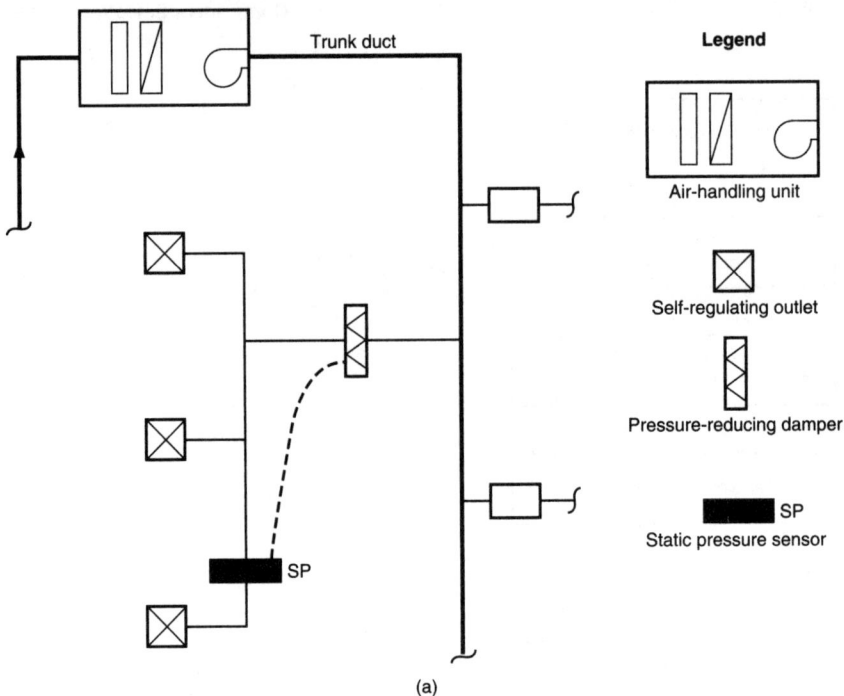

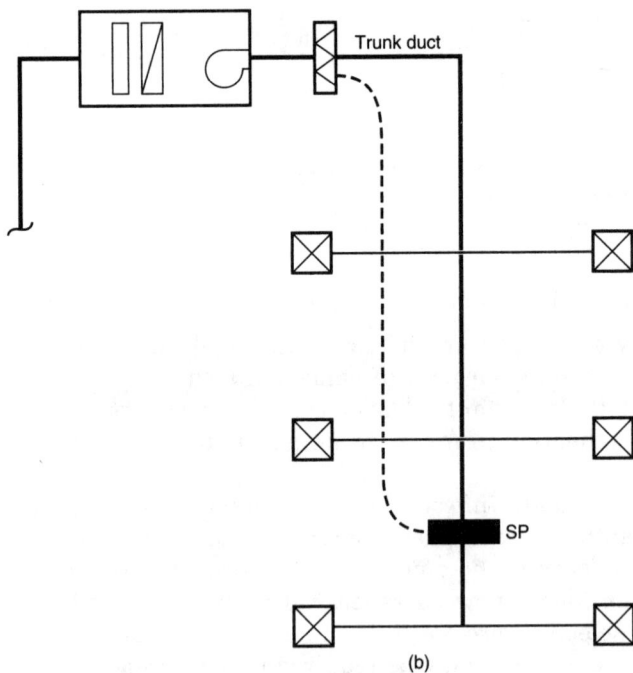

Figure 9.8 Examples of pressure-reducing damper installations. (*a*) High-pressure trunk low-pressure branch duct. (*b*) Low-pressure trunk and branch ducts.

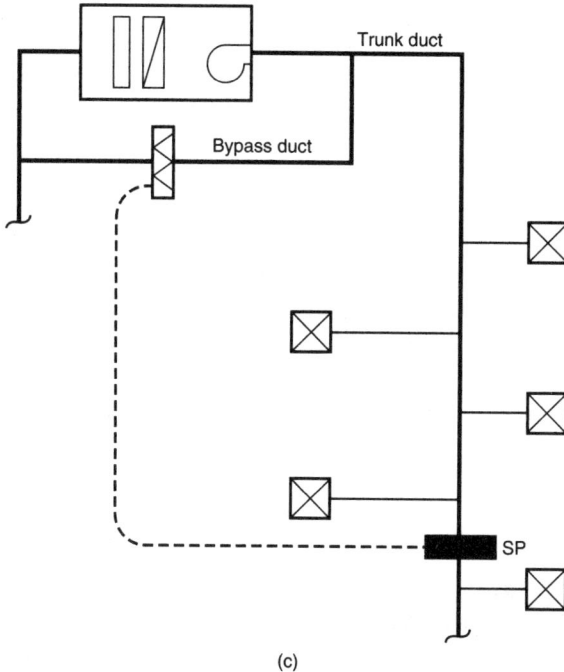

Figure 9.8   (*Continued*) (*c*) Bypassed low-pressure trunk and branch ducts.

9.3.4 Fan-powered terminals

A fan-powered unit is a combination of a fan and a VAV terminal. The fan is used to provide good air circulation in the conditioned space. Figure 9.12*a* shows basic components of a typical constant volume fan-powered unit. Figure 9.12*b* summarizes basic types available and the major features and differences.

Figure 9.12*c* describes the step-by-step procedure of selecting fan-powered terminals. The only difference in the selection of VAV terminals and fan-powered terminals is fan selection. Often, fans require additional attenuation or need to be relocated or reselected to maintain an acceptable acoustical environment in the occupied space.

Fan-powered terminals combine most VAV system advantages (zoned temperature control, operating flexibility, and economy) with constant air circulation in conditioned spaces. With a heating coil, these units provide economical overhead heating as well as cooling and ventilation. The central air-handling unit can be shut off during unoccupied periods while fan-powered-unit fans deliver the required perimeter heating.

The major drawbacks of fan-powered terminals are noise and costs. When selecting these units, the system designer should pay attention

Figure 9.9   VAV terminal. *(Courtesy of Titus Division of Tomkins Industries, Inc.)*

to noise and cost controls throughout the entire selection process. Major design considerations are the number and type of units to be used and where they should be located, the amount of attenuation required mechanically and architecturally, and feasibility of using other types of VAV terminals.

9.3.5   VAV induction terminals

VAV induction terminals are a hybrid of simple VAV terminal and dual-duct concepts. Instead of ducted warm air supply, these terminals make use of internal loads by mixing cold primary air with recirculated ceiling plenum air.

A major advantage of VAV induction terminals is their ability to maintain nearly constant supply air until the primary air is reduced below 50 percent of the design flow. Simple VAV terminals reduce the air supply almost immediately as loads begin to decrease, while induction terminals can maintain relatively high air supply rates unless the primary air drops to a very low level. Figure 9.13a indicates that the total airflow is still 60 percent of design flow when the primary air is reduced to 20 percent of the design flow rate.

The major drawbacks are higher inlet static pressure requirements and pressure limitations for downstream duct design. Usually, the available pressure is less than 0.3 inches w.g. which, in turn, means larger duct sizes. In this sense, the return air grille pressure loss must be kept low and included in the system loss estimate. Also, the system designer should select a low inlet static pressure for both energy and noise reasons.

Overall design considerations for the selection of VAV induction terminals should include the following:

- Inlet pressure required and available outlet pressure are the most important consideration. Lower inlet and outlet pressures should be used whenever practical and possible. An increase of 100 percent inlet pressure increases supply air only 40 percent. Use of small induction terminals by increasing the inlet static pressure may not be able to compensate for the increased operating costs resulting from a high system pressure at the central air-handling unit.

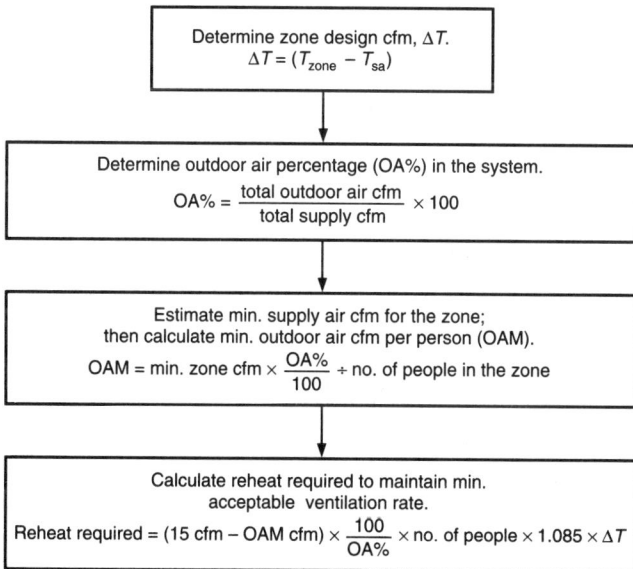

Determine zone design cfm, ΔT.
$$\Delta T = (T_{zone} - T_{sa})$$

Determine outdoor air percentage (OA%) in the system.
$$OA\% = \frac{\text{total outdoor air cfm}}{\text{total supply cfm}} \times 100$$

Estimate min. supply air cfm for the zone; then calculate min. outdoor air cfm per person (OAM).
$$OAM = \text{min. zone cfm} \times \frac{OA\%}{100} \div \text{no. of people in the zone}$$

Calculate reheat required to maintain min. acceptable ventilation rate.
$$\text{Reheat required} = (15\ \text{cfm} - OAM\ \text{cfm}) \times \frac{100}{OA\%} \times \text{no. of people} \times 1.085 \times \Delta T$$

Legend

T_{zone}: Zone temperature (°F)
T_{sa}: VAV terminal supply air temperature (°F)

Note

If necessary, the transmission heat loss needs to be added to the above reheat load.

Figure 9.10 Reheat load calculation for VAV terminals.

Determine design cfm and ΔT.
ΔT = room temp. − SA temp.

Is reheat necessary?
Go through the described procedure
in Fig. 9.10 for reheat calculation.

Reheat requirement determined.

Select terminal type.

Select terminal size and model no.

Select reheat coil if required.

Determine noise level and ΔP.

Provide additional attenuation,
if required.

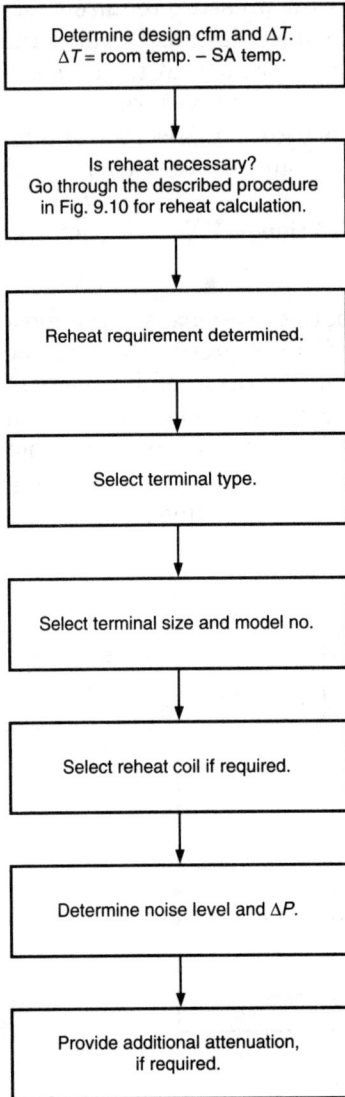

Figure 9.11 VAV terminal selection procedure.

Constant volume unit

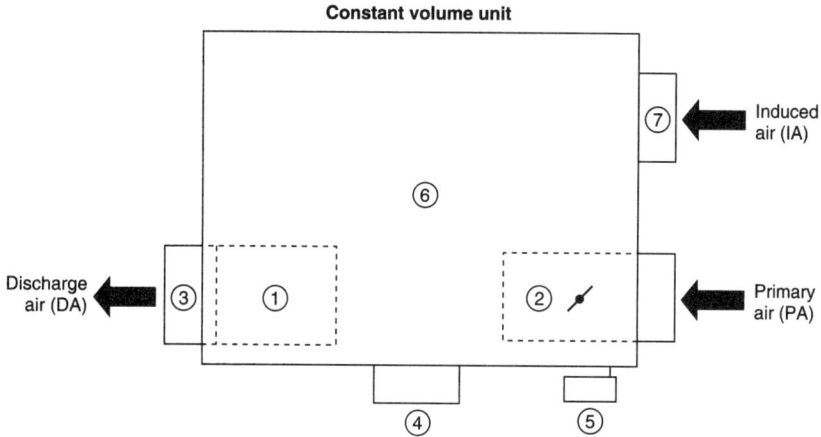

Components

① : Fan

② : Primary airflow measuring and regulating unit (air valve)

③ : Reheat coil (option)

④ : Control box

⑤ : Damper operator

⑥ : Unit casing

⑦ : Filter (option)

(a)

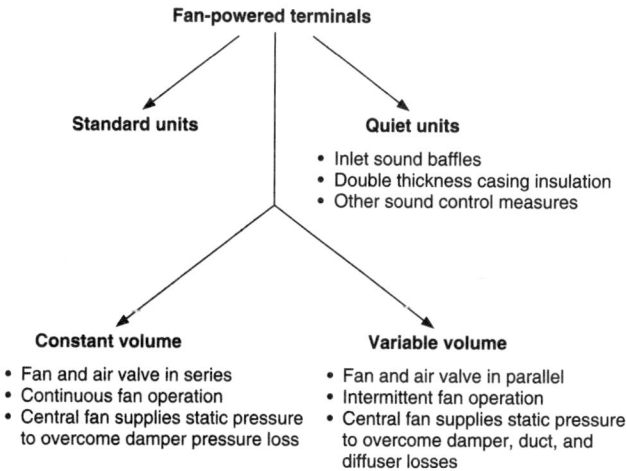

Fan-powered terminals

Standard units

Quiet units
- Inlet sound baffles
- Double thickness casing insulation
- Other sound control measures

Constant volume
- Fan and air valve in series
- Continuous fan operation
- Central fan supplies static pressure to overcome damper pressure loss

Variable volume
- Fan and air valve in parallel
- Intermittent fan operation
- Central fan supplies static pressure to overcome damper, duct, and diffuser losses

(b)

Figure 9.12 Various types of fan-powered terminals and their applications. (*a*) Basic components of fan-powered terminals. (*b*) Classification.

Select type: Constant or variable vol.

Determine design PA cfm and ΔT.
(ΔT = room temperature − PA temperature)

Calculate reheat requirement.
(See Fig. 9.10 for calculation.)

Determine fan cfm.
- Fan cfm = 65% PA cfm
  (variable volume units)
- Fan cfm = space air circulation required

Select unit size/model no.

Select reheat coil.

Determine noise level and ΔP.

Select a quieter unit if required.

(c)

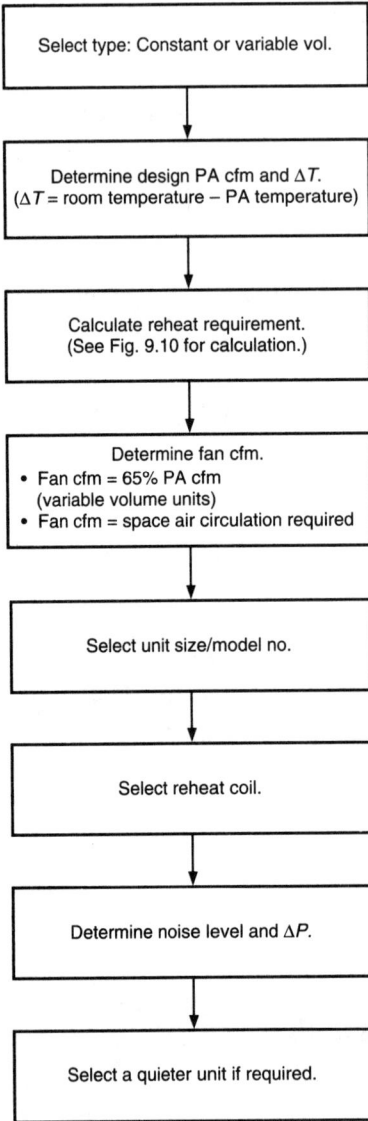

Figure 9.12 (*Continued*) (*c*) Selection procedure.

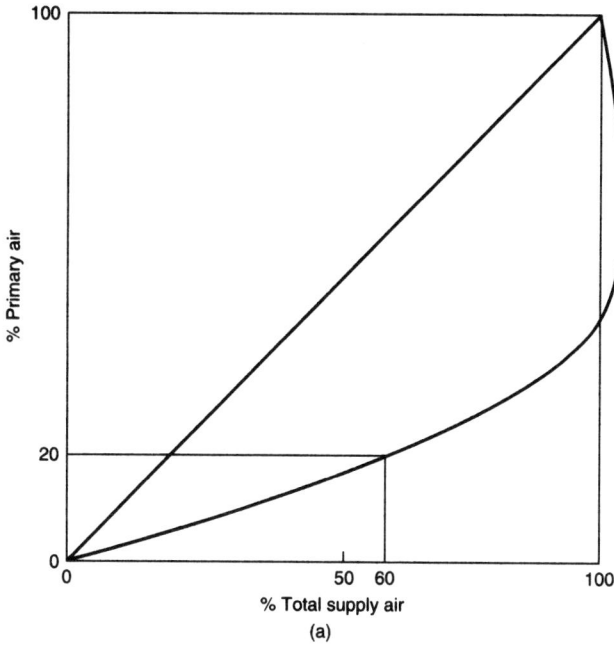

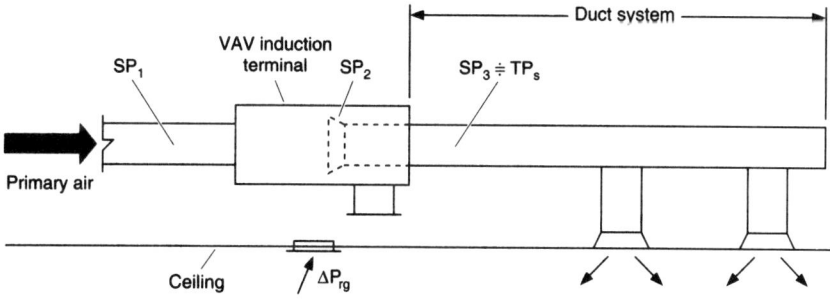

Figure 9.13   Performance of VAV induction terminals. (*a*) Reduction in supply air. (*b*) Required and available static pressures.

- Needs for adding a heating coil in induction terminals should be carefully analyzed. In mild climates, morning warm-up in winter is not a major problem, and it is usually sufficient to circulate warmer primary air for a certain period before the building is fully occupied.

References

1. Honeywell Inc., *Engineering Manual of Automatic Control for Commercial Buildings,* 1988.
2. Belimo Aircontrols (Can.), Inc., *Damper Application Guide 1,* 1993.
3. Johnson Controls, Inc., *Damper Manual.*
4. Titus Division, Tomkins Industries, Inc., Tytus Air Management Products.

10

VAV System Design Procedures

10.1 Design and Construction Phases

Buildings are designed and constructed through several phases (Fig. 10.1). The client initiates and funds a building project. At this phase of the project, a building/environmental program is prepared (Table 10.1). The program describes the client's needs, general space requirements, scope of services, schedules, budgets, and conceptual drawings. It is ideal for the system designer to get involved in the conceptual design of HVAC and other systems at this time to facilitate the integration of HVAC systems during later phases of the project.

The role of design professionals is to design a functional, cost-effective, and high-quality-environment building that meets the client's requirements. This is accomplished through three design stages—schematic, preliminary, and final design—followed by the construction administration of the project during the construction phase (Fig. 10.1, design and construction phases).

Because of the complexity of modern building systems, design professionals, particularly HVAC engineers, are frequently involved in start-up and commissioning activities as well as a variety of post-occupancy services such as IAQ investigations and checking of building and system performances (Fig. 10.1, post-construction phase). As a result of these post-construction activities, documentation of design intent has been increasingly emphasized in the preparation of construction documents.

All in all, it is important for VAV system designers to understand the flow of design information through each phase of a building project and how such information is generated and utilized, and therefore which documents are prepared at various phases of the design process.

Phase	Activity	Flowchart

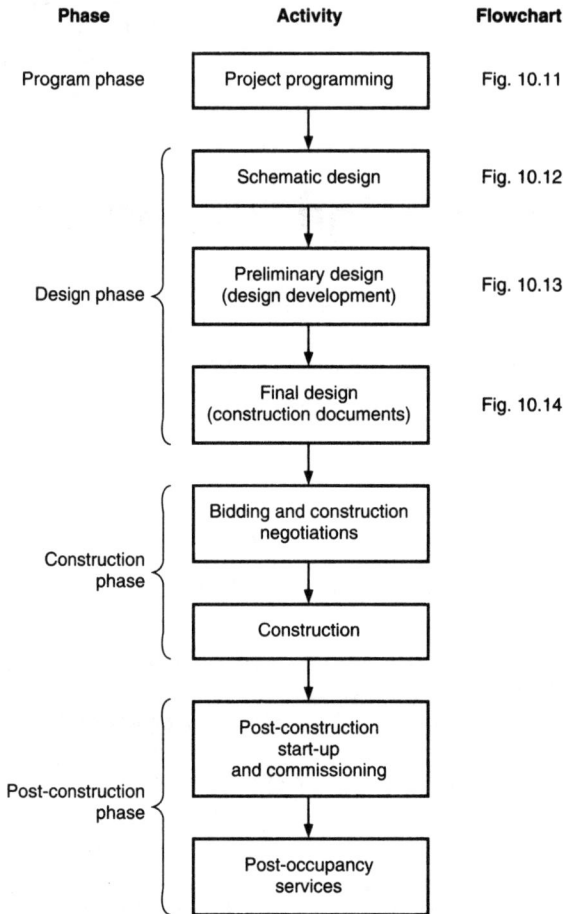

Figure 10.1   Building design and construction phases.

10.2   Information Flow and Document Preparation

The information required to prepare design documents comes from outside design sources or is generated by design professionals. Timing to receive or generate necessary and accurate design information is critical for the successful preparation of design documents. Figure 10.2 shows design information flow and how the outside information is utilized to generate each design discipline's own information which, in turn, is used to prepare design documents required at each design phase.

TABLE 10.1   An Example of Architectural/Environmental Program

General Space Requirements	
Office (open)	200,000 ft²
Office suite	200,000 ft²
Office (private)	40,000 ft²
Commercial/service space	60,000 ft²
Basement (2 levels)	500,000 ft²
	100,000 ft²
Total	600,000 ft²

Scope of Design Services

- Conceptual design
- Schematic design
- Preliminary design (design development)
- Final design (construction documents)
- Construction administration

Requirements of Drawings (Conceptual Design)

- Site plan, circulation diagram, site description, and the following written narratives:
  a. Description of the architectural design concept and structural, mechanical, and electrical systems
  b. Environmental, fire, security, daylight, and work safety design description
  c. Areas of each programmed space
  d. Zoning calculations, including lot coverage, building height, floor-area ratio, number of floors, landscaping ratio and total building area
- Model at 1:400
- Plan and section development
- Main public space, interior perspective, and description of major materials

Scale of the drawings will be 1:200.

Schedule

- The period for the trial production will be 70 days, from May 1, 1994 to July 10, 1994.
- The owner requires the architect to provide several conceptual schemes in 45 days for the owner's and planning department's initial review.

10.2.1 Program phase

During this phase, the architect defines the general characteristics of the building, its space usage and area requirements, as well as a construction budget. The architect also prepares conceptual building drawings and documents general statements on building services requirements. Table 10.1 illustrates an example of an architectural/environmental program for an office building project. At this phase of the project, the environmental system designer is usually asked to provide information on candidate systems for the project, particularly in the areas of major equipment locations, air distribution methods, and space requirements.

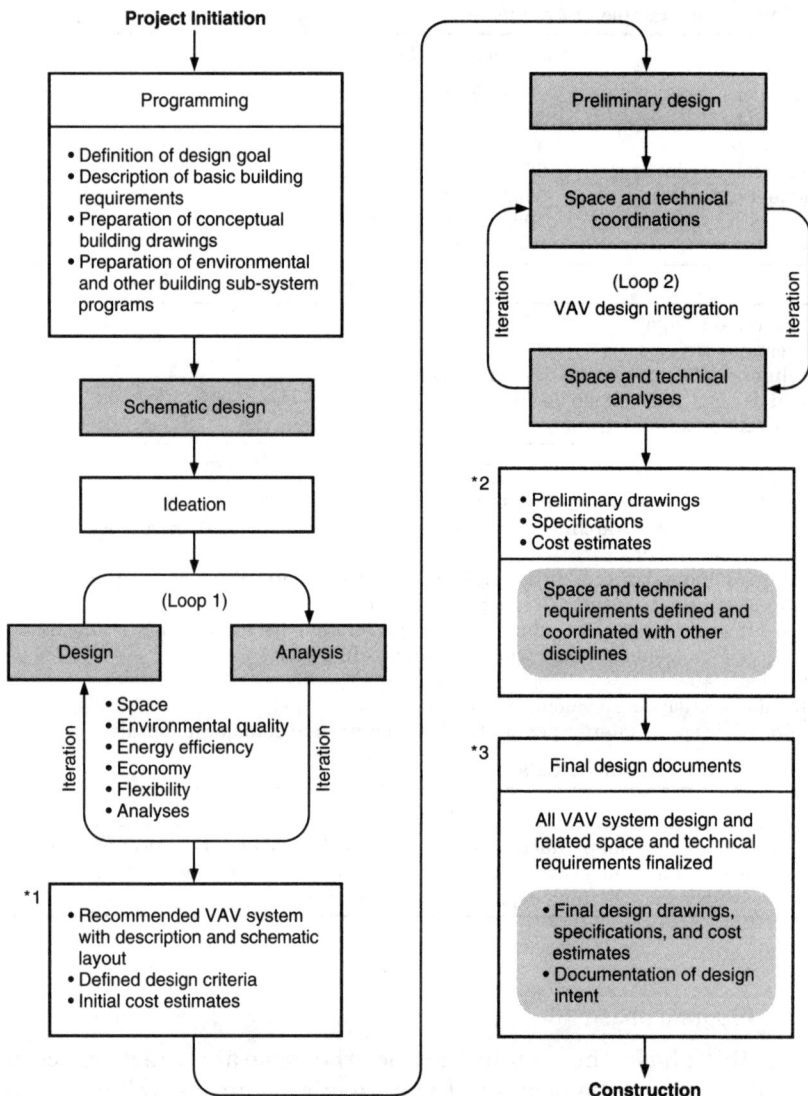

Figure 10.2   Information flow and document preparation for VAV System Design.

10.2.2 Schematic design

Based on the building/environmental program, schematic design identifies candidate systems and establishes design criteria for each system. Then each system is analyzed for its space requirements,
environmental quality (comfort, IAQ, and acoustics), energy efficiency,

economy, and system flexibility. Such analyses are necessarily repetitive, and may become a time-consuming process. However, depending on the size of the project and the budget available, the level and extensiveness of the analyses performed can be adjusted to suit the project requirements. Table 10.2 shows sample environmental design criteria for an office building.

In the actual process of analyses, an iterative design procedure is employed, first identifying new ideas (the process of ideation), then

TABLE 10.2   A Sample Environmental Design Criteria

A. *Codes and Standards*
The HVAC systems will be designed to conform, as a minimum, with the following codes and standards:
- The BOCA National Building Code—1990
- The BOCA Mechanical Code—1990
- National Fire Protection Association (NFPA)
- American Society of Heating, Refrigeration and Air Conditioning Engineers (ASHRAE)
- American Society for Testing and Materials (ASTM)
- American National Standards Institutes (ANSI)
- Sheet Metal and Air Conditioning Contractors' National Association (SMACNA)
- American Society of Mechanical Engineers (ASME)
- Air Conditioning and Refrigeration Institute (ARI)
- Underwriters Laboratory (UL)
- All applicable local codes

B. *Climate*
   The HVAC system's design will be based on the following climate factors:

   Outdoor temperatures.
   Summer: 95°F db 78°F wb (1% ASHRAE)
   Winter: 20°F db

C. *Space Conditions*

	Indoor Design Temperature	
Space	*Summer*	*Winter*
Conference center	75°F db	72°F
Multifunction spaces	75°F db	72°F
Underground parking	—	—
Utility spaces	—	60°F
Offices	75°F db	72°F
Computer center	68°F db	68°F

D. *Ventilation Rate (outdoor air)*

Space	
Conference center	20 cfm/person
Multifunction spaces	20 cfm/person
Underground parking	6 air changes
Utility spaces	1.50 cfm/ft^2
Offices	20 cfm/person
Computer center	20 cfm/person

testing each idea or design against physical laws (the process of design/analysis iteration). A great design idea may not be practical because of its space requirement, environmental quality, cost, or some other reason. In Table 10.3, two air-handling systems are analyzed for their floor space requirements and the results are tabulated for comparison. Usually, several vertical and horizontal distribution schemes are designed and analyzed before the best scheme is selected jointly by the architect, system designer, and structural engineer. This process is iterative as well as interactive; and all architectural, system, and structural ramifications should have been discussed and analyzed before the results were tabulated in Table 10.3.

After a consensus is reached concerning the air-handling system to be used for the project, the system designer prepares the description of the recommended system along with design criteria, a schematic layout, and initial cost estimates as the design documents for schematic design (item 1, schematic design documents in Fig. 10.2). An example of schematic layout for a VAV system is shown in Fig. 10.3.

10.2.3 Preliminary design (design development)

As shown in Fig. 10.1, the preliminary design phase follows immediately after the completion and approval of the schematic design. In this stage, the recommended air-handling system is closely coordinated with the architectural and structural subsystems and, if necessary, also with other disciplines. The coordination at this design stage involves all members of the design team and includes both space and technical coordination.

Space coordination is concerned with the physical space requirements on each floor, above-the-ceiling space, air intakes and exhaust

TABLE 10.3   Space Analysis for Two Air-Handling Schemes

Item of comparison	Scheme 1	Scheme 2
System description	Central plant, floor-by-floor air-handling units (AHUs)	Central plant fan coil units commercial area air-handling units (AHUs)
Central plant rooms (basement)	13,200 ft²	13,200 ft²
Total AHU room	13,600 ft²	3,000 ft²
Vertical shaft (piping)	150 ft² (low temperature)	250 ft² (conventional temperature)
Vertical shaft (outdoor air)	0 ft² (floor-by-floor intake)	2,000 ft²
Vertical shaft (general exhaust)	2,000 ft²	2,000 ft²
Horizontal duct space	18 in	12 in
Total floor area	28,950 ft²	20,450 ft²
Relative space requirement	100%	71%

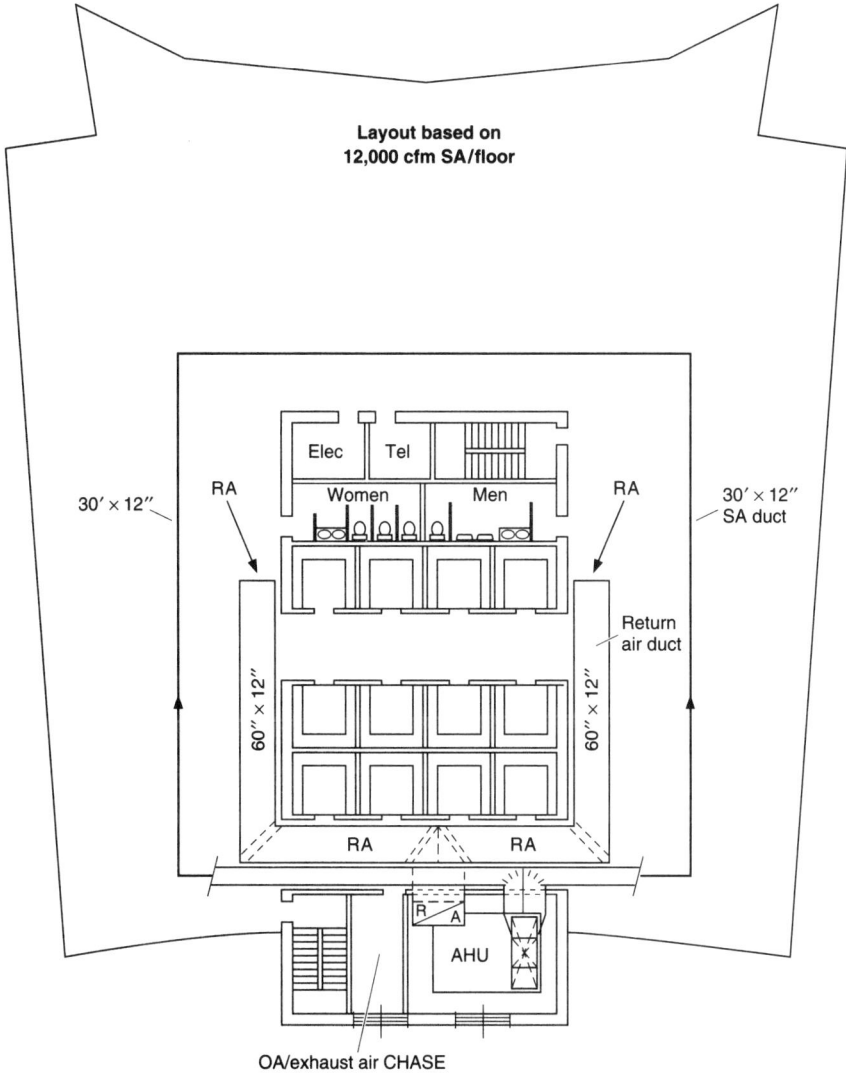

Figure 10.3 An example of schematic design layout.

openings, and the spaces occupied by ductwork and its components, as well as separate rooms and floors required by air-handling equipment (Fig. 10.4). Insufficient space requested by the system designer or provided by the architect may seriously affect the system performance and also produce annoying noise and some other system problems.

Technical coordination, on the other hand, is more concerned with the technical soundness of the system and components selected, code

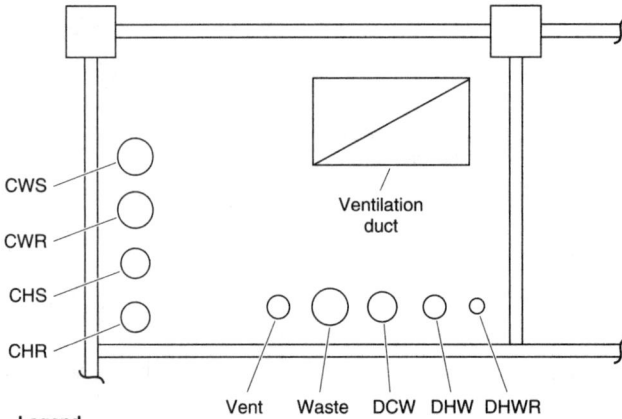

Legend

CWS = Condenser water supply
CWR = Condenser water return
CHS = Chilled water supply
CHR = Chilled water return
DCW = Domestic cold water supply
DHW = Domestic hot water supply
DHWR = Domestic hot water return

(a)

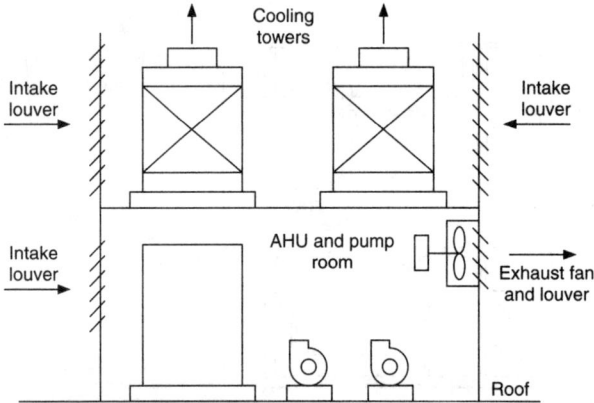

Coordination items

- Intake/exhaust louver size and location
- Intake and exhaust louver separation
- Contamination of intake air
- Noise and air pollution of intake and exhaust air

(b)

Figure 10.4 Examples of space and technical coordination. (a) Vertical shaft space. (b) Air intake and exhaust openings.

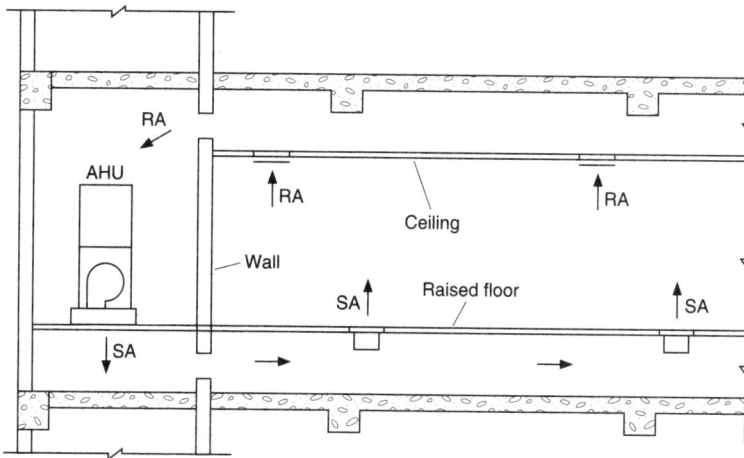

Coordination items
• Ceiling and floor plenum height requirements for proper air supply and return
• Acoustical requirements for walls, floors, and ceilings
• Plenum pressurization and ceiling/floor materials and construction methods

(c)

Figure 10.4 *(Continued)* (c) Above ceiling and below floor space coordination.

compliance, and coordination with other trades. Table 10.4 lists major coordination items for VAV system design at the preliminary design phase. Again, as in the schematic design, an iterative technique is employed to test the interaction between the building and the proposed system, using the updated architectural and structural drawings with greater details. When this phase is completed, a well-coordinated set of drawings and specifications with cost estimates is produced (item 2 in Fig. 10.2) for the continued refinement in the next phase, or final design, of the project. During this phase, only minor changes are usually necessary for space and technical coordination.

10.2.4 Final design

Final design involves finalizing design parameters, satisfying design criteria, and producing databases, drawings, and specifications—in other words, a set of construction documents which include sufficient details to show equipment schedules, plans and sections, diagrams, and control strategies. The budget is also finalized. In addition, the description of design intent is an important part of the documents for post-construction system commissioning, operation, maintenance, and other system-related activities. The databases of final design include

TABLE 10.4   **Major Technical Coordination Items for VAV System Design at the Preliminary Design Phase**

A. *Engineering Coordination*
   - Use whenever possible appropriate mathematical models for prediction of performance.
   - The validation of system and component performance is an iterative process. Use design criteria and system parameters and updated building drawings to refine heating/cooling load calculations, comfort, IAQ, and noise analyses at both full and part load conditions. Apply the results of refined analyses to achieve the close integration with other building subsystems. Try combined building and system solutions, instead of system solutions alone, by mutually adjusting building and system design parameters.
   - Verify system layouts for code compliance and space performance and budget requirements.

B. *Code and Standards Compliance Coordination*
   - BOCA building and mechanical codes
   - NEPA standards
   - ASHRAE and ISO standards
   - State and federal regulations

C. *Coordination with Other Disciplines*
   - Supply electrical load information to the electrical engineer.
   - Interdisciplinary coordination for ceiling-/floor-mounted air outlets, lighting fixture, sprinklers, smoke detectors, and other equipment.
   - Equipment weights for the structural engineer.
   - Supply the cold water and drain locations and capacity requirements to the plumbing engineer.

design criteria, load, air quantity, IAQ, noise, and mathematical models used for design analyses—these constitute an integral part of the final design documents (item 3 in Fig. 10.2).

10.3   Building and VAV System Design Interactions

Building and VAV system design is necessarily an interactive process. It is a teamwork requiring constant interaction among the team members—architects, engineers, and other design professionals. This process starts soon after the project is initiated by the owner. Following the owner's instructions, the architect prepares a building/environmental program. During this initial stage of design interactions, several design options are conceptualized and their feasibility studied. Finally, one or two design schemes are selected for further refinements in the schematic and preliminary design stages. The system designer's interaction with the architect and other team members is rather limited at this stage; nevertheless, sufficient attention should be paid to ensure the proper development of the design concept. (See Sec. 10.2.1, Program Phase.)

TABLE 10.5   VAV System Design Decisions at Schematic Design

A. *Design Criteria Decisions*
- Design temperature and humidity conditions
- Ventilation and IAQ criteria
- Air filtration criteria
- Noise criteria
- Pressurization criteria
- Design duct velocity criteria
- Air intake and exhaust opening design velocity criteria

B. *System Decisions*
- Horizontal air transportation and space air distribution
- Air transportation routing
   1. Space air distribution method, type of outlet, and outlet location
   2. Air-handling options, central versus separate AHU options
   3. Outdoor air intake options, central versus individual supply
   4. Local ventilation and exhaust options
- Vertical air transportation
   1. Supply air shafts
   2. Return air shafts
   3. Exhaust air shafts
   4. Combined air shafts
   5. Smoke exhaust shafts
- Air-handling decisions
   1. Central versus separate AHU options
   2. Multifloor, floor-by-floor, two-floor AHU options
   3. Mixed CAV and VAV air-handling options
   4. Return fans versus exhaust fan options
   5. Single-duct versus dual-duct VAV AHU options
- Duct and VAV operation decisions
   1. Single-duct VAV
   2. Single-duct bypass VAV
   3. Dual-duct VAV
- VAV terminal decisions
   1. Simple VAV terminals
   2. Induction VAV terminals
   3. VAV fan-powered terminals
   4. VAV outlets
- Return air decisions
   1. Ducted return
   2. Plenum return
- Smoke evacuation decisions
   1. Engineered smoke exhaust systems
   2. Dedicated smoke exhaust systems
- Ventilation decisions
   1. ASHRAE ventilation rate or air quality procedure
   2. Central, separate, or independent outdoor air supply
   3. Outdoor air intake size and location
   4. Exhaust air opening size and location (general, kitchen, fume hood)
- Zoning decisions
   1. Fire and smoke zoning
   2. VAV air-handling zoning
   3. VAV terminal zoning
   4. Perimeter heating/cooling zoning
   5. Lighting control zoning
   6. Smoke evacuation zoning
   7. Pressurization zoning
   8. Mechanical ventilation zoning

TABLE 10.5   VAV System Design Decisions at Schematic Design (*Continued*)

- Energy conservation decisions
  1. 100 percent OA or minimum outdoor air
  2. Demand-controlled outdoor air or fixed outdoor air
  3. Exhaust air heat recovery, types and locations
  4. Methods to minimize reheat/recool requirements
  5. Economizer cycle or fixed outdoor air
  6. Conventional or energy-efficient building envelopes

C. *Building Service Bay Decisions* (building mechanical/electrical facility decisions that indirectly affect VAV system design)
  - Central energy plant (chillers, boilers, pumps, etc.)
  - Fan rooms
  - Pump rooms
  - Heat exchanger rooms
  - Transformer/switchgear rooms
  - Electrical/communication closets
  - Automation centers/control rooms

D. *Major Building Decisions Affecting VAV Systems*
  - Mechanical/electrical and other service rooms and floors
  - Floor-to-floor heights and associated ceiling heights
  - Vertical mechanical and other service shafts
  - Architectural noise- and vibration-prevention provision
  - Building envelope design decisions
  - Air intake and exhaust louver locations

10.3.1   Design interactions during the schematic design phase

A series of decisions must be made by the system designer during the schematic phase of VAV system design. As summarized in Table 10.5, the decisions to be made are quite extensive and must be made at the right time. Also certain system-related decisions are highly building-dependent and must be made interactively by the building and system designers. Figure 10.5 shows the interactive and interdependent nature of schematic building and VAV system design. The system designer needs certain architectural inputs before initiating the selection and layout of candidate VAV systems and their components. The building designer, on the other hand, needs mechanical inputs for mechanical room and shaft sizes and locations, as well as other system information. Often, insufficient interactions take place between the building and system designers at this design stage, causing major design problems at the later phases of building design. In this sense, Table 10.5 can serve as a checklist to help the design professionals cover all relevant decision items for schematic design.

Architectural design and generation of information for
schematic VAV system design

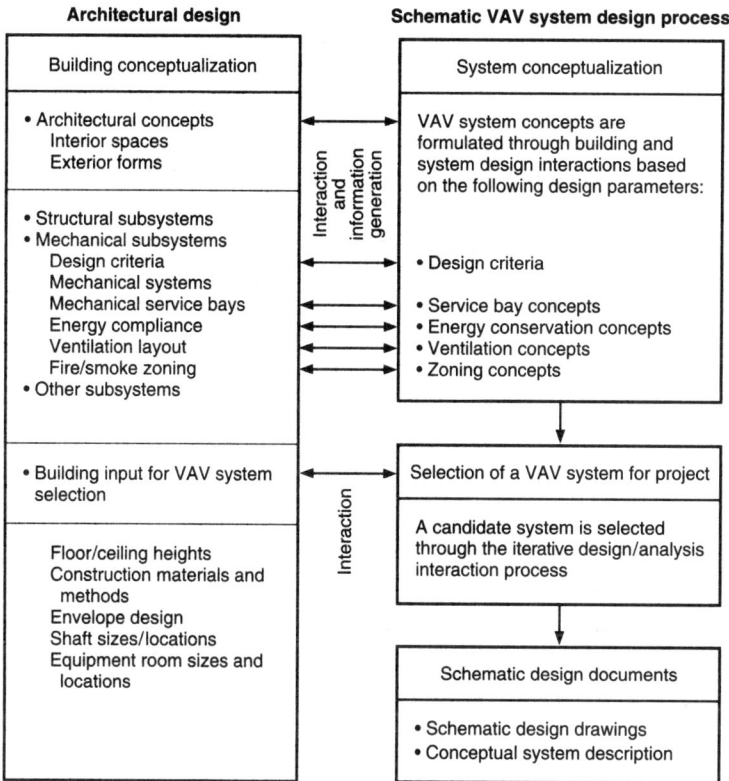

Architectural design	Schematic VAV system design process
Building conceptualization	**System conceptualization**

• Architectural concepts
   Interior spaces
   Exterior forms
	VAV system concepts are
	formulated through building and
	system design interactions based
	on the following design parameters:

• Structural subsystems
• Mechanical subsystems
   Design criteria
   Mechanical systems
   Mechanical service bays
   Energy compliance
   Ventilation layout
   Fire/smoke zoning
• Other subsystems
	• Design criteria
	
	• Service bay concepts
	• Energy conservation concepts
	• Ventilation concepts
	• Zoning concepts

• Building input for VAV system selection	**Selection of a VAV system for project**

Floor/ceiling heights
Construction materials and
   methods
Envelope design
Shaft sizes/locations
Equipment room sizes and
   locations
	A candidate system is selected
	through the iterative design/analysis
	interaction process

	Schematic design documents
	• Schematic design drawings • Conceptual system description

Center labels: *Interaction and information generation* ; *Interaction*

Figure 10.5 Schematic phase design interaction.

10.3.2 Design interactions during the preliminary design phase

Different types of design interactions take place during this design phase. Although two similar design loops are shown in Fig. 10.2 for both the schematic and preliminary design phases, these interactive loops serve entirely different purposes. The loop for schematic design (loop 1) is executed to conceptualize design ideas formulated in the process of coming up with design ideas at the time of ideation. The conceptualization usually takes the form of freehand sketches of system schematics and layouts. The loop for preliminary design (loop 2) is performed to quantify certain physical and technical aspects of the selected system. Its ultimate purpose is *system integration,* to produce design documents with scaled drawings and specifications showing

**Architectural design and generation of information for
preliminary VAV system design**

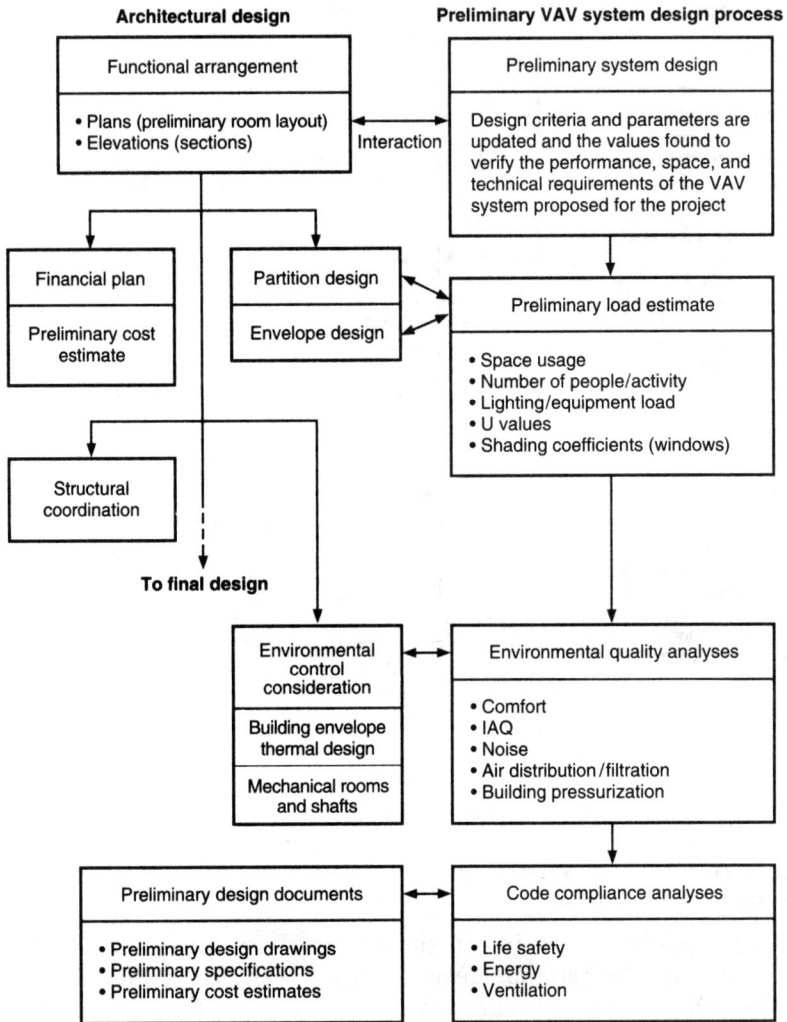

Architectural design

Preliminary VAV system design process

Functional arrangement

• Plans (preliminary room layout)
• Elevations (sections)

Interaction

Preliminary system design

Design criteria and parameters are
updated and the values found to
verify the performance, space, and
technical requirements of the VAV
system proposed for the project

Financial plan

Preliminary cost
estimate

Partition design

Envelope design

Preliminary load estimate

• Space usage
• Number of people/activity
• Lighting/equipment load
• U values
• Shading coefficients (windows)

**Structural
coordination**

To final design

**Environmental
control
consideration**

Building envelope
thermal design

Mechanical rooms
and shafts

Environmental quality analyses

• Comfort
• IAQ
• Noise
• Air distribution / filtration
• Building pressurization

Preliminary design documents

• Preliminary design drawings
• Preliminary specifications
• Preliminary cost estimates

Code compliance analyses

• Life safety
• Energy
• Ventilation

Figure 10.6   Preliminary phase design interactions.

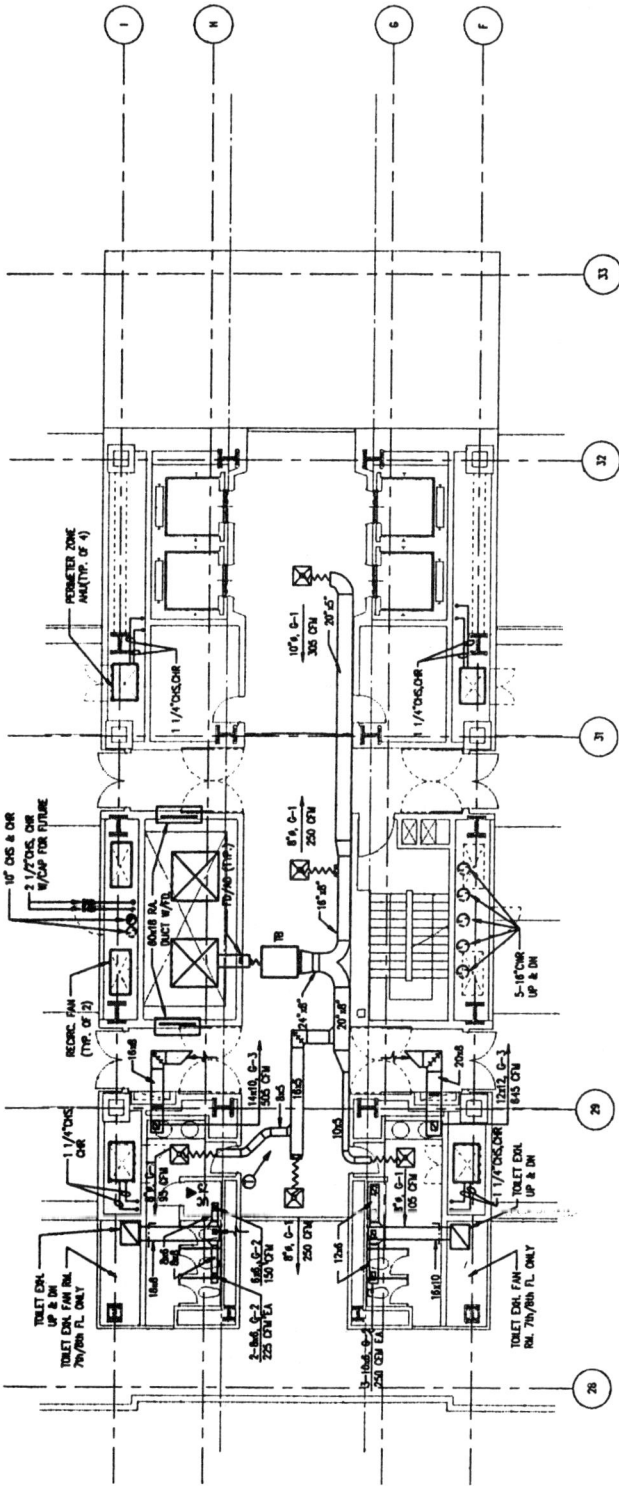

Figure 10.7 An example of preliminary VAV system layout. (*Courtesy of H. C. Yu & Asso. Reprinted with permission.*)

**Architectural design and generation of information for
final VAV system design**

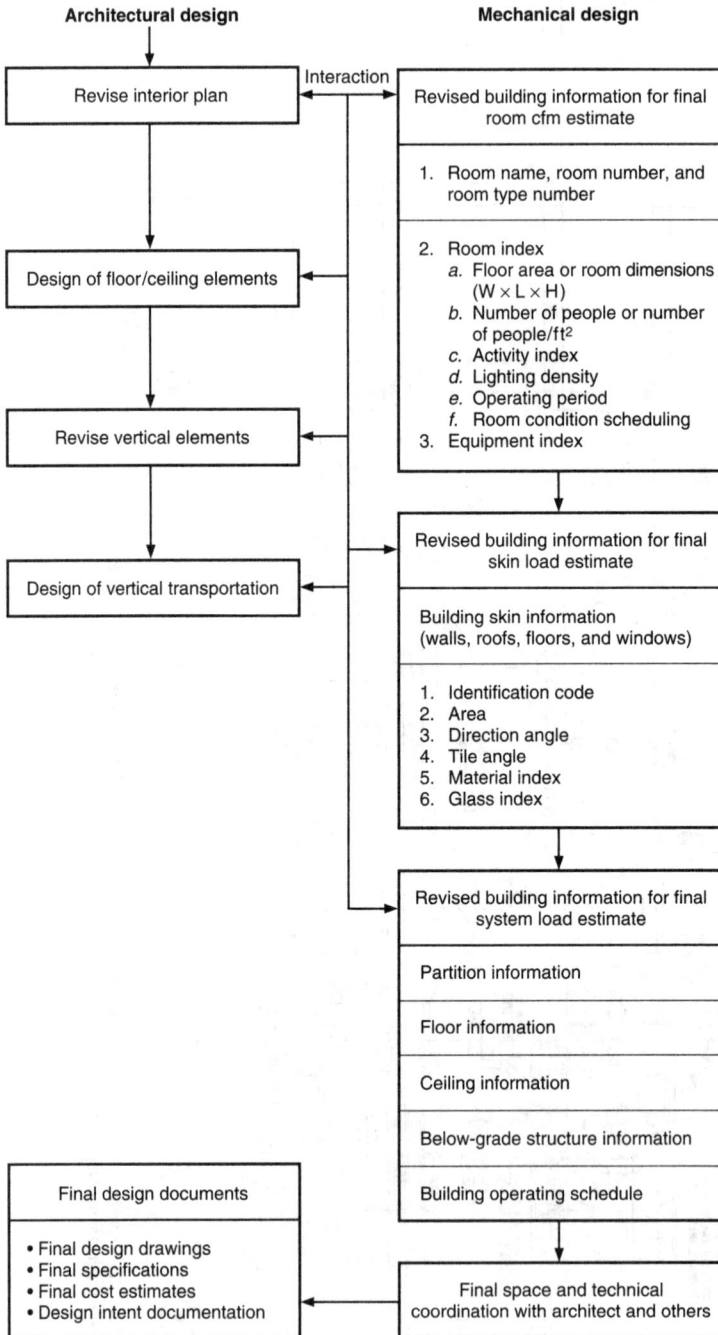

Architectural design	Mechanical design

Architectural design

↓

Revise interior plan

Interaction

Revised building information for final room cfm estimate

1. Room name, room number, and room type number

2. Room index
   a. Floor area or room dimensions (W × L × H)
   b. Number of people or number of people/ft²
   c. Activity index
   d. Lighting density
   e. Operating period
   f. Room condition scheduling
3. Equipment index

Design of floor/ceiling elements

Revise vertical elements

Design of vertical transportation

Revised building information for final skin load estimate

Building skin information (walls, roofs, floors, and windows)

1. Identification code 2. Area 3. Direction angle 4. Tile angle 5. Material index 6. Glass index

Revised building information for final system load estimate

Partition information

Floor information

Ceiling information

Below-grade structure information

Final design documents

Building operating schedule

• Final design drawings • Final specifications • Final cost estimates • Design intent documentation

Final space and technical coordination with architect and others

Figure 10.8 Final phase design interactions.

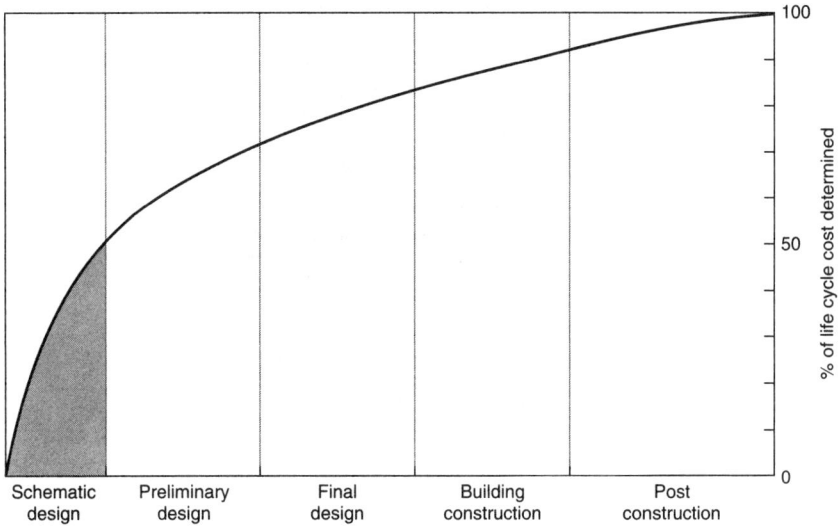

Figure 10.9 Impacts of design decisions on life-cycle costs.

Figure 10.10 Process checking versus product checking quality control.

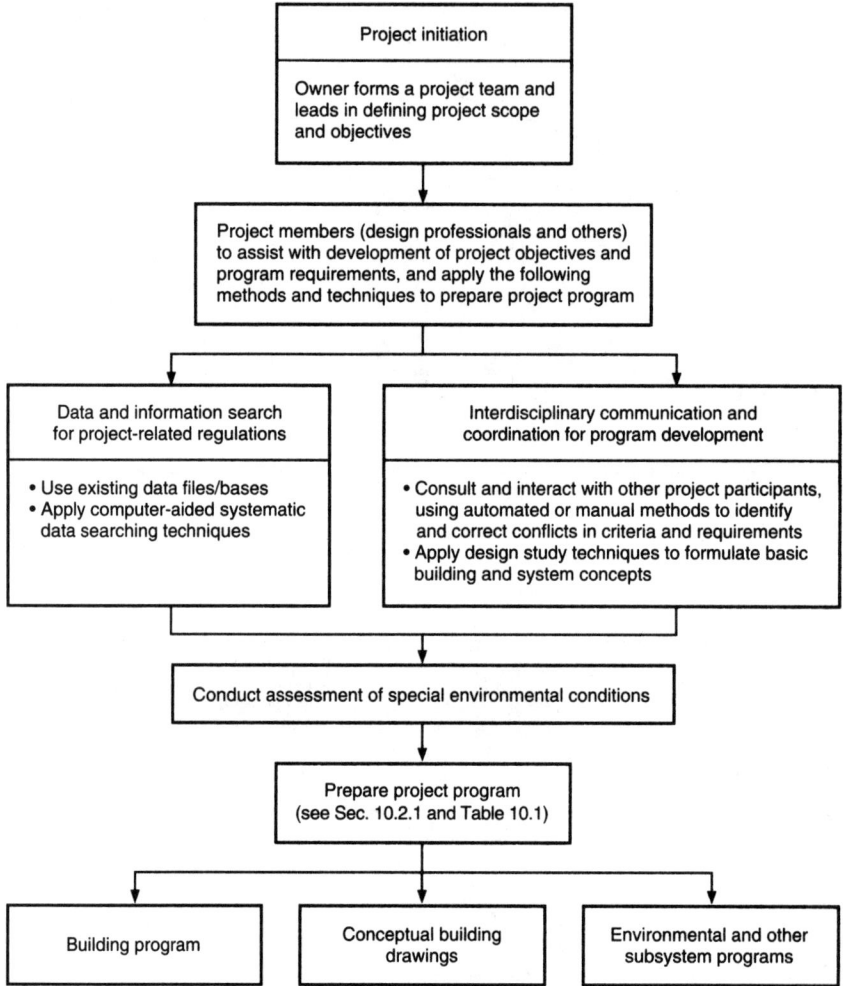

Figure 10.11   Design flowchart 1: preparation of project programs.

and describing the building/system relationships (see Fig. 10.6). Figure 10.7 is an example of preliminary VAV system layout.

10.3.3 Design interactions during the final design phase

The interactions during this stage of system design are mainly refinements and final coordination. As shown in Fig. 10.8, all specific details of building and system information should be reviewed and finalized for technical soundness, material compatibility, and spatial coordina-

Step 1

Ideation and conception
Formulate two to three VAV systems and perform quantitative (space and economy) and qualitative (environmental, energy efficiency, and system flexibility) conceptualization

(See Sec. 10.2.2 and Fig. 10.2, Loop 1)

Step 2

System definition
Determine basic system parameters for each system • Design criteria • AHU location floor area/ceiling height requirements • Basic duct layout • Terminal and outlet choice • Shaft size and location • Air intake and exhaust openings

(See Table 10.2)

Step 3

Interactive analysis
• Conduct analyses for each system using quantitative and qualitative parameters defined for conceptualization in Step 1 • Communicate and interact with other design professionals (mainly with architects) to optimize these parameters for each system

(See Secs. 7.1/7.2 and Figs. 7.1/7.2)

(See Table 10.3)

Step 4

System selection
Select a VAV system for further analysis and evaluation in preliminary design phase

(See Chap. 8)

Step 5

Dooumentation
Prepare documents for the system selected • Design criteria • System description/layout • Initial cost estimates

(See Fig. 10.3)

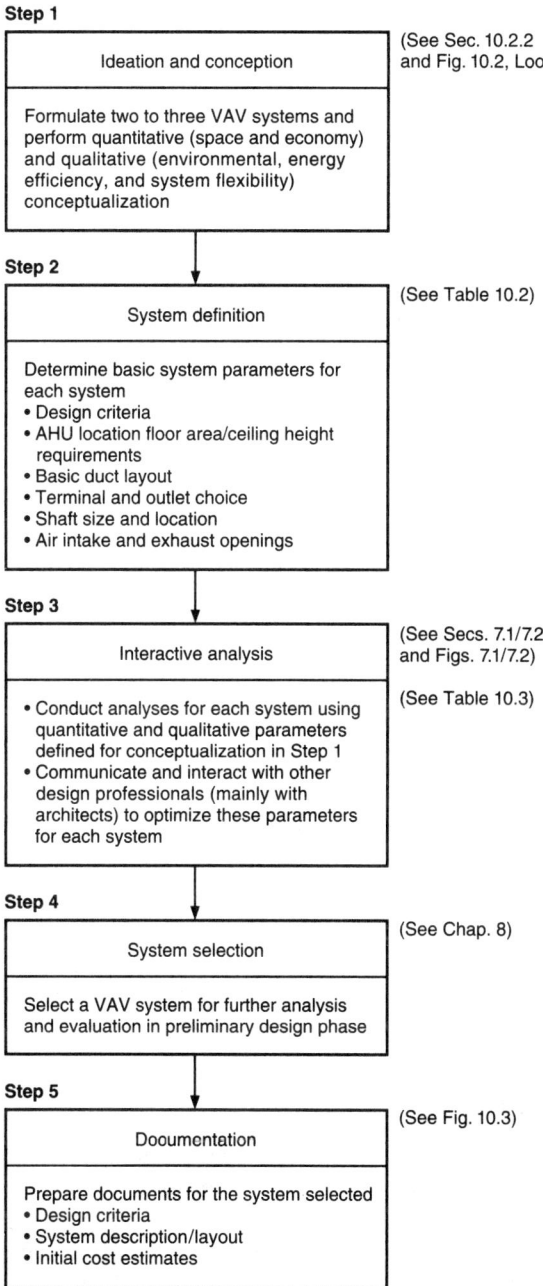

Figure 10.12 Design flowchart 2: schematic design procedure.

Step 1

System refinement and component selection

(See Chap. 9)

- Refine load estimate based on updated architectural information
- Select major system components
- Refine system layout
- Perform environmental quality analysis

(See Fig. 10.7)

Step 2

Physical space coordination		Technical coordination
• Perform space coordination by design interaction	(See Sec. 10.2.3 and Table 10.4)	• Perform technical coordination by shared design information and interactive design simulation

Step 3

Code compliance analysis

- Perform material, method, and quantity analyses for code compliance

Step 4

Preparation of preliminary design documents

- Apply total quality concepts (shared information and design by interactive simulation) to develop preliminary design documents

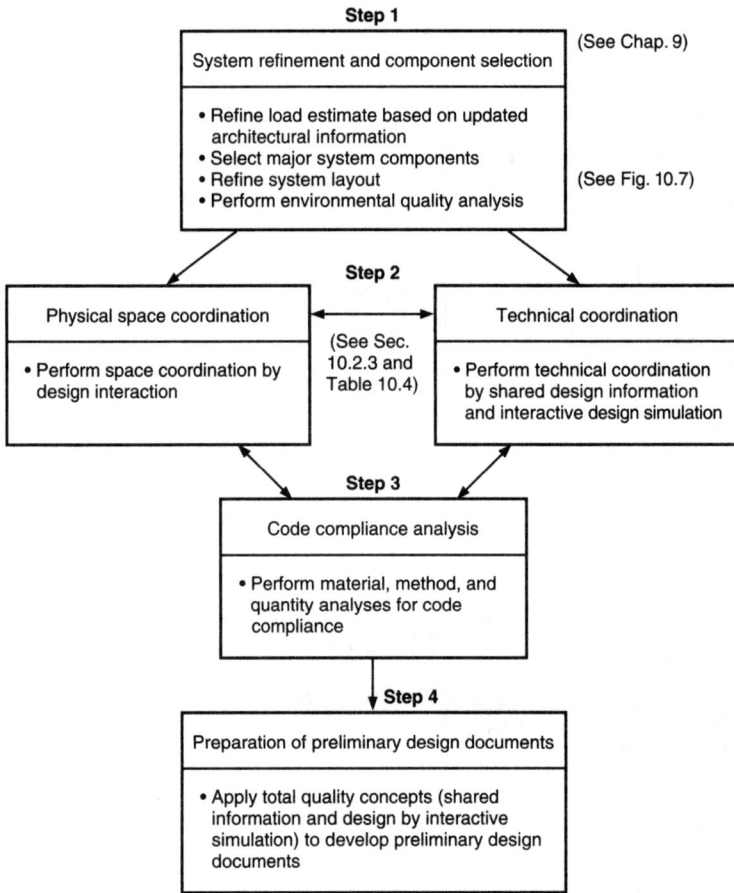

Figure 10.13   Design flowchart 3: preliminary design procedure.

tion. Also, this is the time to finalize load calculations, various environmental quality analyses, and design intent documentation.

10.4   Shared Design Information

Sharing design information among the design team members will greatly enhance communication and achieve high quality in design and construction. As shown in Fig. 10.9, most likely half of the building life cycle cost is determined by the time that the schematic design is completed, and probably more than three-quarters of the cost is set by the time that construction actually begins.

The information listed in Table 10.5 can serve as a checklist for quality decision making for the team members concerned, and the resulting

Step 1

Final design refinement	(See Sec. 10.2.4)

- Finalization of system design and layout
- Finalization of control system design
- Design accommodation to meet the special system requirements
- Finalization of design calculations and analyses

Step 2

Continued space and technical coordination	(See Sec. 10.3.3)

- Continued coordination to ensure total design quality

Step 3

Preparation of final design documents

- Continued application of total quality concepts to develop construction documents

Step 4

Documentation of design intent

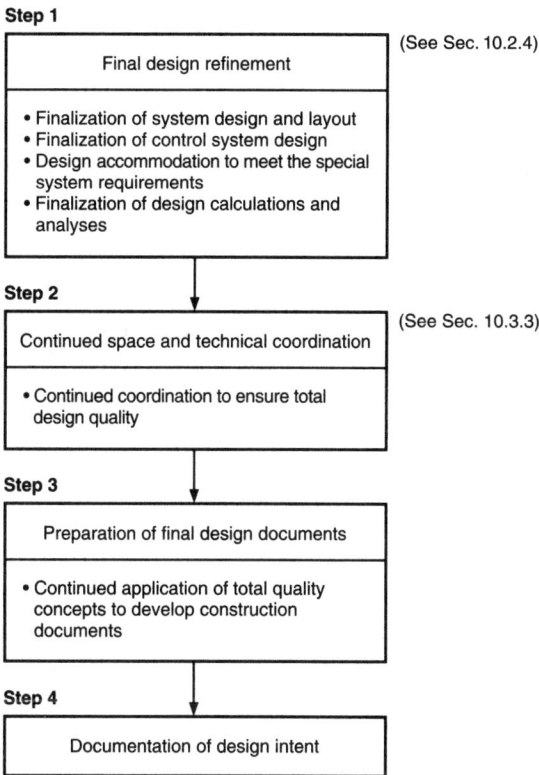

Figure 10.14   Design flowchart 4: final design procedure.

design reports and documents constitute a common data and knowledge base, providing transparent design models for prediction of system performance, its limits, serviceability, and modifiability. Also, the information explaining the critical design interactions that took place during the schematic and preliminary design phases should be included in the shared information package as part of design intent documentation.

Sharing design information at an early stage of system design can serve as an effective quality control means to achieve cost-effective Total Quality Management. This concept is illustrated in Fig. 10.10. The traditional way of checking design data, reports, drawings, and specifications when they are almost completed does not provide total quality control. After the schematic and preliminary design phases are passed, the willingness and practicality to change rapidly diminish, while the difficulties and lack of time grow exponentially. Sharing well-defined and well-documented design information continuously throughout the entire design process is the key to successful VAV or any other HVAC system designs.

10.5 Step-by-Step Design Procedure

Many design and selection considerations as well as specific design methods and techniques need to be included in the design procedure for VAV system designs. Figure 10.1 (design and construction phases) gives an overall design information flow, and Figs. 10.11 through 10.14 show the step-by-step procedure the VAV system designer must follow for quality design of VAV systems. The designer should refer to the appropriate chapters and sections throughout Part 2 and notes on Figs. 10.11 through 10.14 for detailed explanation.

Design Analysis

Analysis is a key element in the system design process. Once a design is conceptualized, it must be tested against physical laws. This verification process is recognized as analysis and further divided into two phases: (1) system testing and (2) performance analysis by simulation. The design and analysis process is necessarily iterative and constitutes an interactive loop between design and analysis. As a result, the process is recursive and may be repeated several times until a new or redefined old system emerges to meet the designer's system requirements.

First, a system is formulated and tested for a predetermined set of design criteria and conditions in the phase 1 process. This system is analyzed in phase 2 by a suitable simulation technique, either manual or automated, for partial and nonstandard conditions. This phase is particularly important for VAV systems because most of environmental and other VAV-related problems occur under these nonstandard conditions. Simulation techniques used in this phase range from simple two-point manual analyses to multipoint computer simulations. It is the designer's responsibility to decide whether a simple analysis based on manual simulation or a more complex computer-simulation-based analysis is required to produce the necessary information (Fig. 11.1). How to make this judgment is best illustrated by practical design situations. Sections 11.5 and 11.10 discuss this subject in detail.

Critical design considerations directly impacting the performance of VAV systems are analyzed in the following sections of this chapter:

- Load analysis (Sec. 11.1)
- Ventilation analysis (Sec. 11.2)
- Thermal comfort analysis (Sec. 11.3)

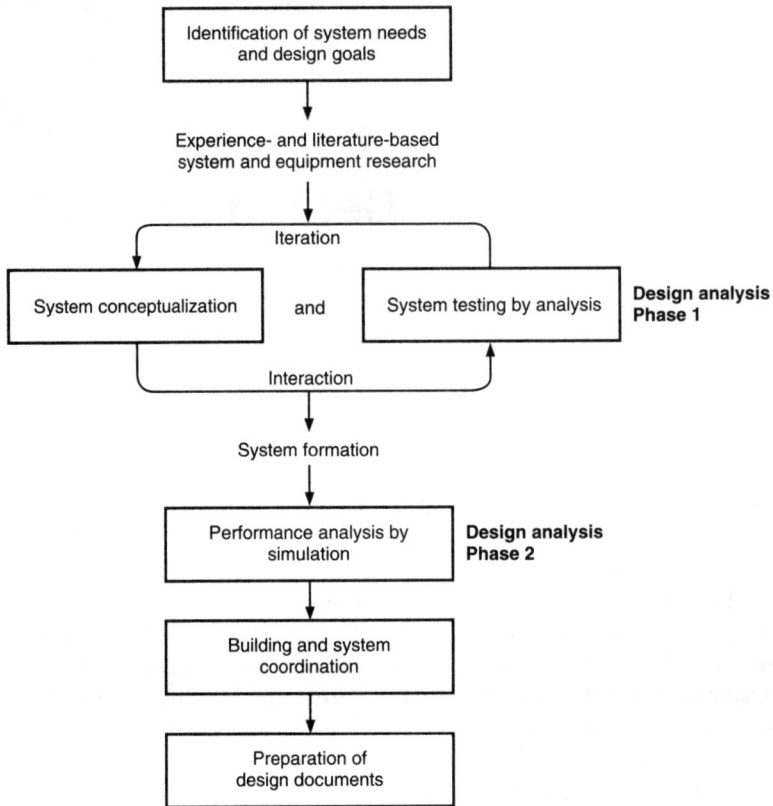

Figure 11.1   System design information flow and analysis.

- IAQ analysis (Sec. 11.4)
- Noise analysis (Sec. 11.5)
- Building pressurization analysis (Sec. 11.6)
- Outdoor air variation analysis (Sec. 11.7)
- Duct static pressure variation analysis (Sec. 11.8)
- System reliability and redundancy analysis (Sec. 11.9)
- Integrated design analysis (Sec. 11.10)

11.1 Load Analysis

11.1.1 Load calculations in general

Building HVAC loads are calculated component by component, then they are grouped together in load categories such as space, zone, coil, and plant loads (Fig. 11.2). This categorization is convenient because

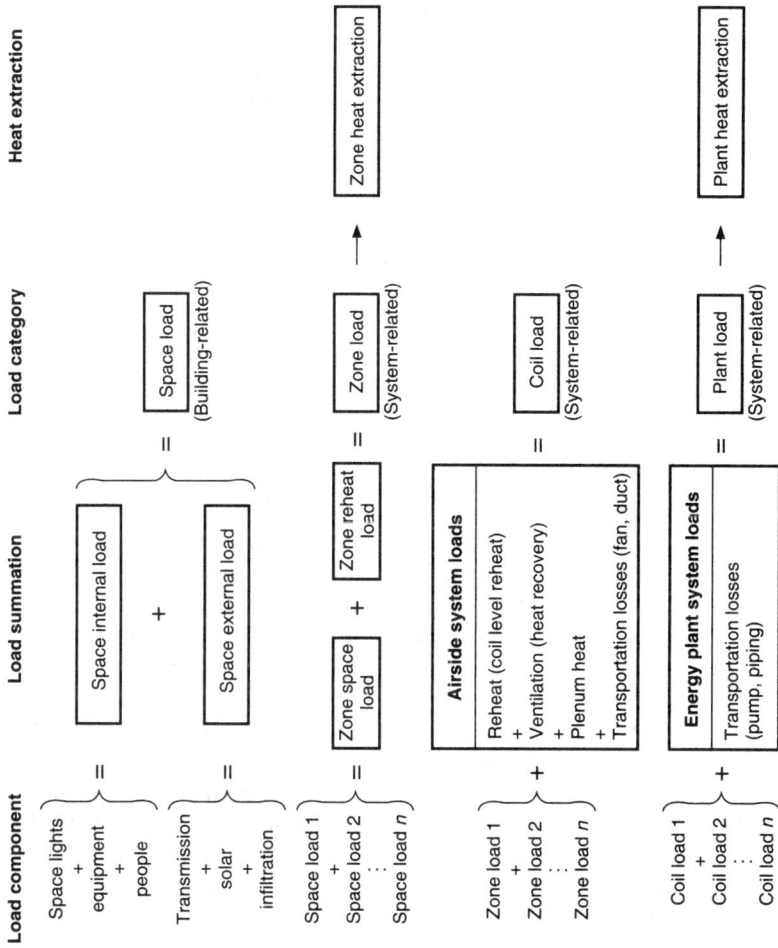

Figure 11.2 VAV system HVAC load and heat extraction calculations.

many system design parameters and operating conditions—for example, reheat, ventilation, plenum heat, transportation losses, thermostat settings, and operating schedules—all affect the magnitude of these load groups. Figure 11.2 schematically illustrates the relationship between load components and load categories. When loads are categorized in this way, it is easier to estimate the impact of system design on load variations and resulting airflow changes and energy use variations.

In VAV systems, zone loads are related to terminal unit selection and layout. Too-large or too-small zone loads may not be practical in selecting terminal units for the intended application. Also, excessive fluctuation in zone loads may cause comfort, IAQ, and noise problems. It is therefore important to analyze the impact of airflow variation on the performance of a VAV system. This is usually performed at the zone level for both peak and minimum airflow conditions. (See Fig. 11.1, Design Analysis Phase 1.)

The details of load component calculation are discussed in the appropriate chapters of ASHRAE Fundamentals and Cooling and Heating Load Calculation Manual (Ref. 1).

11.1.2 VAV system load analysis

Load calculations are divided into two parts, building- and system-related loads. In Fig. 11.2, space loads are building-related and can be computed without system considerations. Zone and coil loads are system-related and must consider the system impact on load variations. This impact may be insignificant and can be ignored, or it may become a major design concern, and its significance must be determined by proper load analyses.

There are two types of system loads—airside and energy plant system loads. In analyzing VAV system loads, the major concern is the airside system loads, which can be further broken into subcategories of zone and coil level reheat loads, ventilation, plenum heat, and airside transportation loads. (See Fig. 11.2, Load Summation column.)

For VAV system design, typical load analyses may include some or all of the analyses listed below:

Building impact analysis (Sec. 11.1.3)
Effect of climate (Sec. 11.1.4)
Ventilation requirements (Sec. 11.1.5)
Reheat load analysis (Sec. 11.1.6)

11.1.3 Building impact analysis

Building design influences cooling and heating loads in many different ways. Building usage, occupancy, lights, equipment and internal loads, building configuration and orientation, number of floors, mechanical space size and location, envelope area, and construction materials all increase or decrease cooling and heating loads (Fig. 11.3). Some of these factors may affect loads and therefore air quantities, which in turn affect VAV system design. The system designer should identify the magnitude of these changes at an early stage of design, preferably at the schematic design phase. This usually can be accomplished by a simple two-point or peak and minimum load point manual calculation.

11.1.4 Effect of climate

The major impact of climate on VAV systems is skin load variation. However, this impact can be analyzed only by associating it with the building design. For example, if the building design specifies a wall construction with 4-inch insulation, 20 percent window area, and a double-pane construction, the skin load may not be significant to justify a separate perimeter heating system even in a cold northern climate. Again, the system designer needs to exercise proper judgment to determine whether a skin load analysis is justifiable for the given building design.

11.1.5 Ventilation requirements

Ventilation is the most important consideration in IAQ-conscious VAV system design. The ASHRAE Standard 62-89, Indoor Air Quality Procedure, specifies 15 cfm/person as a minimum level of ventilation required. As the actual amount of outdoor air supply to each VAV zone is a pure function of zone loads and not related to type of occupancy and its ventilation requirement, it is mandatory to perform a ventilation analysis for various load and occupancy combinations (Table 11.1). Again, the system designer's experience and personal judgment must be applied to determine whether a complex simulation or a simple manual analysis is sufficient for the building and system under consideration.

11.1.6 Reheat load analysis

In a pure VAV concept, reheat is not required and should not be used, because the quantity of supply air is varied according to the space load fluctuation and should be reduced to zero when there is no load in the space. But certain practical considerations may not allow the air quantity to be reduced to zero, and reheat must be applied to create an artificial load to maintain the airflow required. As shown in Fig. 11.4, both internal and external factors cause this situation, and reheat becomes greatest when both the factors coexist.

Generally, reheat load analyses are performed for those VAV zones where the cooling load fluctuates or changes substantially while the

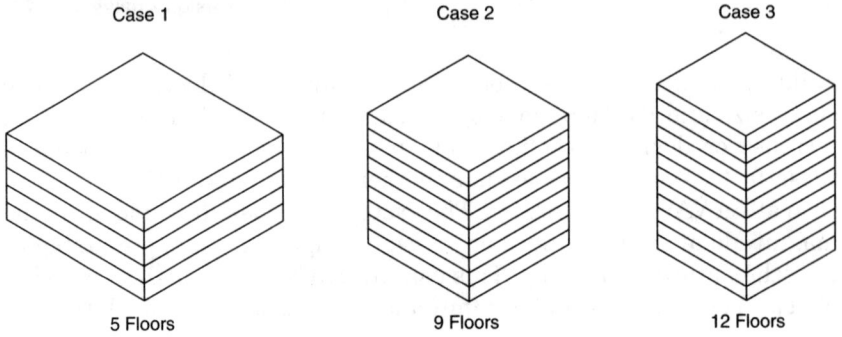

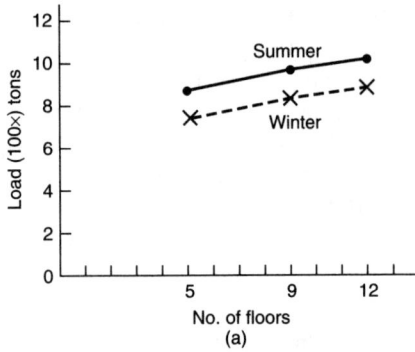

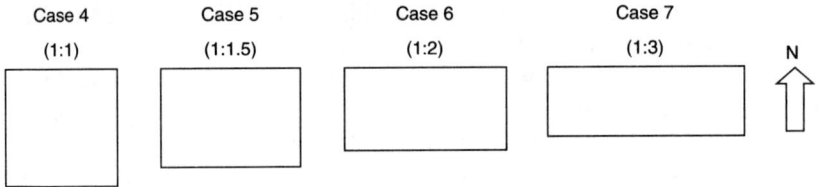

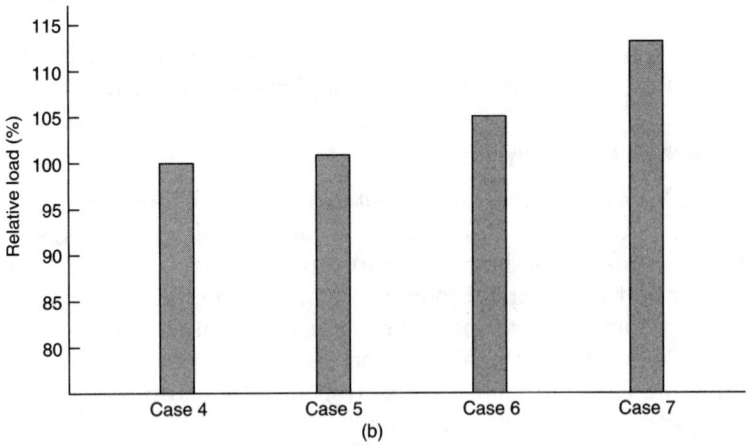

Figure 11.3 Impact of building design on load variations. (*a*) Impact of building heights; (*b*) Impact of building plans.

TABLE 11.1 Ventilation Analysis for Various Load/Occupancy Combinations

Load/Occupancy Combination	Urgency of Ventilation Analysis
Case A	
Large office areas combined with high occupancy density, high ventilation rate, but relatively low load density areas	This situation almost always assures the need of a ventilation analysis. It reveals the seriousness of ventilation imbalance quantitatively.
Examples	Degree of Urgency
1. Offices with a smoking lounge 2. Offices with an auditorium 3. Offices with a dining room 4. Offices with a conference room 5. Offices with a class room	Most urgent Least urgent
Case B	
Load/occupancy situation is similar to Case A above, except high occupancy density, high ventilation rate areas also have a high load density (cfm/ft^2 floor area).	Ventilation imbalance is less severe in this case. The results of a ventilation analysis reveal whether simple reheat or more drastic measures are required to solve the imbalance problem.
Examples	Degree of Urgency
Same as Case A above except load densities are high and fluctuate.	With repeated ventilation analyses a VAV designer can develop design limitations for combining various load/occupancy areas and avoid excessive imbalance of ventilation.

occupancy density remains relatively constant. Typical examples are building perimeter zones or internal zones with high-density occupancy (e.g., classrooms and auditoriums). Sometimes, reheat load analyses are performed to make certain that the room humidity is maintained below a maximum acceptable level, usually 60 percent RH under all load conditions.

11.2 Ventilation Analysis

11.2.1 Introduction

The purpose of ventilation in buildings is the dilution, capture, and removal of indoor pollutants inside the building to maintain acceptable indoor air quality. Here, *ventilation* refers only to the outdoor air introduced into the spaces and not to the total air supplied to the spaces. Thus, in a pure sense of ventilation, outdoor air is introduced and exhausted outdoors after diluting indoor contaminants (Fig. 11.5a). Ventilation is also accomplished by the supply air which contains a certain percentage of outdoor air. Figure 11.5b illustrates two cases of this type of ventilation: (1) dilution ventilation by the outdoor air portion of supply air, and (2) capturing and removing of local contaminants by supply air.

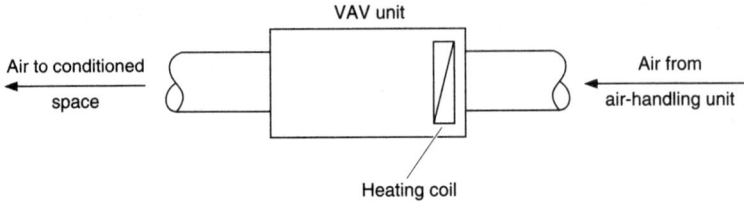

VAV unit

Air to conditioned space

Heating coil

Air from air-handling unit

Situation A
To maintain a minimum air supply
to compensate the effect of external factors
(skin heating loads and outdoor air infiltration
at a lower temperature)

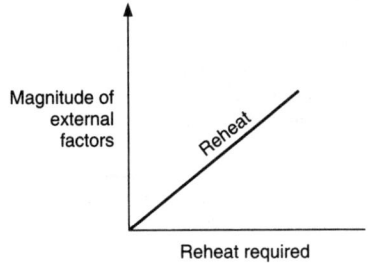

Magnitude of external factors

Reheat

Reheat required

Situation B
To maintain a minimum air supply and compensate
load reduction due to internal factors
(lights, equipment, and people loads)

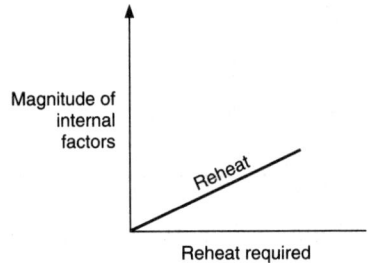

Magnitude of internal factors

Reheat

Reheat required

Situation C
Combination of Situations A and B

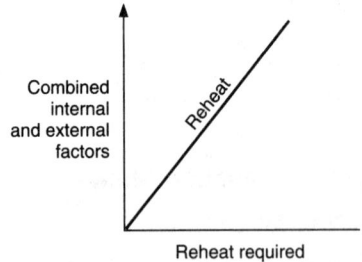

Combined internal and external factors

Reheat

Reheat required

Figure 11.4 Reheating of conditioned supply air.

For most VAV systems, ventilation is performed as shown in Fig. 11.5b. However, there is a distinctive difference between the ventilation by VAV and CAV (constant air volume) systems. In VAV systems, the supply and outdoor air supplied to each VAV zone, $V_{s,1}$, $V_{s,2}$, $V_{o,1}$, and $V_{o,2}$ are all variables continuously changing their values (Fig. 11.5c), while these same values are all constant for CAV systems. Obviously this causes a ventilation problem unless the corresponding indoor contaminants also change their concentrations proportionally. Since this

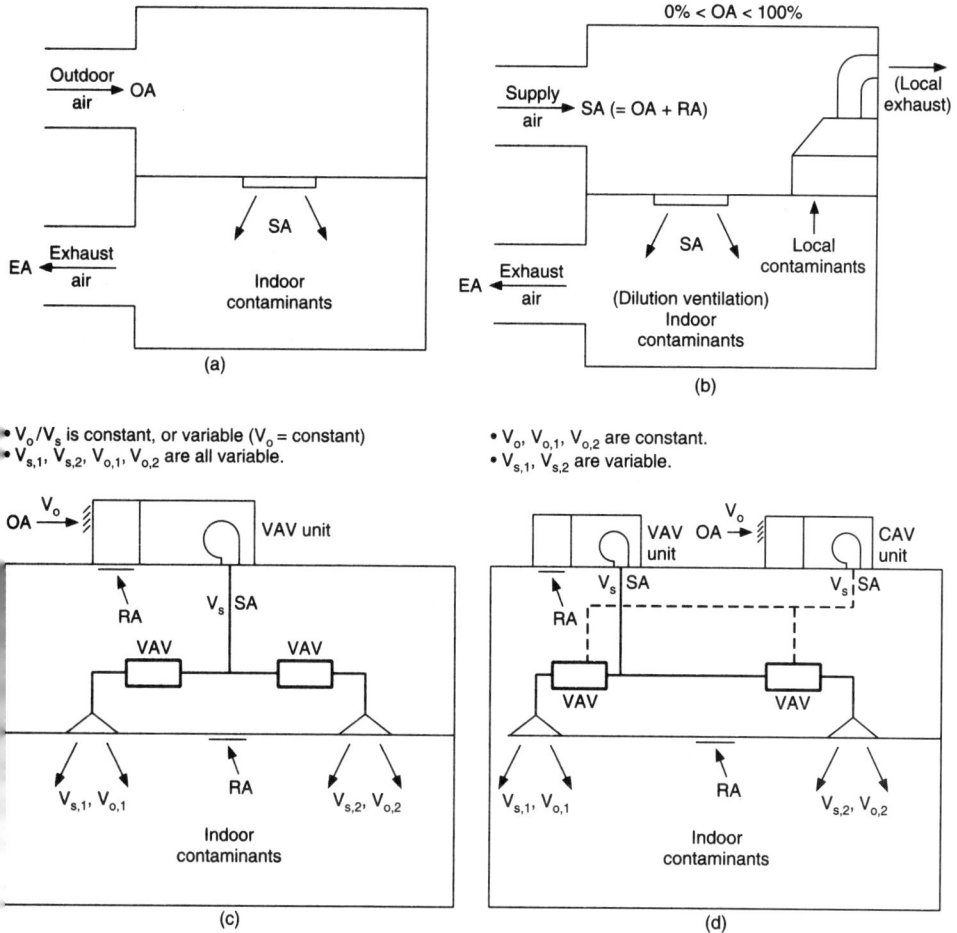

Figure 11.5 Principles of ventilation and VAV system ventilation. (a) Ventilation by dilution of contaminant concentrations; (b) Ventilation by capture and removal of contaminants; (c) VAV system with a common duct supplying OA for ventilation; V_o/V_s is constant, or variable (V_o = constant); $V_{s,1}$, $V_{s,2}$, $V_{o,1}$, $V_{o,2}$, are all variable; (d) VAV system with a separate system supplying OA to each VAV terminal for ventilation; V_o, $V_{o,1}$, $V_{o,2}$, are constant; $V_{s,1}$, $V_{s,2}$, are variable.

may or may not happen in a real VAV system operation, each VAV system must be analyzed for the amount of outdoor air introduced per occupant under both design and part load conditions, as well as the ventilation effectiveness of supply air in diluting, capturing, and removing indoor contaminants.

The preceding ventilation problem can be avoided by installing a separate and independent outdoor air system supplying a constant amount of outdoor air to each VAV zone irrespective of the variation in supply air. Figure 11.5d illustrates this situation, where V_o, $V_{o,1}$, $V_{o,2}$ are constant while V_s, $V_{s,1}$, $V_{s,2}$ are all variable. Since most VAV systems rarely utilize this type of system layout and operation, the following sections concentrate on the ventilation analysis of conventional VAV systems with a common air-handling and duct system supplying outdoor air to all VAV zones.

11.2.2 Outdoor air supply to each space

Ventilation codes and standards usually specify a minimum outdoor flow rate to each space in terms of air changes per hour and/or cfm per person. Table 11.2 lists part of the ASHRAE standard 62-1989 ventilation requirements (Ref. 2). While codes and standards specify minimum outdoor flow rates, the quantity of outdoor air actually supplied to each space is determined independently by the cooling load in each VAV zone. This often creates incompatibility between outdoor airflow rates and ventilation requirements. The following case study analyzes this incompatibility and how it affects the ventilation of VAV systems.

Design parameters affecting ventilation for acceptable indoor air quality are (1) floor area of each space (ft^2), (2) space supply air density (cfm/ft^2 floor area), (3) space people density ($person/ft^2$), (4) code/standard specified minimum ventilation rates ($cfm/person$). If all these parameters are in equilibrium with ventilation, the design of a ventilation system would be very simple and straightforward, as illustrated in Fig. 11.6. When these parameters differ considerably from space to space and from time to time for the same space, it may become necessary to modify the ventilation system to maintain an acceptable level of indoor air quality in the conditioned spaces. Figures 11.7 through 11.9 illustrate various combinations of design parameters mentioned above and their impacts on outdoor air supply to each VAV zone.

TABLE 11.2 Outdoor Air Requirements for Ventilation

(A Partial List of ASHRAE 62-1989, Table 2)

Application	Max. occupancy ft^2/person	Outdoor air requirements	
		cfm/person	cfm/ft^2 OA
Office	143	20	0.14
Conference rooms	20	20	1.0
Smoking lounge	14	60	4.3
Malls	50	—	0.2
Spectator rooms	7	15	2.2
Swimming pools	—	—	0.5
Auditorium	7	15	2.2
Waiting rooms (transportation)	10	15	1.5
Dining rooms	14	20	1.4
Meat processing (workrooms)	100	15	0.15
Classrooms	20	15	0.75

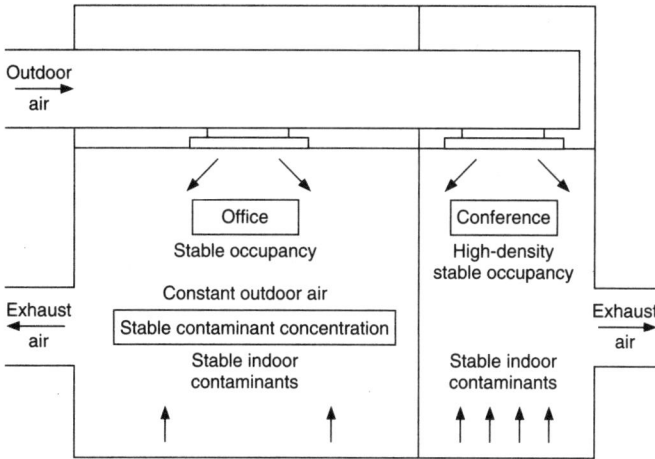

Major sources of space pollutants

- Building-oriented: Wall and floor finishes, furniture, etc.
- System- and equipment-oriented: Air-handling systems, copying equipment
- Human-oriented: Smoking and bioeffluents

Figure 11.6 Ventilation and indoor contaminants in equilibrium.

The outdoor air required per square foot of floor area varies considerably from application to application. In Table 11.2 the highest outdoor air density is 4.3 cfm/ft^2 in smoking lounges, and the lowest is only 0.14 cfm/ft^2 in general offices. It is clear from the table that certain applications are not suitable for VAV systems because these applications require constant and high rates of ventilation which goes contrary to the operating principle of most VAV systems, namely, the continuous variation of the conditioned supply air and the outdoor air contained in it.

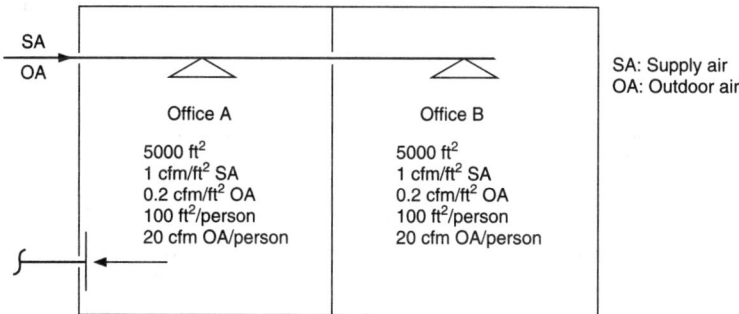

Total outdoor air supply required = 5000/100 × 20 + 5000/100 × 20 = 2000 cfm
% OA supply = 2000/(5000 + 5000) × 100 = 20% (OA/SA × 100)
cfm OA/person = 20 cfm for each office area

Figure 11.7 Uniform occupancy.

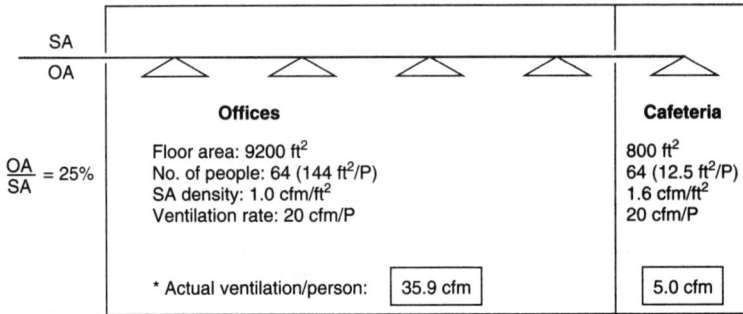

Figure 11.8 92 percent office area and 8 percent cafeteria.

Even for those applications with low ventilation rates, the same ventilation problem occurs if the cooling load in one part of the system drops drastically while the loads in other areas remain relatively stable. Figure 11.7 is a case in point. In this case, the design supply air and people densities are exactly the same, 1 cfm/ft^2 supply air and 100 ft^2/person. When supplied with 2000 cfm of outdoor air, this system meets the ventilation rate of 20 cfm/person at the design condition of 1 cfm/ft^2 of supply air, but not at nondesign, part load conditions.

For example, if the supply air volume is reduced to 0.2 cfm/ft^2, or 1000 cfm for office B and if the total outdoor air of 2000 cfm is maintained, the actual outdoor air supply to office B will be reduced to 6.7 cfm/person, while 33.3 cfm/person instead of 20 cfm/person is now supplied to office A. The situation becomes more complex as other parameters affecting ventilation are introduced, as demonstrated in the following case studies. Figures 11.8 and 11.9 show the impact of supply air and people densities on actual ventilation rate, or cfm/person.

Analysis A: system I versus system II. Although the people and supply air density are the same for both systems I and II, an increase in floor

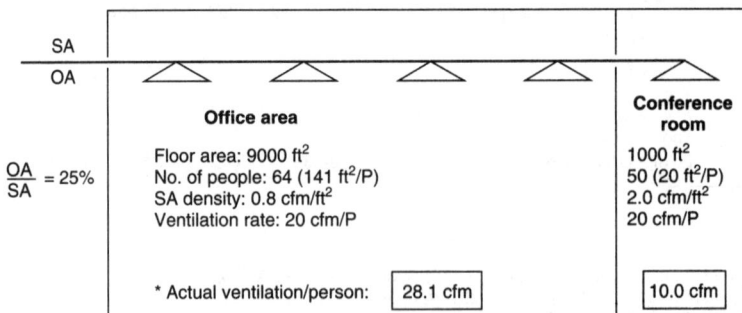

Figure 11.9 90 percent office area and 10 percent conference.

area has improved the ventilation in the conference room for system II from 6.6 to 13.8 cfm/person. This is because a larger floor area for the conference room for system II has brought in more supply air (from 1500 to 7500 cfm), which contains a large amount of outdoor air (from 22 to 46 percent). (See Figs. 11.10 and 11.11.)

Analysis B. Increasing the outdoor air contained in the total system supply air and also increasing the supply air amount, or cfm/ft^2 for a high-people-density space (such as a conference room), both improve ventilation in high-people-density spaces. However, which is better would depend on several design parameters. Where the annual cost of conditioning outdoor air is low, increasing outdoor air would be preferable. On the other hand, if the climate is humid and has a long summer season, it may be cheaper to increase the supply air to high-people-density areas and apply reheat whenever it is necessary to maintain an adequate space temperature. Figure 11.12 shows the impact of these design parameters on outdoor air/person to office and conference spaces.

For example, when 2 cfm/ft^2 is supplied to the conference room, 50 percent OA is required in order to supply 20 cfm of outdoor air per person. But for 4 cfm/ft^2 supplied to the conference room, the percent of outdoor air can be reduced to 25 percent and still meet the 20 cfm per

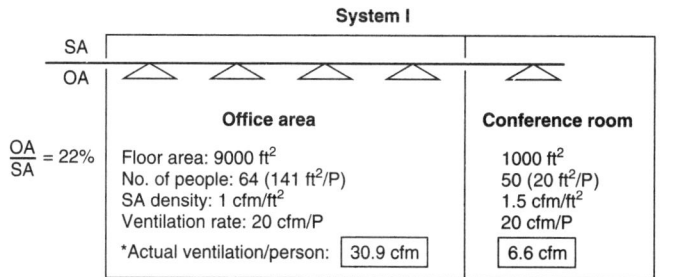

System I

	Office area	Conference room
SA		
OA		
$\frac{OA}{SA} = 22\%$	Floor area: 9000 ft^2 No. of people: 64 (141 ft^2/P) SA density: 1 cfm/ft^2 Ventilation rate: 20 cfm/P	1000 ft^2 50 (20 ft^2/P) 1.5 cfm/ft^2 20 cfm/P
	*Actual ventilation/person:	30.9 cfm	6.6 cfm

Figure 11.10 Large office area (90%) and a small conference room (10%) supplied by the same VAV HVAC system.

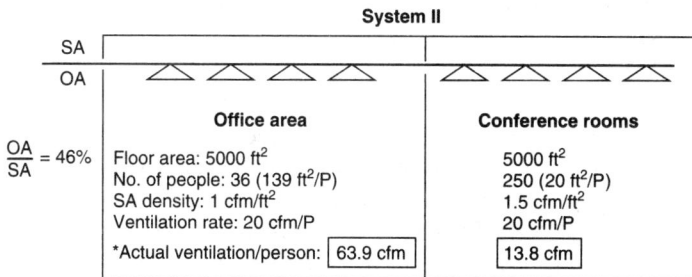

System II

	Office area	Conference rooms
SA		
OA		
$\frac{OA}{SA} = 46\%$	Floor area: 5000 ft^2 No. of people: 36 (139 ft^2/P) SA density: 1 cfm/ft^2 Ventilation rate: 20 cfm/P	5000 ft^2 250 (20 ft^2/P) 1.5 cfm/ft^2 20 cfm/P
	*Actual ventilation/person:	63.9 cfm	13.8 cfm

Figure 11.11 Equal floor area offices and conference rooms.

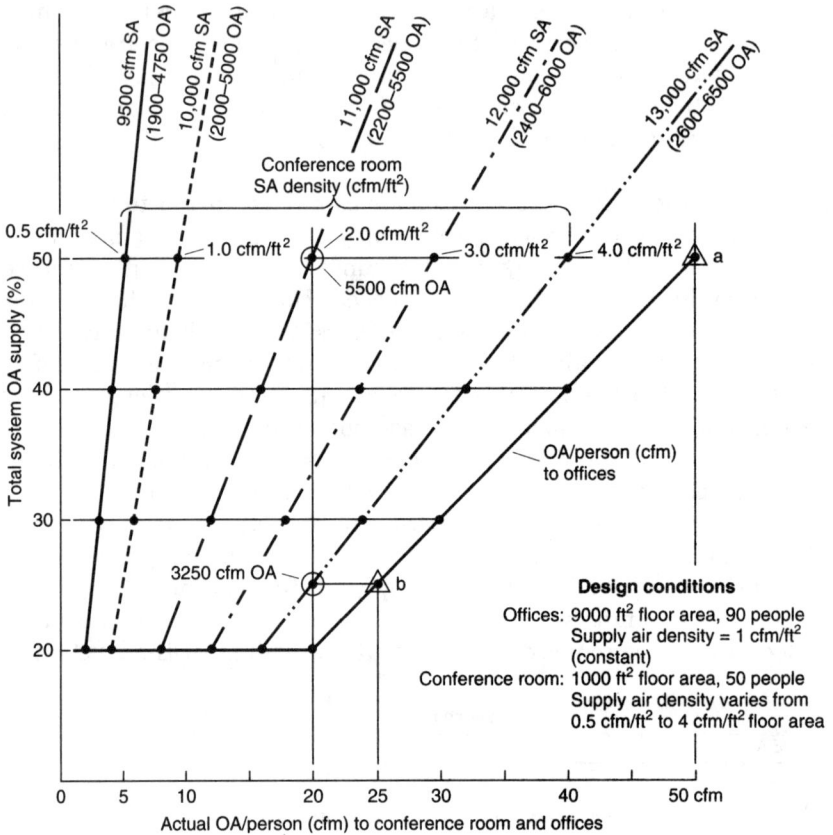

Figure 11.12 Impact of total OA supply and supply air density on outdoor air supply to spaces.

person ASHRAE standard requirement. In this case, the corresponding outdoor air supply will be reduced from 50 (point a in **Fig. 11.12**) to 25 cfm/person in the office area (point b).

11.2.3 Application of ventilation reduction allowance described in ASHRAE Standard 62-1989

From the preceding analyses, it is obvious that when high- and low-occupancy-density areas are combined and the required ventilation air is supplied by a VAV system with a common outdoor intake and supply duct, either oversupply or undersupply of ventilation air is unavoidable. In this situation the ASHRAE Standard 62-1989 specifies the use of Eq. (11.1) to calculate a minimum outdoor air quantity for an air-handling unit serving zones with different occupancy densities. (See Ref. 2.)

$$Y = \frac{X}{1 + X - Z} \qquad (11.1)$$

where $Y = V_{ot}/V_{st}$ = corrected fraction of outdoor air in system supply
  $X = V_{on}/V_{st}$ = uncorrected fraction of outdoor air in system supply
  $Z = V_{oc}/V_{sc}$ = fraction of outdoor air in critical space (the critical space is that space with the greatest required fraction of outdoor air in the supply to this space)
  V_{ot} = corrected total outdoor airflow rate
  V_{st} = total supply flow rate (i.e., the sum of all supply for all branches of the system)
  V_{on} = sum of outdoor airflow rates for all branches on system
  V_{oc} = outdoor airflow rate required in critical spaces
  V_{sc} = supply flow rate in critical space

The system outdoor air supply calculated by this equation is larger than the system outdoor air supply based on the summation of the minimum outdoor air supply per person to each zone in the system. However, it is smaller than the system outdoor air supply designed to meet the minimum outdoor air per person requirement for all zones in the system. For further explanation, see conclusion of analysis.

The following case studies summarize actual outdoor air supplied to each space using Eq. (11.1) under various building usage applications.

Case 1: 90 percent office area, 10 percent conference area.

Area	Floor area	People density	No. of people	Supply air	Min. cfm per person	Actual OA cfm per person	Min. OA req'd per ft^2
Office	9,000	140 ft^2/p	64	1 cfm/ft^2	20	55.5	0.142
Conference	1,000	20 ft^2/p	50	1.5 cfm/ft^2	20	11.9	1.0

$$X = \frac{9,000 \times 0.142 + 1,000 \times 1.0}{9,000 \times 1.0 + 1,000 \times 1.5} = \frac{1,280 + 1,000}{9,000 + 1,500} = \frac{2,280}{10,500} = 0.217$$

$$Z = \frac{50 \times 20}{1,000 \times 1.5} = 0.667$$

$$Y = 0.217/(1 + 0.217 - 0.667) = 0.395 \text{ or } 0.395 \times 10,500 = 4,148 \text{ cfm}$$

In this case, actual OA supply would be as follows:

$$\text{Conference: } 0.395 \times 1{,}000 \times 1.5 = 593 \text{ cfm or } \frac{593}{50} = 11.9 \text{ cfm/p}$$

$$\text{Office: } 0.395 \times 9{,}000 \times 1.0 = 3{,}555 \text{ cfm or } 55.5 \text{ cfm/p}$$

Thus, the office spaces would receive 55.5 cfm/p outdoor air, while the conference room would receive only 11.9 cfm/p.

Case 2: 50 percent office area, 50 percent conference area.

Area	Floor area	People density	No. of people	Supply air	Min. cfm per person	Actual OA cfm per person	Min. OA req'd per ft²
Office	5,000	140 ft²/p	36	1 cfm/ft²	20	80.4	0.144
Conference	5,000	20 ft²/p	250	1.5 cfm/ft²	20	17.4	1.0

$$X = \frac{5{,}000 \times 0.144 + 5{,}000 \times 1.0}{5{,}000 \times 1.0 + 5{,}000 \times 1.5} = \frac{5{,}720}{12{,}500} = 0.4576$$

$$Z = \frac{5{,}000}{7{,}500} = 0.667$$

$$Y = 0.4576/(1 + 0.4576 - 0.667) = 0.5788$$

In this case, the office spaces would receive 80.4 cfm/p outdoor air, while the conference rooms would receive approximately 17 cfm/p and percentage of system outdoor air supply would be increased from 39.5 percent for Case 1 to 57.9 percent for Case 2, but average outdoor air per person would be reduced from 36.4 cfm/p for Case 1 to 25.3 cfm/p for Case 2.

Case 3: 92 percent office area, 8 percent cafeteria space.

Area	Floor area	People density	No. of people	Supply air	Min. cfm per person	Actual OA cfm per person	Min. OA req'd per ft²
Office	9,200	140 ft²/p	64	1 cfm/ft²	20	143.8	0.139
Conference	800	12.5 ft²/p	64	1.6 cfm/ft²	20	20.0	1.6

$$X = \frac{9{,}200 \times 0.139 + 800 \times 1.6}{9{,}200 \times 1.0 + 800 \times 1.6} = \frac{2{,}560}{10{,}480} = 0.244$$

$$Z = \frac{64 \times 20}{800 \times 1.6} = 1$$

$$Y = 0.244/(1 + 0.244 - 1.0) = 1.0$$

In this case, the classroom would receive exactly 20 cfm/p, while the offices would receive in excess 143.8 cfm/p outdoor air, and the average outdoor air per person would be 81.9 cfm/p, far greater than the average outdoor air (36.4 cfm/p) for Case 1.

Case 4: All conditions same as Case 1, except supply air quantities changed from 1 to 0.8 cfm/ft² in offices and increased from 1.5 to 2.0 cfm/ft² in the conference room.

Area	Floor area	People density	No. of people	Supply air	Min. cfm per person	Actual OA cfm per person	Min. OA req'd per ft²
Office	9,000	140 ft²/p	64	0.8 cfm/ft²	20	37.4	0.142
Conference	1,000	20 ft²/p	50	2.0 cfm/ft²	20	13.3	1.0

$$X = \frac{9{,}000 \times 0.142 + 1{,}000 \times 1.0}{9{,}000 \times 0.8 + 1{,}000 \times 2.0} = \frac{2{,}278}{9{,}200} = 0.248$$

$$Z = \frac{50 \times 20}{1{,}000 \times 2} = 0.5$$

$$Y = 0.248/(1 + 0.248 - 0.5) = 0.332$$

In this case, the actual OA supply would be modified as:

Conference: $0.332 \times 1{,}000 \times 2.0 = 664$ cfm or $\dfrac{664}{50} = 13.3$ cfm/p

Office: $0.332 \times 9{,}000 \times 0.8 = 2{,}390$ cfm or $\dfrac{2{,}390}{64} = 37.4$ cfm/p

Compared to Case 1, the outdoor air supply would be increased 12 percent for the conference but would be decreased 32.6 percent for the offices.

11.2.4   Effectiveness of ASHRAE ventilation reduction method

Outdoor air calculations without applying ventilation reduction allowance described in ASHRAE standard 62-89

Case 1A: 90 percent office area, 10 percent conference area.

Area	Floor area	People density	No. of people	Supply air	Min. cfm per person	Actual OA cfm per person	Min. OA req'd per ft²
Office	9,000	140 ft²/p	64	1 cfm/ft²	20	30.5	0.142
Conference	1,000	20 ft²/p	50	1.5 cfm/ft²	20	6.5	1.0

Total outdoor air required $= (64 + 50) \times 20 = 2{,}280$ cfm

$\%\ \text{OA} = 2{,}280/(9{,}000 \times 1 + 1{,}000 \times 1.5) = 21.7\%$

Actual OA for offices $= 0.217 \times 9{,}000 \times 1.0 = 1{,}953$ cfm

Actual OA for conference $= 0.217 \times 1{,}500 = 326$ cfm

Comparison of Case 1 and Case 1A

Case No.	Total OA cfm	OA cfm per person	
		Office	Conference
1	4,148	55.5	11.9
1A	2,280	30.5	6.5

Case 2A: 50 percent office area, 50 percent conference area.

Area	Floor area	People density	No. of people	Supply air	Min. cfm per person	Actual OA cfm per person	Min. OA req'd per ft²
Office	5,000	140 ft²/p	36	1 cfm/ft²	20	63.6	0.144
Conference	5,000	20 ft²/p	250	1.5 cfm/ft²	20	13.7	1.0

Total outdoor air required $= (36 + 250) \times 20 = 5{,}720$ cfm

$\%\ \text{OA} = 5{,}720/(5{,}000 \times 1 + 5{,}000 \times 1.5) = 45.76\%$

Actual OA for offices = $0.4576 \times 5{,}000 \times 1.0 = 2{,}288$ cfm

Actual OA for conference = $0.4576 \times 5{,}000 \times 1.5 = 3{,}432$ cfm

Case 2B: 50 percent office area, 50 percent conference area, same as Case 2A except supply cfm to conference reduced from 1.5 cfm to 1 cfm/ft^2.

Area	Floor area	People density	No. of people	Supply air	Min. cfm per person	Actual OA cfm per person	Min. OA req'd per ft^2
Office	5,000	140 ft^2/p	36	1 cfm/ft^2	20	79.4	0.144
Conference	5,000	20 ft^2/p	250	1 cfm/ft^2	20	11.4	1.0

Total outdoor air required = $(36 + 250) \times 20 = 5{,}720$ cfm

% OA = $5{,}720/(5{,}000 \times 1 + 5{,}000 \times 1) = 57.2\%$

Actual OA for offices = $0.572 \times 5{,}000 \times 1.0 = 2{,}860$ cfm

Actual OA for conference = $0.572 \times 5{,}000 \times 1.0 = 2{,}860$ cfm

Comparison of Cases 2, 2A, and 2B

Case No.	Total OA cfm	Office	Conference
2	7,235	80.4	17.4
2A	5,720	63.6	13.7
2B	5,720	79.4	11.4

Case 3A: Same as Case 3 except total OA supply cfm is based on number of people without applying $Y = X/(1 + X - Z)$ equation.

Area	Floor area	People density	No. of people	Supply air	Min. cfm per person	Actual OA cfm per person	Min. OA req'd per ft^2
Office	9,200	140 ft^2/p	64	1 cfm/ft^2	20	35.1	0.139
Classroom	800	12.5 ft^2/p	64	1.6 cfm/ft^2	20	4.9	1.6

$$\text{Total outdoor required} = (64 + 64) \times 20 = 2{,}560 \text{ cfm}$$

$$\% \text{ OA} = 2{,}560/10{,}480 = 24.43\%$$

$$\text{Actual OA for offices} = 0.2443 \times 9{,}200 \times 1.0 = 2{,}247 \text{ cfm}$$

$$\text{Actual OA for classroom} = 0.2443 \times 800 \times 1.6 = 313 \text{ cfm}$$

Comparison of Case 3 and Case 3A

Case No.	Total OA cfm	OA cfm per person	
		Office	Classroom
3	10,480	143.8	20
3A	2,560	35.1	4.9

Case 4A: Same as Case 4 except total OA supply cfm is based on number of people without applying $Y = X/(1 + X - Z)$ equation.

Area	Floor area	People density	No. of people	Supply air	Min. cfm per person	Actual cfm per person	Min. OA req'd per ft^2
Office	9,000	140 ft^2/p	64	0.8 cfm/ft^2	20	27.9	0.142
Conference	1,000	20 ft^2/p	50	2.0 cfm/ft^2	20	9.9	1.0

$$\text{Total outdoor required} = (64 + 50) \times 20 = 2{,}280 \text{ cfm}$$

$$\% \text{ OA} = 2{,}280/9{,}200 = 24.78\%$$

$$\text{Actual OA for offices} = 0.2478 \times 9{,}000 \times .8 = 1{,}784 \text{ cfm}$$

$$\text{Actual OA for conference} = 0.2478 \times 1{,}000 \times 2.0 = 496 \text{ cfm}$$

Comparison of Case 4 and Case 4A

Case no.	Total OA cfm	OA cfm per person	
		Office	Conference
4	3,054	37.4	13.3
4A	2,280	27.9	9.9

				OA (cfm/person)		
		Total	System OA			
Case no.	Total SA	OA cfm	(% OA)	Zone office	Conference	Cafeteria
1	10,500	4,148	39.5	55.5	11.9	
1A	10,500	2,280	21.7	30.5	6.5	
2	12,500	7,235	57.9	80.4	17.4	
2A	12,500	5,720	45.8	63.6	13.7	
2B	10,000	5,720	57.2	79.4	11.4	
3	10,480	10,480	100	143.8		20.0
3A	10,480	2,560	24.4	35.1		4.9
4	9,200	3,054	33.2	37.4	13.3	
4A	9,200	2,280	24.8	27.9	9.9	

Summary of Analysis

Conclusion of analysis. The ASHRAE equation $Y = X/(1 + X - Z)$ is effective in increasing cfm/person outdoor air supply in the high-occupancy areas. However, its effectiveness is dependent upon the floor area, occupant density, and supply air in the high-occupancy areas.

For example, as summarized in the table below, Case 1 substantially improves the outdoor air supply per person to the conference room (from 6.5 to 11.9 cfm/person) while still limiting the system outdoor supply to 4,148 cfm, far less than 7,035 cfm required to supply 20 cfm/person to the conference (ASHRAE specified minimum outdoor air per person). However, there is no reduction in the system outdoor air supply for Case 3. Since the conference room is small (800 ft^2) compared to the office zone (9,200 ft^2), it would be far more economical in this case to increase the total supply air from 10,480 to 13,040 cfm than to increase the system outdoor air supply from 3,260 to 10,480 cfm.

Case No.	Total SA	System OA	Conference OA/person	Note
1	10,500	4,148	11.9	1
1A	10,500	2,280	6.5	2
1A-1	10,500	7,035	20	3
3	10,480	10,480	20	1
3A	10,480	2,560	4.9	2
3A-1	10,480	10,480	20	3
3A-2	13,040	3,260	15	4

1. System OA based on ASHRAE Eq. (11.1).
2. System OA = summation of zone minimum OA/person.
3. System OA meets the minimum OA/person in conference room.
4. Supply air to conference room increased from 1,280 to 3,840 cfm, and system OA is maintained at 3,260 cfm.

11.3 Thermal Comfort Analysis

11.3.1 The ISO 7730 model for comfort prediction

Comfort analysis tests system performance against parameters affecting comfort. It requires the quantification of the relationship between comfort parameters and predicted comfort. The ISO 7730 model (Ref. 3) is one of the mathematical models that can be used by VAV system designers to quantify these relationships.

The major parameters affecting comfort are:

- Metabolic rate resulting from a certain activity
- External work (equal to zero for most activities)
- Thermal resistance of clothing
- Mean radiant temperature
- Air temperature
- Relative air velocity
- Partial water vapor pressure in ambient air

As shown in Fig. 11.13, the ISO 7730 PMV/PPD prediction model (Ref. 3) calculates PMV and PPD values for a given set of the preceding comfort parameters. Table 11.3 defines predicted mean vote (PMV) and predicted percentage dissatisfied (PPD). Their relationship is shown in Fig. 11.14. Using this ISO model, the VAV system designer can manipulate certain comfort parameters such as air temperatures, air motion, and vapor pressure through system design, thus improving the perfor-

TABLE 11.3 PMV and PPD Calculations

Comfort parameter	Metabolism (M, w/m²)				Input/ output
	81	70	70	70	
Clothing (I_{cl}, M²·°C/W)	0.116 (.75 clo)	0.155 (1.0 clo)	0.078 (0.5 clo)	0.124 (.8 clo)	Input variable
Air temperature (T_a, °C)	30.0	22.0	25.0	24.0	Input variable
Mean radiant temperature ($\bar{t}_r$, °C)	30.0	22.0	25.0	22.0	Input variable
Relative air velocity (V_{ar}, m/s)	0.10	0.0	0.0	0.15	Input variable
Partial water vapor pressure, (P_a, Pa)	1,000	1,000	2,000	1,500	Input variable
PMV	1.6	0.1	0.2	–0.0	Output
PPD, %	56.8	5.1	6.1	5.0	Output

SOURCE: International Standard ISO 7730 (Ref. 3).

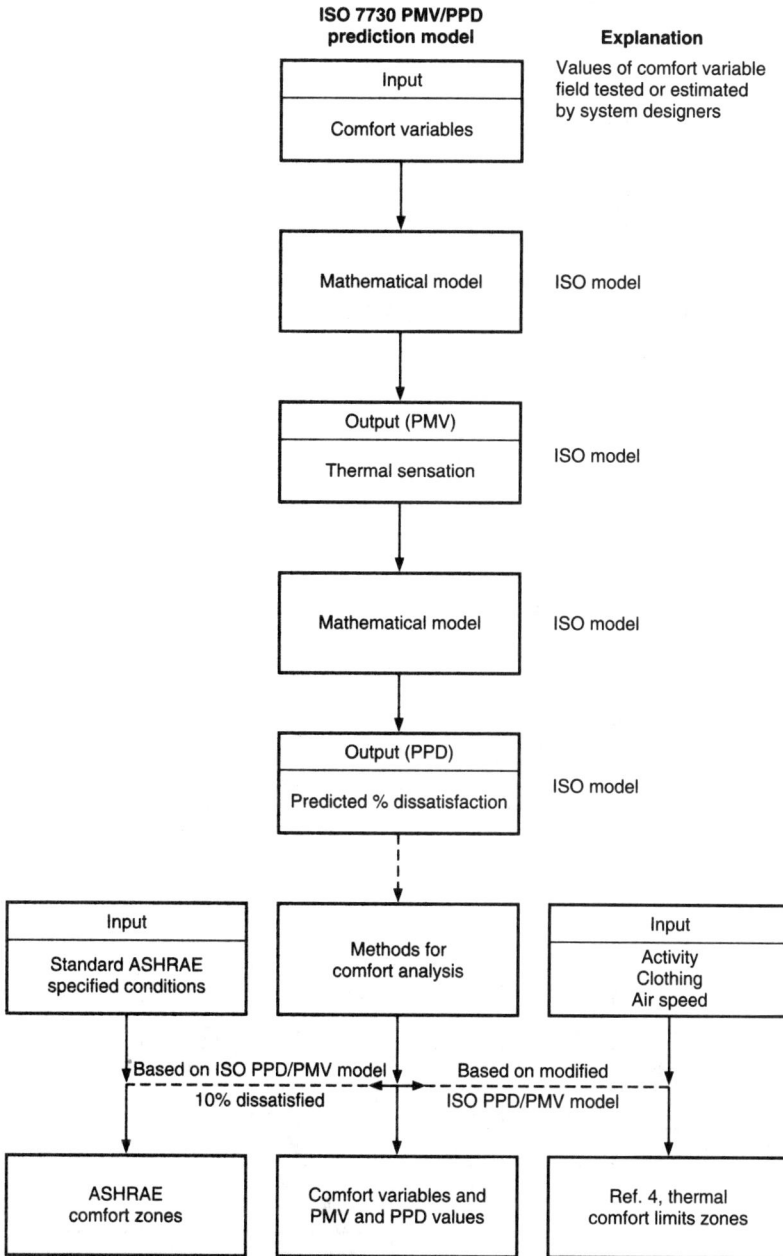

Figure 11.13 Process of comfort analysis.

mance of the system being designed, especially at part load conditions for VAV systems.

11.3.2 Other methods for comfort prediction

Using the ASHRAE comfort chart for sedentary occupants. The comfort chart in Fig. 11.15 shows the winter and summer comfort zones for the activity, air movement, and clothing specified. These zones are based on 90 percent acceptance, or 10 percent dissatisfied, using the ISO PPD/PMV program. This chart is therefore more restrictive than the method using the ISO program directly.

7-Point Thermal Sensation Scale

+ 3	hot
+ 2	warm
+ 1	slightly warm
0	neutral comfort
− 1	slightly cool
− 2	cool
− 3	cold

- Predicted mean vote (PMV) represents the mean vote of a large group of people on the above scale.

- Predicted percentage dissatisfied (PPD) is a normalized curve representing the percentage of people thermally dissatisfied.

The PPD predicts the percentage of a large group of people likely to feel thermally uncomfortable, that is, voting warm (+2), slightly warm (+1), slightly cool (−1) and cool (−2). Thus it establishes a quantitative prediction of the number of thermally dissatisfied people for a given PMV value.

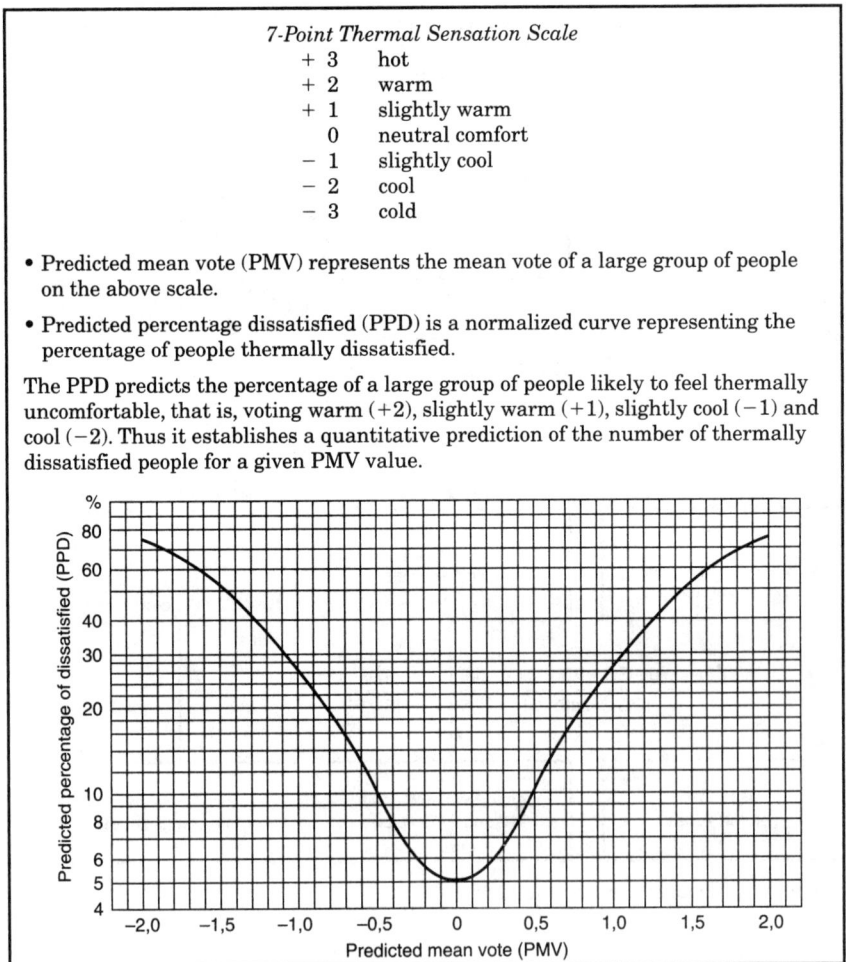

Figure 11.14 PPD as a function of PMV.

Using thermal comfort limits charts.   The thermal comfort limits (Ref. 4) shown in Fig. 11.16 is produced by a computer program which is based on the ISO program, but which includes the effective temperature ET in place of the operative temperature. Also, the terms which were not described fully in the ISO model have been modified and defined. (See Ref. 4 for details.)

In addition to the ISO model method, either of the two preceding methods can be used to develop realistic limitations for VAV system design. Comfort analysis then depends on whether the system designer can come up with a VAV system that can stay within the recommended design limits for its entire operating range. The following section illus-

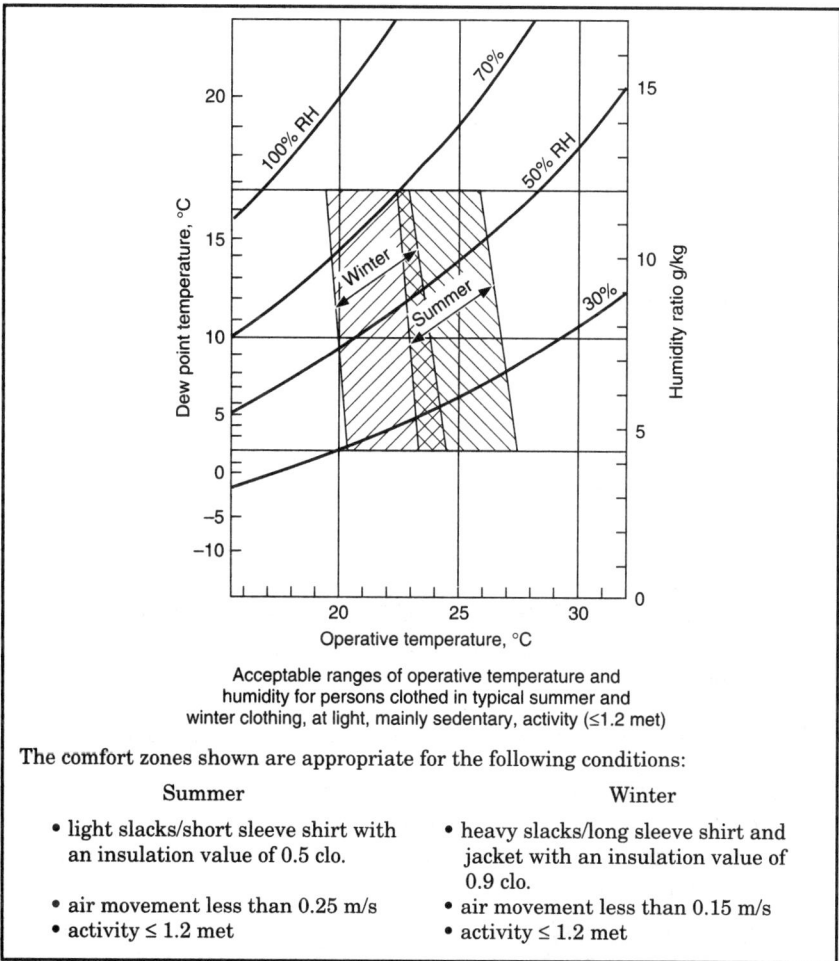

Acceptable ranges of operative temperature and
humidity for persons clothed in typical summer and
winter clothing, at light, mainly sedentary, activity (≤1.2 met)

The comfort zones shown are appropriate for the following conditions:

Summer	Winter
• light slacks/short sleeve shirt with an insulation value of 0.5 clo.	• heavy slacks/long sleeve shirt and jacket with an insulation value of 0.9 clo.
• air movement less than 0.25 m/s	• air movement less than 0.15 m/s
• activity ≤ 1.2 met	• activity ≤ 1.2 met

Figure 11.15   ASHRAE standard 55-1981 comfort zones for sedentary occupants.

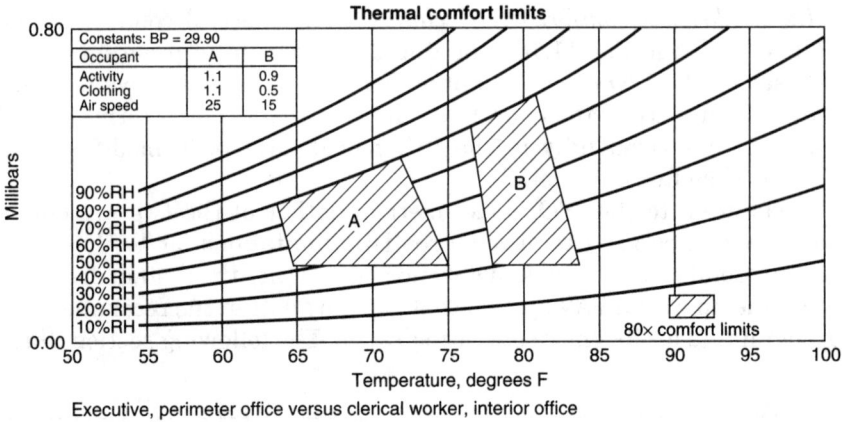

Figure 11.16 Thermal comfort limits. *(Source: Carrier Corp., Ref. 4. Reprinted with permission.)*

trates the process of developing such design limitations and analyzing their impacts on VAV system design.

11.3.3 Comfort limits and analysis of their impacts on VAV design

Using the ISO method

1. Enter various combinations of six comfort parameters and construct a comfort parameter, PMV, and PPD table (Table 11.3).

2. Compare the tabulated values with the performance of the VAV system being designed. For example, if the system is designed to maintain a space air temperature of 24°C (75.2°F) with 0.15 m/s (30 fpm) air movement, only 5 percent of the occupants with 0.8 clo clothing will feel thermally uncomfortable. This is a satisfactory design. On the other hand, Table 11.3 also lists a combination of six parameters that will produce a PPD value of 56.8 percent (column 1). This is clearly not acceptable, and no VAV system should be designed to perform under these conditions.

Using the ASHRAE comfort chart for sedentary occupants.

1. Select an air temperature and humidity combination from the design limits shown in the ASHRAE comfort chart (Fig. 11.15) for system design. This will ensure a comfortable operation of the system being designed.

2. If a VAV system needs to be designed to operate outside the design limits (or comfort zones), then increase air movement to see whether a higher air velocity can improve the comfort level.

3. If not, the VAV system must be redesigned to operate at a different set of conditions to meet the comfort requirements. This situation typically occurs for the air-conditioning of kitchens, laundries, and other facilities where high internal loads prevail and it is not economical to design HVAC systems for conventional air temperature, humidity, and movement. Usually a special combination of space temperature and air movement, such as spot cooling or high-velocity local air supply, will provide reasonable comfort and can be predicted by the comfort chart.

Using thermal comfort limits charts. As demonstrated in Fig. 11.16, this method (Ref. 4) provides realistic design limitations for space temperature and humidity for a given level of occupant activity, clothing, and ambient air movement. The VAV system designer can use these limits to establish design and operating parameters for the system being designed. For example, in Fig. 11.16 a design condition of 75°F and 50 percent RH will not satisfy both occupants A and B. Occupant A may feel warm while occupant B may complain of drafty air. Possible design solutions include two-zone temperature control, personalized task air-conditioning systems, or modification of clothing.

11.3.4 Local thermal discomfort

Thermal neutrality, or PMV = 0, which is the major objective of comfort design for VAV systems, is not the only condition for thermal comfort. Space occupants may not feel comfortable if one part of the body is warm and another cold. Such localized discomfort is caused by a variety of reasons. Major ones are:

1. Asymmetric radiant field resulting in lower mean radiant temperatures. *Examples:* cold wall and window surfaces.

2. Local convective cooling, or draft.

3. Contact with a cold or warm floor.

4. Uneven vertical temperature distributions in the occupied zone from the floor to the 6-ft level.

Most of these items can and should be controlled by the careful design of VAV systems. The impact of item 1 is addressed by the ISO model as one of its input, mean radiant temperature. However, its value must be preestimated before entering it as an input. The ASHRAE Standard 55-1992 covers the subject of draft in detail under 5.1.6.4 Draft. Items 3 and 4 are addressed by Standard 55, as well as the ISO 7730 standards in the following sections:

- *Floor temperatures:*   Sec. 5.1.6.3, Standard 55; Annex A, ISO 7730
- *Vertical temperature distribution:*   Sec. 5.1.6.1, Standard 55; Annex A, ISO 7730

These factors causing local discomfort should also be considered when the thermal comfort of a VAV system design is analyzed.

11.4   IAQ Analysis

11.4.1   Introduction

Many factors influence indoor air quality (IAQ) in buildings. Major ones include:

- Outdoor air quality
- Indoor pollutants caused by building occupants, materials, and equipment, including metabolic gases, tobacco smoke, formaldehyde, volatile organic compounds, and microorganisms
- Variable occupancy
- Outdoor air rates for ventilation
- Ventilation effectiveness
- Maintenance of ventilation system

TABLE 11.4   Calculation of Ventilation Air Requirements

Based on *European Guidelines for Ventilation Requirements in Buildings* Published by Commission of the European Communities and ASHRAE 62-89

Unit Conversion: 1 l/s = 2.119 cfm, $\epsilon_v = C_e/C_i$, $Q_c = 10 \cdot G/(C_i - C_o) \cdot 1/\epsilon_v$
Examples using both European Guidelines and USA Standards

Design conditions: Perceived Outdoor Air Quality (C_o) = 0.1 Decipol
Standard Indoor Air Quality (C_i) = 1.4 Decipol
No smoking, or 1 olf/occupant (20 cfm/p)   G = Sensory pollution load
Building olf load = 0.3 olf/m² floor   = 1 × 0.07 + 0.3 = 0.37
Occupancy: 0.07 occupant/m² floor (100 ft²/p)
Ventilation effectiveness (ϵ_v): 0.9 (overhead) 1.3 (displacement)

European Guidelines	ASHRAE 62-89
Ex. 1: ϵ_v = 0.9 (overhead vent.) $Q_c = 10 \cdot 0.37/(1.4 - 0.1) \cdot 1/0.9$ = 3.16 l/s (m²) = 0.62 cfm/ft²	Ex. 1: 0.7 person/100 ft² floor $Q_c = 0.7 \times 20/100 = 0.14$ cfm/ft²
Ex. 2: ϵ_v = 1.3 (displacment vent.) $Q_c = 10 \cdot 0.37/(1.4 - 0.1) \cdot 1/1.3$ = 2.19 l/s (m²) = 0.43 cfm/ft²	Ex. 2: 1 person/100 ft² $Q_c = 1 \times 20/100 = 0.2$ cfm/ft²
	Ex. 3: 2 person/100 ft² $Q_c = 2 \times 20/100 = 0.4$ cfm/ft²

The system designer can directly or indirectly influence all these factors. The purpose of IAQ analysis is to assess the impact of VAV system design and selection on these factors quantitatively and qualitatively, and determine what the system designer should do to provide acceptable indoor air quality for the entire life cycle of the VAV system.

11.4.2 Definition of indoor air quality (IAQ)

ASHRAE Standard 62-89 (Ref. 2) defines acceptable indoor air quality as "air in which there are no known contaminants at harmful concentrations as determined by cognizant authorities and with which a substantial majority (80 percent or more) of the people exposed do not express dissatisfaction." Report number 11 *Guidelines for Ventilation Requirements in Buildings* published by the European Communities, on the other hand, divides indoor air quality into two parts. First, the

TABLE 11.5 Outdoor Air Quality Standards as Set by the U.S. Environmental Protection Agency

Contaminant	Long term Concentration ug/m^3	Long term ppm	Long term Averaging	Short term Concentration ug/m^3	Short term ppm	Short term Averaging
Sulfur dioxide	80	0.03	1 year	365	0.14	24 hours
Total Particulate	75[a]	—	1 year	260	—	24 hours
Carbon monoxide				40,000	35	1 hour
Carbon monoxide				10,000	9	8 hours
Oxidants (ozone)				235[b]	0.12[b]	1 hour
Nitrogen dioxide	100	0.055	1 year			
Lead	1.5	—	3 months[c]			

[a] Arithmetic mean
[b] Standard is attained when expected number of days per calendar year with maximal hourly average concentrations above 0.12 ppm (235 ug/m^3) is equal to or less than 1, as determined by Appendix H to subchapter C, 40 CFR 50
[c] Three-month period is a calendar quarter.

TABLE 11.6 Outdoor Levels of Air Quality

	Perceived air quality decipol	Air pollutants Carbon dioxide mg/m^3	Air pollutants Carbon monoxide mg/m^3	Air pollutants Nitrogen dioxide µg/m^3	Air pollutants Sulfur dioxide µg/m^3
At sea	0	680	0–0.2	2	1
In towns, good air quality	<0.1	700	1–2	5–20	5–20
In towns, poor air quality	>0.5	700–800	4–6	50–80	50–100

* The values for the perceived air quality are typical daily average values. The values for the four air pollutants are annual average concentrations.
SOURCE: Table 5 of Report No. 11, Guidelines for Ventilation Requirements in Buildings prepared by Commission of the European Communities.

health risks of breathing the indoor air should be negligible. Second, the indoor air should be perceived as fresh and pleasant. The required ventilation rates are calculated using these criteria.

Although stated in different ways, the ASHRAE and European definitions of indoor air quality are very similar in substance. However, the ASHRAE standard and the European guidelines produce considerably different results in ventilation rate requirements. Also, the ASHRAE standard is more application-conscious and allows the system designer to modify ventilation rates if one or more of the following conditions apply: contaminants are present, ventilation effectiveness is other than unity, occupancy is intermittent or variable, or multiple spaces are ventilated by a common system. Because of this and other reasons, the ASHRAE standard and European guidelines result in considerably different ventilation requirements. This is illustrated in Table 11.4. Still, both definitions consider the dilution of outdoor contaminants by outdoor air ventilation as the major source of indoor air quality control.

TABLE 11.7   Exhaust Air Recirculation Cost-Benefit Analysis Projections for VA Medical Center, Cincinnati, Ohio

Exhaust air cleaned and recirculated	22,000 cfm
Refrigeration ton-hr saved	91,216 ton-hr per year
Heating fuel saved	59,923 therms per year
Humidification fuel saved	10,973 therms per year
	Annual Operating Costs and Savings:
Cooling cost saved	$4,925.68
Water and chemical cost saved	0.00
Heating cost saved	47,938.18
Humidification cost saved	8,778.11
Extra fan energy cost	(431.31)
Gas phase filtration medium and labor costs	(3,900.00)
Net operating cost savings	$57,311.00
	Capital Investment Costs and Savings:
Chiller capacity reduction	72 tons
Boiler capacity reduction	2,163 MBtuh
Chiller cost savings	$0.00
Boiler cost savings	$0.00
Other capital savings	$0.00
Gas-phase filtration equipment (less medium) cost	($13,000.00)
Extra ductwork, cost	($1,300.00)
Extra controls, cost	($2,600.00)
Net capital savings (cost if negative)	($16,900.00)
Simple Payback	0.29 years*

* Exclusive of air filtration system installation costs.
SOURCE: *Heating/Piping/Air Conditioning*, November, 1991. Reprinted with permission.

11.4.3   Analysis of other IAQ factors

The first three factors listed in Sec. 11.4.1 are not directly related to VAV systems, but they equally affect the quality of indoor air and should be analyzed qualitatively and, if possible, also quantitatively.

Outdoor air quality.   Tables 11.5 and 11.6 list outdoor air quality standards. The values shown should be used as the basis for IAQ analysis. If the actual quality of outdoor air is less than the required minimum, then the system designer needs to consider one or more design options, including location of outdoor air openings, to prevent intake of outdoor air contaminants, reduction of outdoor air intake, and treatment of both outdoor and recirculation air. A cost-benefit analysis should be performed to determine which design options should be used for an optimum result.

Indoor contaminants.   Most indoor contaminants are difficult to control directly by VAV systems. However, there are several design options available to capture, separate, and remove such contaminants.

1. *Local exhaust systems.*   If the generation of pollutants is local and concentrated in a relatively small area, this method is most effective and should be used whenever practical. Still, the supply and exhaust air balance should be carefully analyzed to ensure the proper space pressurization for the entire supply air flow range of the VAV system.

2. *Nighttime or preoccupancy flashing of the air inside the building.* It is often observed in actual buildings that the IAQ is the freshest in the morning and gradually worsens toward the end of the day. As building flashing is most effective with maximum circulation of 100 percent outdoor air, the system designer should analyze the feasibility of designing a VAV system capable of introducing and exhausting a large amount of outdoor air.

3. *Displacement air flow for tobacco smoke control.*   The ASHRAE IAQ standard (Ref. 2) specifies 4.2 cfm/ft^2 floor area of ventilation air supply for smoking lounges. This air quantity is not practical for conventional VAV systems. However, the system designer should analyze the feasibility of applying displacement airflow concepts to improve ventilation effectiveness, thus reducing the ventilation air requirement, and making it an integral part of the VAV system being designed.

4. *Airborne contaminant control by air filtration.*   Airborne contaminants are divided into solid, liquid, and gaseous types and come in a variety of sizes. Depending upon their characteristics and sizes, these contaminants (except for carbon monoxide and carbon dioxide) can be effectively removed either by particle removal or gas-

phase filtration. A computer program is available (Ref. 5) to analyze the feasibility of cleaning and recirculating normally exhausted air. The analysis is based on a statistically sound method of accurately projecting the cost of heating and cooling outdoor air in a given location. An example of such an analysis is shown in Table 11.7.

11.4.4   System maintenance and operation

System maintenance and operation are not direct design parameters considered in VAV system design. However, these parameters seriously affect the quality of indoor air in many buildings. The system designer should therefore design VAV systems that can facilitate the system maintenance and operation for good indoor air quality.

- Provide sufficient maintenance space around the air-handling equipment for proper housekeeping.

- Specify easy-to-replace air filters and provide a sufficient space for filter removal.

- Provide a sufficient space without obstructions in front of ventilation supply or return outlets.

- Document design intent clearly, particularly in connection with IAQ control.

11.5   Noise Analysis

11.5.1   Introduction

Guidelines and checklists for noise control are always useful and should be followed by system designers in designing quiet VAV systems. In this design approach, system designers simply follow the rules of dos and don'ts for acoustic design. The process is straightforward and uncomplicated. Yet these rules do not provide quantitative information to define the quality of the acoustic design; following, compromising, or sometimes ignoring the rules are decisions the designer must make. The installation of VAV terminals, especially fan-powered terminals in an occupied space, is a case in point. Usually, one of "don'ts" rules recommends not to install fan-powered terminals directly above a space where the required sound rating is RC(NC)40 or less. However, this rule is often unenforceable in many design situations, and system designers need to know the answers to what-if questions, such as "what would be the room sound level if I place a fan-powered terminal here?" or "would the sound level be reduced if the ceiling is acoustically treated even though I still place the fan-powered terminal at the same location?" Manual or computer-aided noise analyses provide answers to these questions and should be performed for all VAV system designs (Fig. 11.17).

Table 11.8 shows the pros and cons of the manual and computer-aided techniques. In fact, the two techniques supplement each other and can be used interchangeably. However, manual techniques are more suitable for a preliminary noise estimate at the conceptual design stage, or a quick estimate of the alternation effect by adding or replacing an acoustic element. On the other hand, computer-aided methods are ideal for multipath noise analyses, applying a what-if technique for

Schematic Stage Design Considerations	Noise Analysis
• Consider placing fan powered terminals outside the conference room. If not practical, make a manual analysis to estimate the resulting noise level at 1000 Hz band.	• Conduct a single band noise analysis. Use manufacturer's typical sound data and average construction and noise absorption data for room surfaces.
• If the results are not acceptable; 1) Try a new terminal layout 2) Find architectural solutions, or 3) Use combined mechanical/architectural solutions.	• Based on design considerations 1), 2) & 3), repeat one band analysis. • Determine the best course of action for the next stage of building design.

(a)

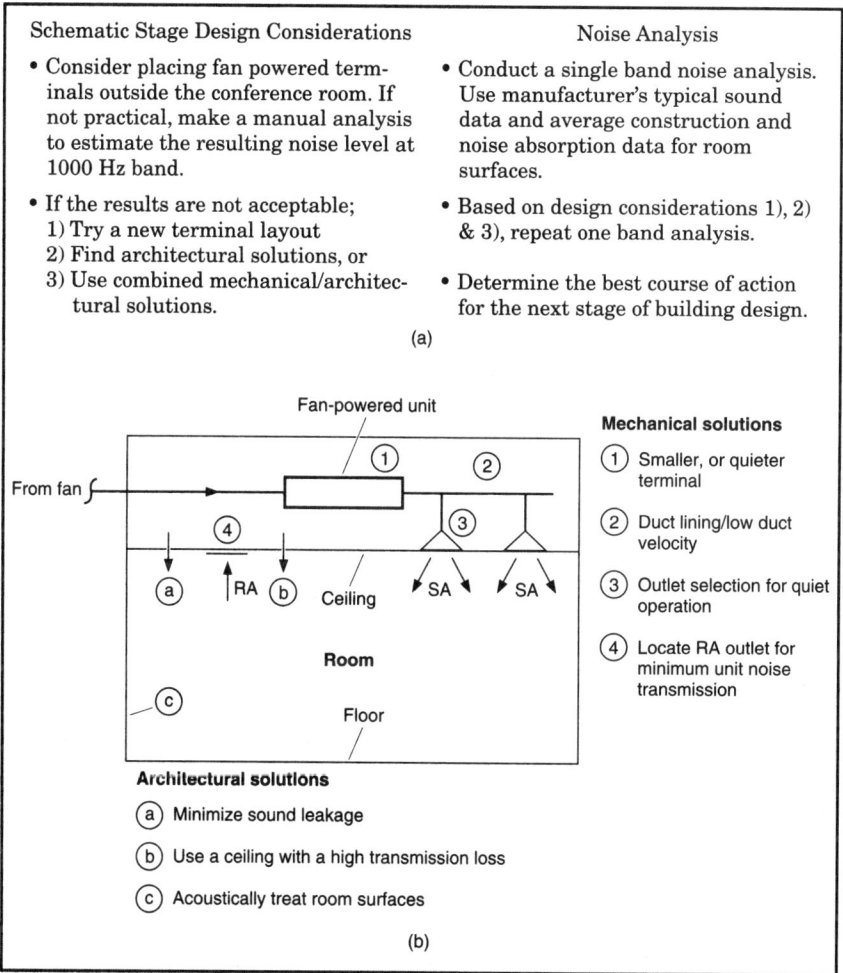

(b)

Figure 11.17 Design interaction and manual noise analysis. (*a*) Placing a fan-powered terminal directly above a conference room; (*b*) Architectural/mechanical solutions.

fine-tuning of acoustic design at the more advanced design stages, justification of certain acoustical treatments, and final documentation of VAV system design data.

11.5.2 Manual methods of noise analysis

The ASHRAE Handbook (Ref. 6) offers a manual method for HVAC designers who estimate the space noise level by manual calculations. Reference 7 also provides a similar manual technique but in a much easier to use table format. The manual approach is generally adequate for a design situation involving a single-source, single-path noise transmission without too many acoustic elements. Figure 11.18 is an example of such a situation. In this case the fan is the major noise source, and all other potential noise sources such as VAV terminals, outlets, and duct fittings are not considered in the analysis. As most of design situations involve more than one source and path, the manual approach has become extremely cumbersome for meeting today's stringent demand for quality acoustic design. There are many reasons for this: (1) ever-shrinking design time and budget; (2) frequent architectural changes; (3) severe space limitations; (4) increasing use of packaged equipment, VAV, and fan-powered terminals which require more acoustical considerations; and above all, (5) difficulty of using manual methods to try different acoustic elements quickly and flexibly to see the impact of such changes on the space sound spectrum. To cope with

TABLE 11.8   Manual versus Computer Method of Noise Analysis

Item	Manual	Computer-aided
Input preparation	Flexible, user can make judgmental modifications of data	More rigid, must follow certain input criteria
Computation	Can be very time consuming	Automatic, by computer
Graphic presentation	Manual, usually somewhat time-consuming	Instant, by computer
Indication of resulting acoustical quality of the space	Must plot the noise spectrum first	Automatic, by computer
Multipath noise analyses and subtraction (attenuation) all by computer	Tedious and time-consuming	Automatic noise addition
Regenerated noise calculations	Table lookup or through tedious manual calculation	Automatic, by computer
Space/noise acoustical models	User dependent but unlimited	Limited to certain models built into the program

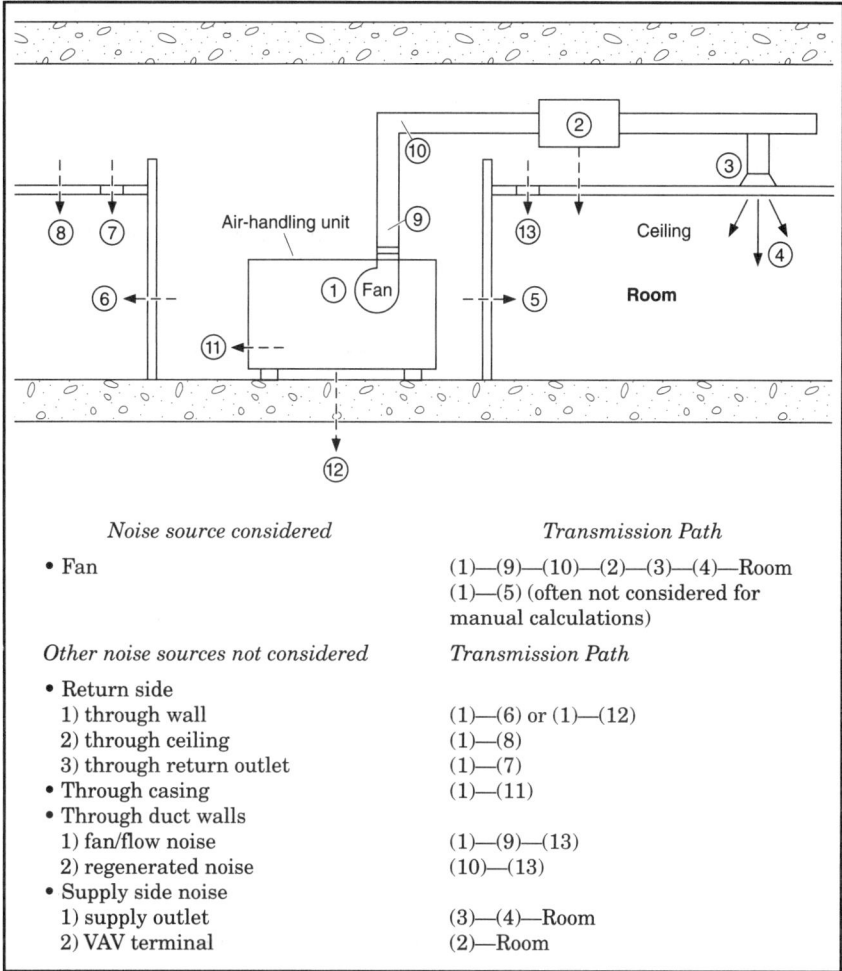

Figure 11.18   An example of manual noise analysis.

this new design trend, system designers need a computer-based design tool to analyze multisource, multipath noise problems.

11.5.3   Computerized multisource, multipath noise analysis

Currently, computer programs (Refs. 8 and 9) are available for multi-source, multipath noise analyses. These programs are based on the ASHRAE algorithms and can be very effective in providing a quality acoustic design by integrating both HVAC and architectural requirements and finding necessary mechanical and architectural solutions for noise control.

In a conventional design approach, acoustic design is routinely limited to specifying standard length of silencers on the supply side and sometimes also on the return side of each fan, and adding acoustically lined ductwork at predetermined locations irrespective of their effectiveness in providing the required attenuation. This simplistic approach is increasingly becoming inadequate for the reasons previously explained.

Path 1: Penthouse supply fan

Acoustic element

1. Supply fan (sound source)
2. Silencer
3. Elbow unvaned
4. Duct unlined
5. Branch division
6. Duct unlined
7. Duct unlined
8. Elbow unvaned
9. Branch division
10. Duct unlined
11. Elbow unvaned
12. VAV terminal (sound source)
13. Duct unlined
14. Tee
15. Duct unlined
16. Branch division
17. Duct unlined
18. Floor outlet (sound source)
19. End reflection
20. Space effect

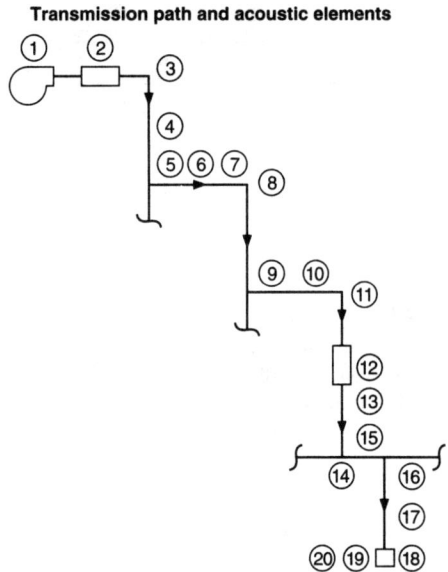

Space L_p (receiver)

Path 2: Penthouse return fan

Acoustic element

1. Return fan (sound source)
2. Silencer
3. Duct unlined
4. Elbow unvaned
5. Elbow unvaned
6. Branch division
7. Duct unlined
8. End reflection
9. Plenum/ceiling effect
10. Ceiling
11. Space effect

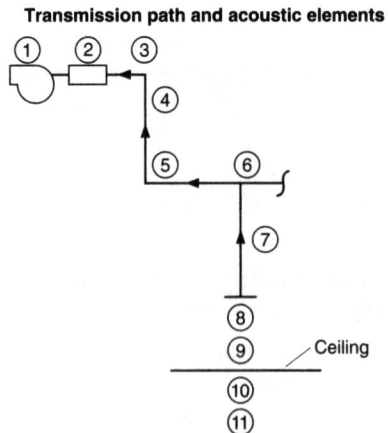

Space L_p (receiver)

Transmission path and acoustic elements

Figure 11.19 Acoustic system description.

Once a VAV system is adequately described acoustically (Fig. 11.19), a computer analysis can provide a quick noise prediction with a complete breakdown of the attenuation as well as noise regeneration of each acoustical element. Then the designer reviews the performance of each acoustic element and, if necessary, modifies or replaces certain acoustic elements in the system to improve the noise level and quality of the resulting space sound spectrum. When used with an appropriate

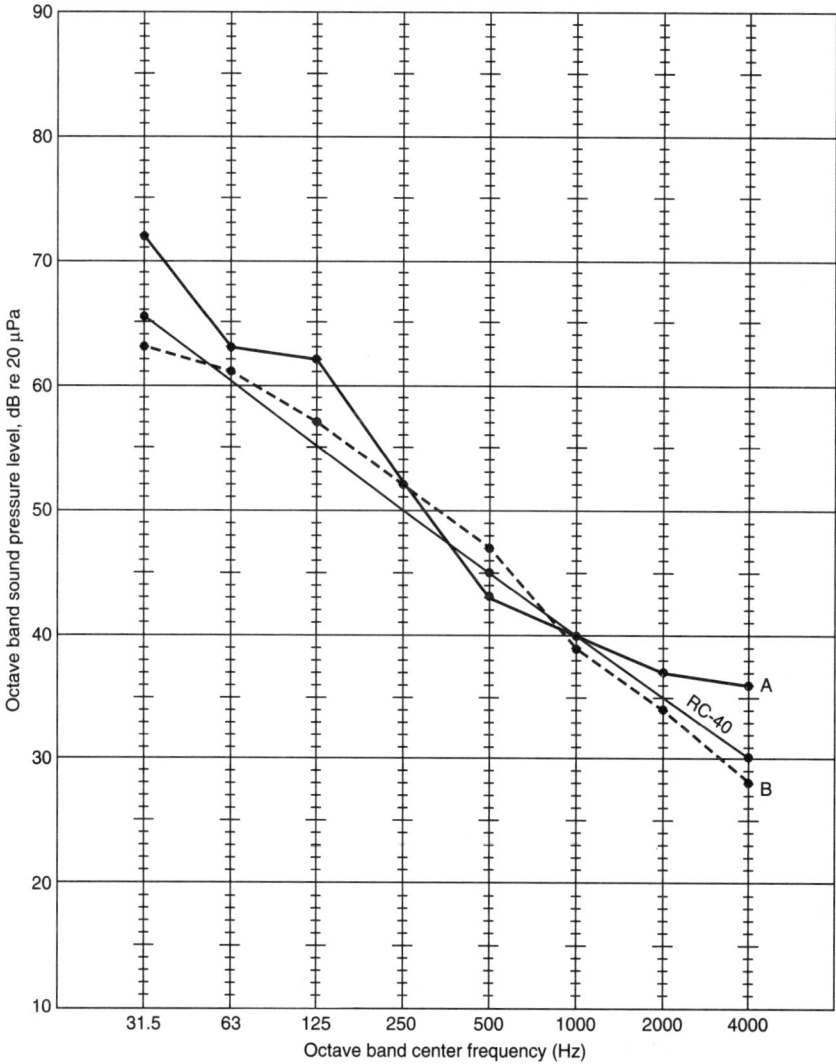

Figure 11.20   Quality and levels of space sound spectra.

what-if technique, this approach can readily result in an optimum acoustic design.

Needs for a what-if technique for computer-based noise analysis. The ultimate goal of acoustic design is to produce an acceptable "neutral" background sound level in occupied spaces. The quality as well as the level of the space sound spectrum are equally important. In Fig. 11.20, the sound spectra A and B are both at RC-40. But while spectrum B is acoustically neutral, spectrum A is not, because it is hissy and also has a tone at the 125-Hz band. Therefore curve A is not a quality spectrum meeting the design goal of an acceptable acoustical environment for general office applications. Generally, the what-if trial must be repeated a few times before reaching a neutral sound spectrum. This process is cumbersome even with a computer, and a new technique needs to be developed for systematic and logical what-if trials.

11.6   Building Pressurization Analysis

11.6.1   Factors affecting building pressurization

Building pressurization is a means to control infiltration and exfiltration to and from buildings. Build-up pressure is affected by both natural and mechanical forces. Wind and stack effect are the main natural forces, while mechanical forces are usually applied by ventilation or outdoor air supply systems. To maintain a suitable pressure inside the building, a mechanical ventilation system is employed to pump outdoor air into the building. How much air is required for proper pressurization depends on many factors: wind velocity and direction, building height, indoor and outdoor temperature difference, crack and opening dimensions, their locations, and the amount of pressurization desired.

As these factors are all variables that are different from building to building, suitable mathematical models or equations should be used to estimate wind pressures and stack-effect-induced pressure differences (Table 11.9) and the resulting infiltration and exfiltration airflow (Fig. 11.21).

11.6.2   Building pressure and airflow analysis

Building pressurization analysis is best understood by practical application examples. Figure 11.22 and Table 11.10 show pressures and airflows due to both natural and mechanical forces for a large industrial plant of 1,000,000 ft² floor.

TABLE 11.9 Wind- and Stack-Effect-Induced Pressures

Wind Pressure	Pressure Difference due to Stack Effect
The wind pressure or velocity head is given by Bernoulli's equation assuming no height change or head losses:	Stack effect pressure, when measured at a height h is:

$$\Delta P_s = (\rho_o - \rho_i)g(h - h_{NPL})$$

$$= (cf)B(h - h_{NPL})\left(\frac{1}{T_o} - \frac{1}{T_i} \right)$$

$$\Delta P_v = (cf)C_p \, \tfrac{1}{2}\rho V^2$$

$$= \rho_i g(h - h_{NPL})(T_i - T_o)/T_o$$

where

ΔP_v = wind pressure relative to the undisturbed flow, in. H$_2$O

ρ = air density, lb/ft^3

v = wind speed, mph

C_p = surface pressure coefficient

cf = conversion factor 0.0129, assuming an air density of 0.075 lb/ft^3

where

ΔP_s = pressure difference due to stack effect, in. H$_2$O

B = atmospheric pressure, 14.7 psi

ρ = air density, 0.075 lb/ft^2

g = gravitational constant, 32.2 ft/s^2

h = height of observation point, ft

h_{NPL} = height of neutral pressure level, ft

T = absolute temperature, °R

cf = conversion factor, 0.52

Subscripts
i = inside
o = outside

SOURCE: *ASHRAE Handbook 1985, Fundamentals,* chap. 22. Reprinted with permission.

Analysis for the building described in Fig. 11.22

1. The building is a relatively airtight structure. It has a total crack and opening area of 74 ft^2 for a total surface area of 1,155,750 ft^2, or a porosity of 0.0064 percent, and only 66,230 cfm or 0.066 cfm/ft^2 floor area of outdoor air is required to maintain a building pressure of 0.05 inches w.g.

2. For more porous buildings, maintaining a 0.05-inch-w.g. building pressure may not be practical or even feasible because of a large quantity of preheated outdoor air required.

3. The example is a low-height structure, and therefore stack effect is not a problem. However, for high-rise buildings, it may become a significant factor for inducing infiltration airflow.

11.6.3 High-rise building infiltration analysis

Two things are distinctive for high-rise buildings: stack effect and a higher wall-to-floor-area ratio. The wall-to-floor-area ratio for the plant described in Fig. 11.22 is 15.6 percent, while the same for the high-rise

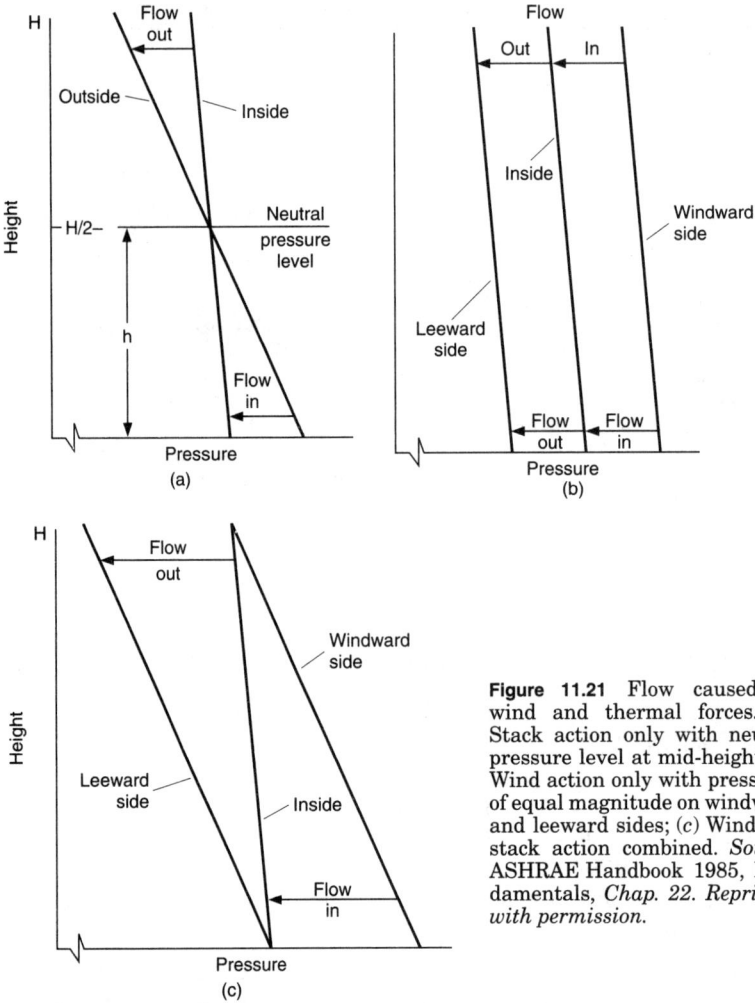

Figure 11.21 Flow caused by wind and thermal forces. (a) Stack action only with neutral pressure level at mid-height; (b) Wind action only with pressures of equal magnitude on windward and leeward sides; (c) Wind and stack action combined. *Source:* ASHRAE Handbook 1985, Fundamentals, *Chap. 22. Reprinted with permission.*

building in Table 11.11 is 33.3 percent. Items B, C, and D in Table 11.11 summarize infiltration for the walls, doors, and buildings under various temperature, wind, and pressurization conditions.

Analysis for the high-rise office building described in Table 11.11.

1. The wall is an average-fitting curtain wall construction. Infiltration due to stack effect is only 6300 cfm. The combined wind and stack infiltration is 6550 cfm. The change is insignificant. However, there is a considerable redistribution in leakage because of wind pressure. In this example, infiltration will increase on the wind side of the lower portion of the building, but decrease on the upper portion.

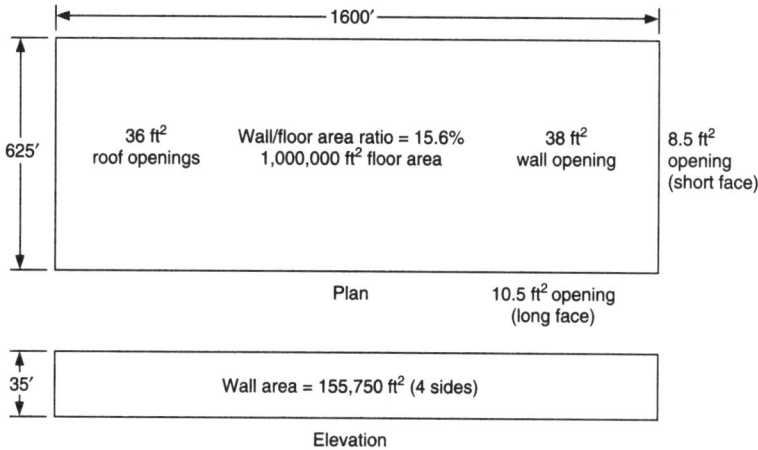

Figure 11.22   Building description. *(Source: Ref. 10. Reprinted with permission.)*

TABLE 11.10   Pressures and Flows by Natural and Mechanical Forces

Table 11.10a—Velocity Pressures

Velocity, FPM	Pressure, in w.g.
895	0.05
801	0.04
693	0.03
566	0.02
400	0.01

Table 11.10b—Wind Velocity Pressures

Wind, mph	Upwind, in w.g.	Downwind, in w.g.
30	+0.348	−0.174
15	+0.087	−0.043
10	+0.039	−0.019
7.5	+0.022	−0.011

Table 11.10c—Outdoor Air Required for Building Pressurization and Prevention of Infiltration

Wind Speed (mph)	Required air volume to prevent infiltration (CFM)	Equivalent Velocity (FPM)	Equivalent Pressure (in. w.g.)
0	26,000	351	0.0076
0	66,230	895	0.05
15	60,000	810	0.041
30	107,000	1,446	0.13

SOURCE: Ref. 10. Reprinted with permission.

TABLE 11.11   High-rise Building Infiltration

(From Examples 5.4, 5.6, 5.9 in Ref. 1)

A. Building description
   20-story office building, floor height 10 ft
   Floor area = 150 × 100 ft/floor or 300,000 ft² total floor area
   Indoor temp = 72°F, outdoor air = 5°F
   Thermal draft coefficient = 0.80
   Wall/floor-area ratio = 33.3%

B. Infiltration through wall
   Infiltration = 6,300 cfm (no wind, no pressurization)
   Infiltration = 1,260 cfm (0.1″ pressurization, no wind)
   Infiltration = 6,550 cfm (15 mph wind, no pressurization)
   6,550 ÷ (300,000) = 0.0218 cfm infiltration/ft² floor area at 15 mph and no
     pressurization
   6,550 ÷ (300,000) × 1 = 0.0218 or 2.18% of supply air (1 cfm/ft² assumed)

C. Infiltration through doors
   Two-door-vestibule-type entrance on windward side
   One single-bank swinging doors on leeward side
   Door size: 3 ft × 7 ft, ⅛″ perimeter gap around each door
   One door carries no traffic; other doors handle 900 people/hour
   Infiltration due to closed door leakage and due to traffic = 8,400 cfm

D. Infiltration for building
   Wall and door infiltration = 6550 + 8400 =14,950 cfm (without pressurization)
   Same, but with 0.1″ pressurization = 23,000 cfm (0.1 cfm/ft² floor area)

2. Door infiltration is significant in this building. Only three active doors provide 8400 cfm infiltration airflow through the building. This quantity is larger than the wall infiltration for the entire building (6550 cfm).

3. The combined wall and door infiltration is 14,950 cfm. This figure will be increased to 23,000 cfm, or roughly 0.1 cfm per ft² of floor area, if the building is maintained at 0.1 inch positive pressure. Generally, building pressurization increases the cooling and heating loads because of increased outdoor air supply required. The common practice of arbitrarily specifying a building pressure should be carefully analyzed to make certain that the resulting outdoor air requirement does not exceed the outdoor air needed for ventilation. If the outdoor air requirement for pressurization becomes excessive, one or more of the following design options should be considered:

   a. Reduce the level of pressurization.

   b. Adopt a tighter building construction.

   c. Use better-constructed doors, doors with a vestibule, or revolving doors.

   d. Limit outdoor air supply to a level sufficient only to neutralize infiltration.

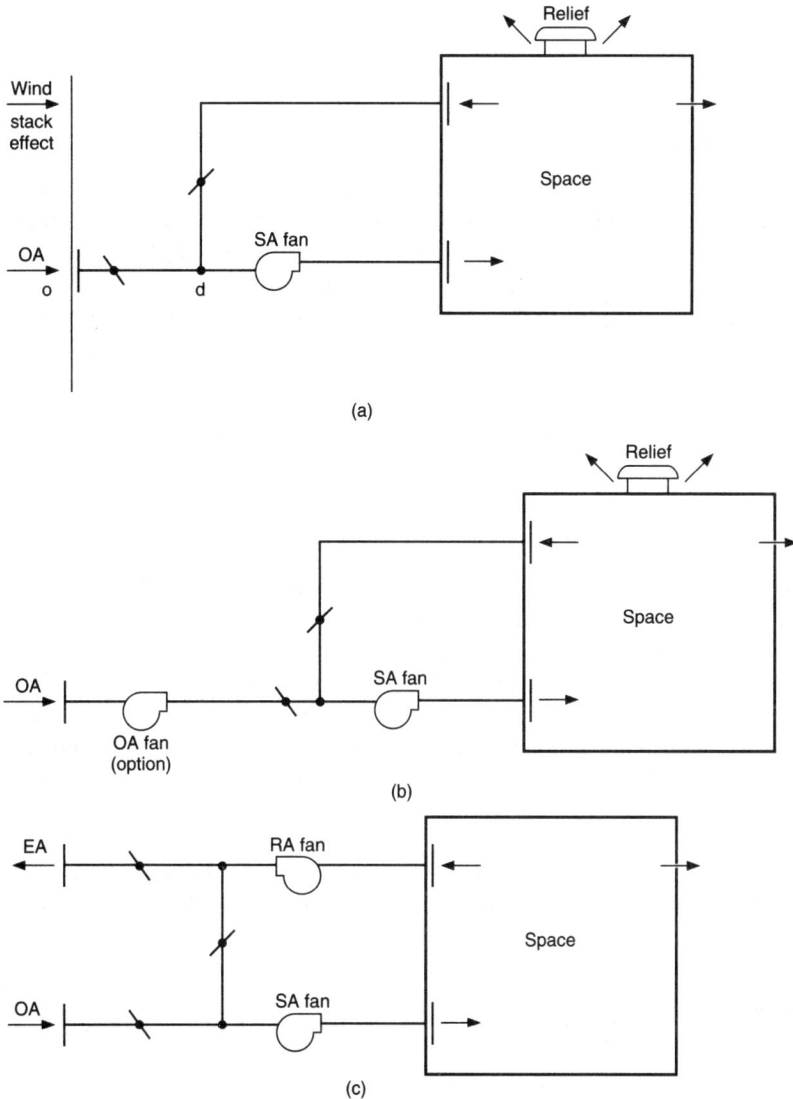

Figure 11.23 Three modes of outdoor air control. (*a*) Constant (minimum) OA supply; (*b*) Demand-controlled OA supply; (*c*) Outdoor air for free cooling (economizer cycle).

11.7 Outdoor Air Variation Analysis

Three modes of outdoor air control are used for VAV systems: constant minimum outdoor air, demand controlled outdoor air, and outdoor air for free cooling (economizer cycle). Each mode employs different control software, hardware, and system components (Fig. 11.23). The problems associated with each control mode are also different and should be analyzed individually.

11.7.1 Constant minimum outdoor air supply analysis

Many different control schemes are possible for this control mode. Figure 11.23a shows one of such control schemes. The objective is to maintain a constant outdoor air intake either by the direct measurement of outdoor airflow or by measuring and maintaining a constant pressure difference between points o and d regardless of the variation in supply air, Q_{sa}. The problems that need to be analyzed in this scheme are as follows:

1. Intake pressure fluctuation at point o due to wind and stack effect

2. Pressure variation at point d due to change in Q_{sa}

3. Relief required to maintain an adequate space pressure

For floor-by-floor VAV systems in high-rise buildings, wind and stack effect may become a serious design consideration when outdoor air is introduced through the louver directly mounted on the external wall.

For the example shown in Fig. 11.24, the outdoor air duct is oversized to avoid drawing rain and snow into the building. This may result in a substantial increase of outdoor air from 4000 to 5470 cfm when there is a 15-mph wind blowing against the exterior wall where the intake louver is located.

If the wind velocity is increased to 30 mph, the outdoor air entering the building will exceed 8000 cfm or almost double the specified minimum air quantity. The simplest way to stabilize the outdoor air supply is

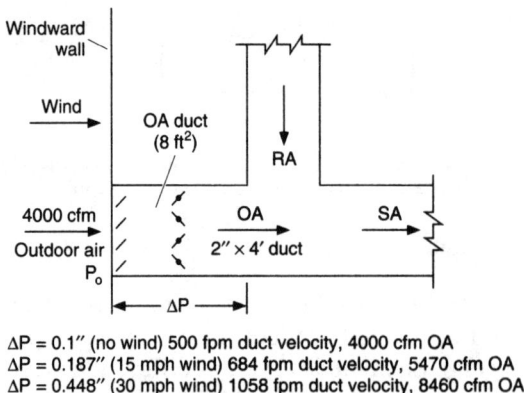

$\Delta P = 0.1''$ (no wind) 500 fpm duct velocity, 4000 cfm OA
$\Delta P = 0.187''$ (15 mph wind) 684 fpm duct velocity, 5470 cfm OA
$\Delta P = 0.448''$ (30 mph wind) 1058 fpm duct velocity, 8460 cfm OA

Figure 11.24 Outdoor air intake and wind velocities.

to add a substantial resistance to the outdoor duct. For example, if the outdoor air duct resistance is increased from 0.1 to 0.3 inch w.g., the outdoor air entering the building will be increased only 543 cfm at 15 mph wind velocity or 14 percent of the specified outdoor air supply. More sophisticated techniques may be employed to stabilize the outdoor air supply. However, such efforts may not be justifiable unless substantial energy savings can be achieved. The prevailing reason for maintaining a consistent outdoor air supply is to prevent the deterioration of indoor air quality resulting from an excessive reduction in outdoor air supply during low-load, low-supply-air conditions. This situation is avoided and considerable energy saving achieved by demand-controlled ventilation.

11.7.2 Demand-controlled ventilation analysis

In demand-controlled ventilation, the amount of outdoor air for ventilation is elaborately modulated to maintain acceptable levels of indoor air quality. Carbon dioxide (CO_2) is usually used as a convenient measure for the adequacy of air supply—usually a mixture of outdoor air and recirculated air to conditioned spaces. The cost-effectiveness of this ventilation, or outdoor air control, scheme is best simulated and analyzed by a suitable mathematical model that interacts with building occupancy, people density, and activity levels to produce hour-by-hour ventilation load calculations for a given level of indoor CO_2 concentration.

Reference 11 shows a schematic layout of a VAV system with demand-controlled ventilation (Fig. 11.25) and the relationships between outdoor ventilation air and CO_2 concentration (Fig. 11.26) for the VAV system described in Ref. 11. The reduction in outdoor air supply is quite obvious when alternatives 2 and 3 are compared with alternative 1. The reduction in outdoor ventilation air supply also means a reduction in energy consumption. Table 11.12 summarizes annual energy savings and payback periods for five U.S. cities comparing alternatives 2 and 3 with alternative 1.

Alternative 1 supplies constant 3600 cfm outdoor air for the entire occupied period.

Alternative 2 supplies various amounts of outdoor air with demand controller set to maintain the indoor CO_2 level of 800 ppm.

Alternative 3 is similar to alternative 2, but demand controller is set to maintain a 920 CO_2 concentration level.

Analysis of the example illustrated in Ref. 11 (Fig. 11.25)

1. This method of outdoor air control is particularly effective in climates with long, hot, and humid summer periods or long, cold win-

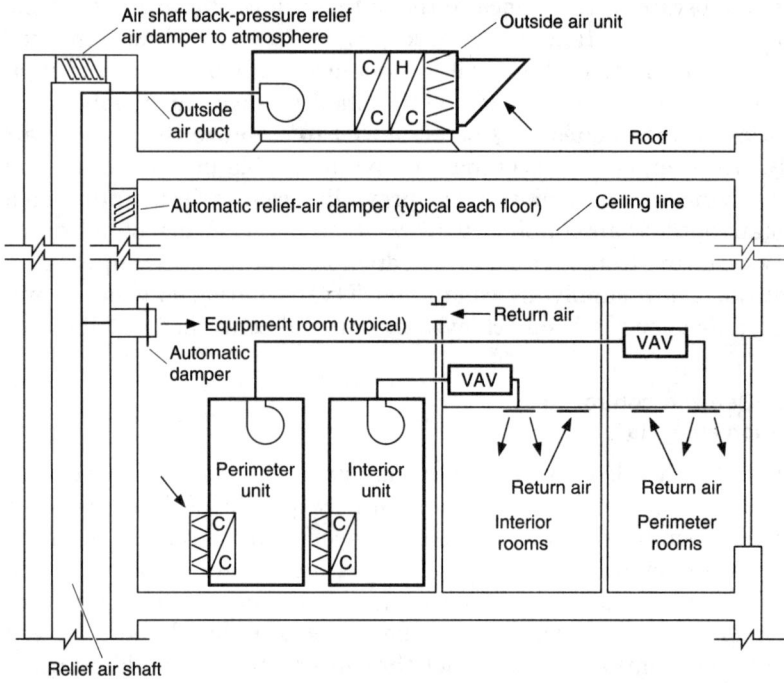

Figure 11.25 Demand-controlled ventilation system. *Source: Ref. 11. Reprinted with permission.*

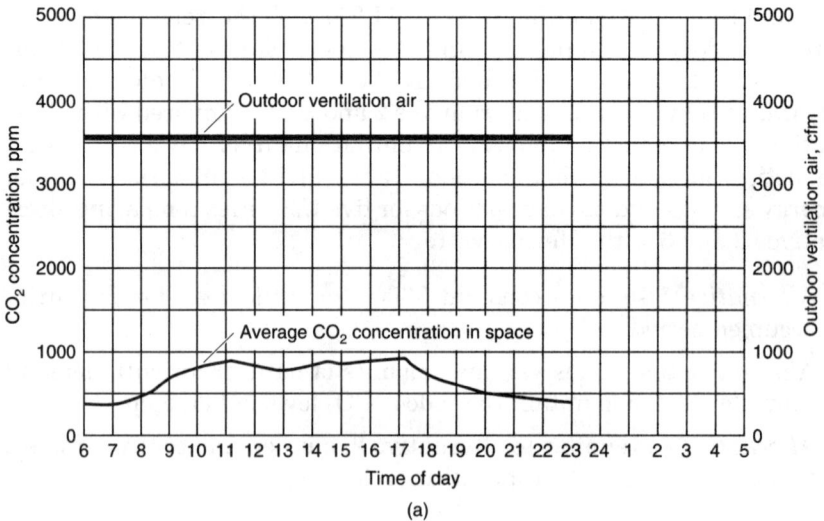

Figure 11.26 Outdoor ventilation air versus CO_2 concentrations. *(Source: Ref. 11. Reprinted with permission. (a)* Alternative 1: calculated hourly carbon dioxide concentration at a constant outdoor airflow of 20 cfm per person (a typical floor).

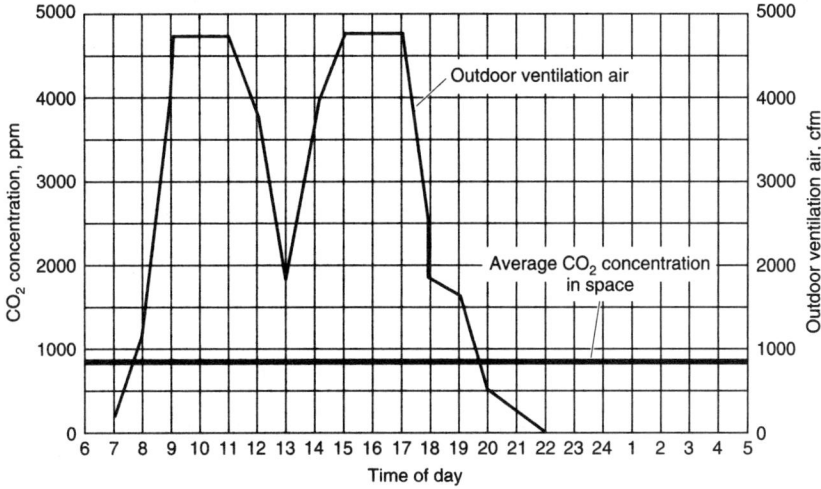

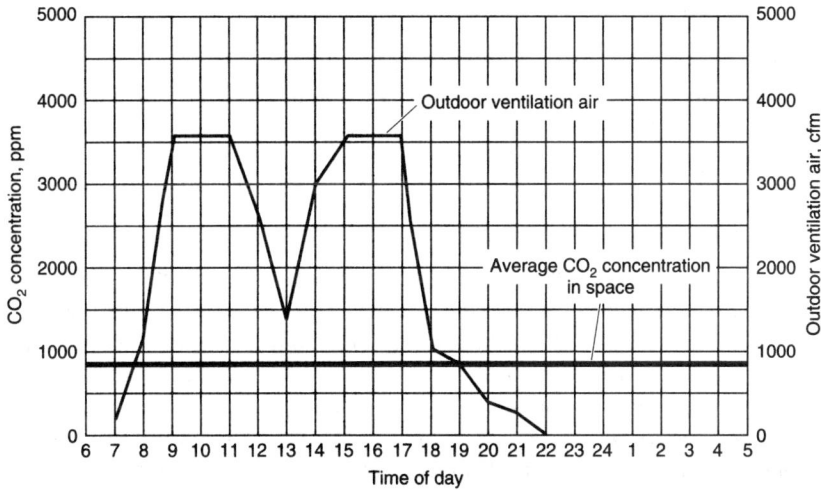

Figure 11.26 *(Continued)* *(b)* Alternative 2: calculated hourly outdoor airflow rate at a preset carbon dioxide concentration of 800 ppm (a typical floor); *(c)* Alternative 3: calculated hourly outdoor airflow rate at a preset carbon dioxide concentration of 920 ppm (a typical floor).

ter periods or a combination of both. Examples are Miami and New York (Table 11.8).

2. On the other hand, the method is not so effective in saving cooling energy in climates with a long free cooling period. New York is a good example (Table 11.8).

TABLE 11.12   Estimated Annual Energy Savings and Payback Periods for Five U.S. Cities—Comparison of Alternatives 2 and 3 with Alternative 1.

	Miami		Atlanta		Washington, D.C.		New York		Chicago	
	Alt. 2	Alt. 3	Alt. 2	Alt. 3	Alt. 2	Alt. 3	Alt. 2	Alt. 3	Alt. 2	Alt. 3
Electricity, kwh	15,907	24,004	15,862	24,012	9,260	13,597	2,277	558	8,765	11,369
Gas, therms	4,372	4,903	4,370	4,900	6,048	6,719	6,769	7,468	8,965	10,285
Total $	4,964	5,934	4,033	5,042	4,256	4,913	5,865	6,337	4,543	5,291
Payback period, yr.	1.81	1.52	2.23	1.79	2.11	1.83	1.53	1.42	1.98	1.70

SOURCE: Ref. 11. Reprinted with permission.

General observations on demand-controlled ventilation

1. Acceptable CO_2 levels do not necessarily indicate that there are no other IAQ problems. In fact, other indoor pollutants, pollen, VOCs, and other contaminants do not have the same rates of pollutant generation. Thus, low CO_2 levels may not necessarily represent a high-quality indoor air environment.

2. Demand-controlled ventilation should be used only when the quality of outdoor air is poor and the use of an economizer cycle is neither practical nor feasible. Whenever feasible, an economizer cycle should be employed for both IAQ and energy-saving purposes.

11.7.3   Economizer cycle analysis

An economizer cycle maximizes the cooling potential of outdoor air. The simplest form of economizer cycles is a dry bulb temperature economizer (Fig. 11.27). In this scheme, whenever the outdoor air temperature drops below a predetermined dry bulb temperature, it begins to introduce 100 percent outdoor air. As the outdoor air temperature decreases, the cycle will reduce the amount of outdoor air to maintain a constant supply air temperature t_s at conditioned spaces. This is accomplished by mixing recirculated air at temperature t_r with outdoor air at temperature t_o and continuously adjusting the outdoor air quantity from 100 percent outdoor air supply to a predetermined minimum for ventilation.

As shown in Fig. 11.27, the concepts of constant and demand-controlled outdoor air supply can be easily incorporated into an economizer cycle. To what extent the outdoor air quantity should be varied, constant minimum, controlled minimum for indoor air quality, or maximum introduction of outdoor air depends on several factors:

1. Number of hours outdoor air temperature in the free cooling zone

2. Severeness of summer (hot and humid) and winter (cold and dry) conditions outside the free cooling zone in a given building location

A. Operating Principle of Dry Bulb Temperature Economizer
B. Application 1: Dry bulb temperature economizer plus constant minimum AO supply.
   1. Min. AO supply whenever OA temperature $> t_s$.
   2. 100% OA supply at OA temp. $= t_s$.
   3. Outdoor air and recirculated air mixing whenever $t_m \leq$ OA temperature $\leq t_x$.
   4. Minimum fixed OA temperature or $t_o < t_m$.
C. Application 2: Dry bulb temperature economizer plus demand-controlled ventilation.
   1. Outdoor air quantity is controlled strictly by CO_2 or other IAQ sensors which are set to maintain a predetermined containment level whenever OA temperature $> t_s$.
   2. 100% OA supply at OA temperature $= t_s$.
   3. Outdoor air and recirculated air mixing whenever $t_m <$ OA temp $\leq t_s$.
   4. Outdoor air quantity controlled by CO_2 or other IAQ sensors for $t_o < t_m$.

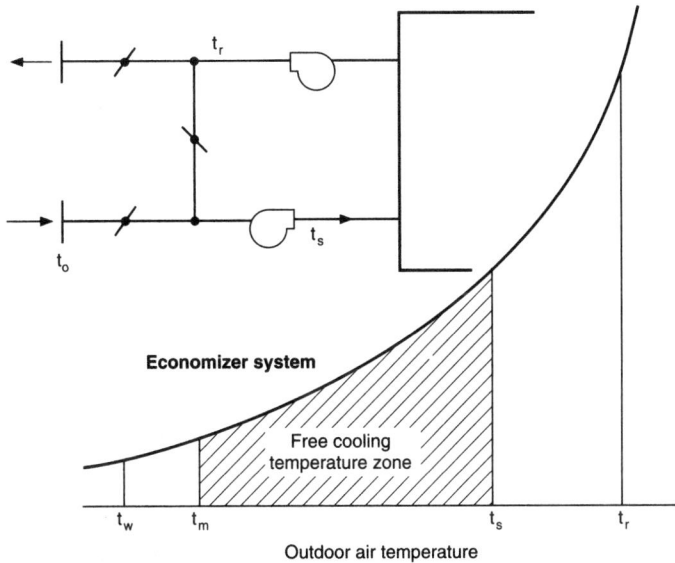

t_s = VAV system SA temperature

t_w = Winter design OA temperature

t_r = Recirculated air temperature

t_m = Outdoor air temperature
corresponding to minimum outdoor
air required for ventilation

t_o = Outdoor air temperature

Figure 11.27   VAV economizer cycle—principles and applications.

3. System design supply air temperature, normal or cold air supply

4. Type of building, occupancy density, and variation

These factors are all related to building and system design and should be analyzed at an early stage of design. A computer-based energy program with hour-by-hour analysis capability and sufficient sensitivity to weather and occupancy variations should be used for this type of analysis. By changing the inputs for items 1 through 4 (Fig. 11.27), simulating for a variety of ventilation schemes, and analyzing the results, the VAV system designer can readily identify the most cost-effective outdoor air control scheme for a given design situation.

Many ventilation schemes are possible. Representative ones are as follows:

- Simple dry bulb temperature economizer

- Enthalpy economizer and other economizer variations

- Demand-controlled ventilation

- Certain combinations of these schemes

Which outdoor air ventilation scheme to choose is a basic design question and highly dependent on each designer's personal preference, experience, and ability to analyze the pros and cons of various control schemes. However, a knowledge base for IAQ-oriented, cost-effective ventilation schemes can be developed for each locality by simulating standardized ventilation systems for typical building usage and occupancy.

11.8   Duct Static Pressure Variation Analysis

Air pressure in VAV duct systems changes constantly and continuously. It is the nature of VAV systems and the result of airflow modulation through VAV terminals as well as the need to control outdoor air for ventilation and free cooling while the system airflow varies over a wide range. However, the real problem with VAV duct systems is not pressure variation but the magnitude of variation and its impacts on system components such as VAV terminals, fans, and system dampers.

11.8.1   Two examples of pressure variation analysis

In example 1, Fig. 11.28, the duct pressure variation has different impacts on VAV terminals VAV1 and VAV3. At VAV3, the entering air pressure is maintained at 1 inch static pressure regardless of the airflow through VAV3, while the entering air pressure at VAV1 fluctuates from

1.574 to 1.109 in w.g. for two values of airflow through VAV terminals shown in Table A3 in Fig. 11.28. The operation of VAV1 is probably less stable and possibly noisy, especially when the terminal is oversized. Reference 12 discusses the problems caused by oversized VAV terminals.

The problem of excessive fluctuation in duct pressure can be lessened by more generous duct sizing and duct layout and configuration techniques such as duct looping, static regain, and uniform cross-section-area duct sizing.

In example 2, Fig. 11.29, the situation is somewhat different. When the system airflow changes from 12,500 cfm maximum to 3400 cfm minimum, the duct pressure at point (3) fluctuates considerably, affecting the outdoor air supply, which needs to be maintained at 2000 cfm constant by design. The outdoor duct size and configuration are such that 0.25 inch of static pressure is required to introduce 2000 cfm of outdoor air. This means that the deficiency in static pressure on the return air side must be compensated by a higher pressure drop through the return air damper. The pressure drops to be supplied by the damper are 0.076 inch w.g. at maximum flow and 0.235 inch w.g. at minimum flow. The corresponding angle of rotation required by the parallel-blade return air damper would be 3° at maximum flow and 60° at minimum flow. Figure 11.30 indicates that the return damper selected is adequate for the application and will be operating in the stable zone where the linearity of control is maintained, although this statement is somewhat less accurate when the rotation angle exceeds 50°.

The preceding example clearly illustrates that control stability and duct pressure fluctuation for outdoor-air/return-air applications are highly dependent on many design parameters—such as maximum-to-minimum VAV turndown ratio, damper selection, duct sizing methods, and layout techniques, as well as actual duct layout, configuration, and sizes. A quantitative analysis will reveal the adequacy of the design in selecting these parameters. What-if trials using a suitable duct analysis program will further reveal the interaction among these design parameters and greatly improve the quality of design. Reference 13 is a duct analysis computer program based on the ASHRAE T-method.

11.8.2   Duct design methods and techniques

A variety of duct design methods and techniques are being used for VAV duct systems. Those commonly used are as follows:

Methods	Techniques
Equal friction method (Ref. 14)	Duct looping (Ref. 15)
Static regain method (Refs. 14, 16)	Equal-size trunk duct
ASHRAE T-method (Ref. 14)	Simulation

Duct design methods have considerable influence over duct pressure fluctuation, especially when ductwork is extensive in length with many high-pressure-loss duct fittings, and duct velocities used are relatively high. If the design situation is in this category, then the static regain method is preferable over other methods (Refs. 14 and 16). However, for relatively small, uncomplicated VAV systems, the equal friction method, possibly combined with duct looping technique (Ref. 15) is usually adequate to provide a sound basis for VAV terminal selection and control design.

The equal-size trunk duct technique provides the most stable environment for VAV systems by using a single-size duct layout for the entire system. The duct size is determined by the maximum design flow and duct velocity. The technique is most appropriate for smaller

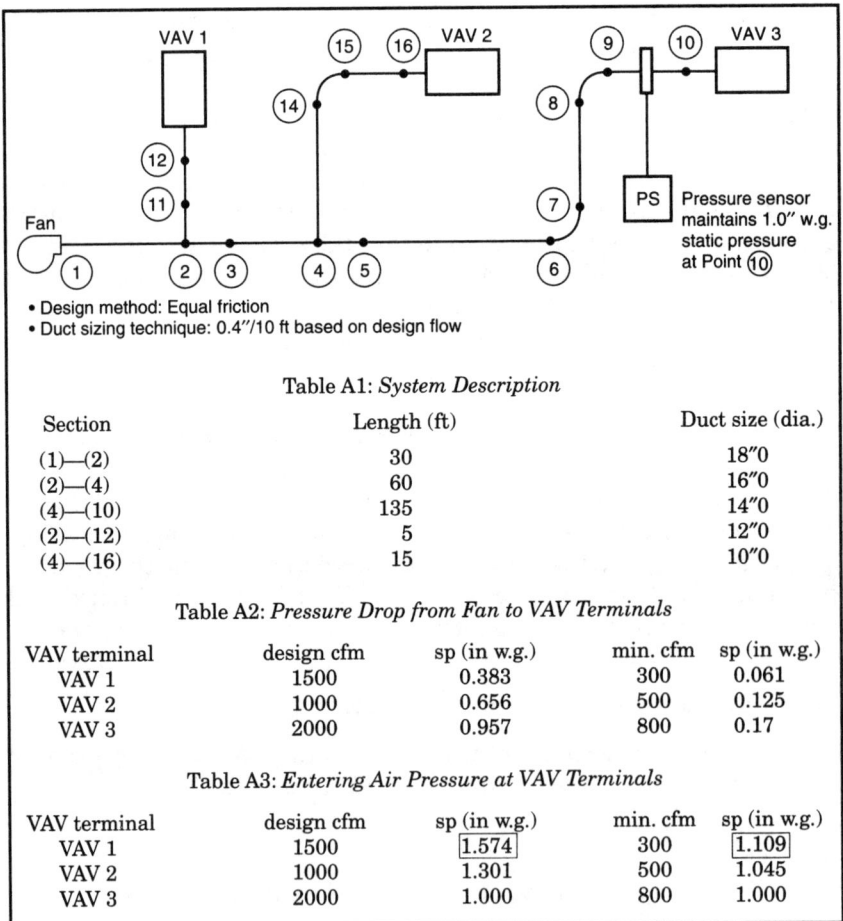

• Design method: Equal friction
• Duct sizing technique: 0.4"/10 ft based on design flow

Table A1: *System Description*

Section	Length (ft)	Duct size (dia.)
(1)—(2)	30	18"0
(2)—(4)	60	16"0
(4)—(10)	135	14"0
(2)—(12)	5	12"0
(4)—(16)	15	10"0

Table A2: *Pressure Drop from Fan to VAV Terminals*

VAV terminal	design cfm	sp (in w.g.)	min. cfm	sp (in w.g.)
VAV 1	1500	0.383	300	0.061
VAV 2	1000	0.656	500	0.125
VAV 3	2000	0.957	800	0.17

Table A3: *Entering Air Pressure at VAV Terminals*

VAV terminal	design cfm	sp (in w.g.)	min. cfm	sp (in w.g.)
VAV 1	1500	1.574	300	1.109
VAV 2	1000	1.301	500	1.045
VAV 3	2000	1.000	800	1.000

Figure 11.28   Duct pressure variation due to VAV terminal operation.

OA and RA ducts

RA damper

OA   OA damper

OA

1

2000 cfm constant
minimum OA
supply to AHU

OA damper

RA   RA duct

OA duct
(12″ × 24″)

2

3

Vertical OA shaft
connected to other
AHU's

RA duct to AHU

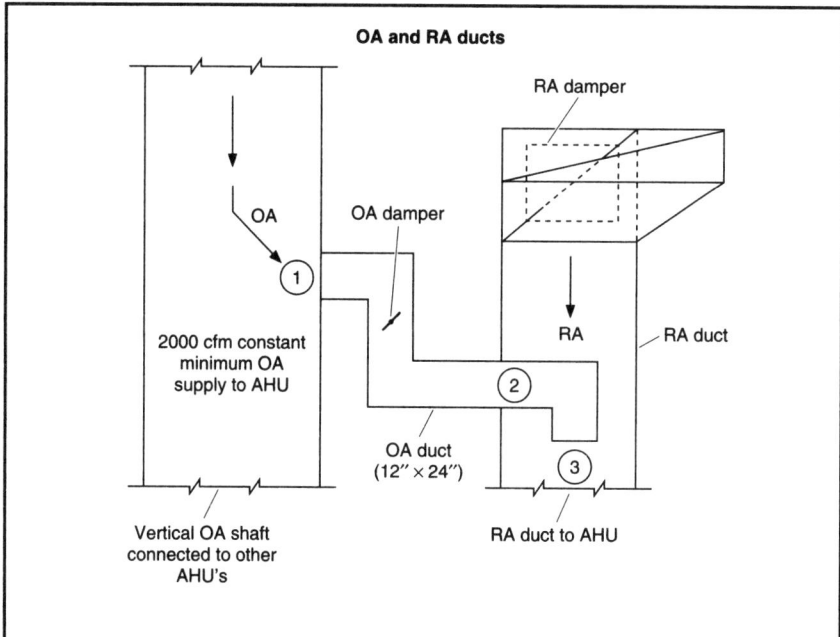

RA Duct Pressure Loss (in w.g.) Calculations

Duct Element	System Airflow	
	Maximum Flow	Minimum Flow
• 56″ × 14″ duct (40 ft long)	0.032″	0.0024″
• 36″ × 30″ duct (10 ft long)	0.0112″	0.0001″
• 56″ × 14″ elbow	0.0033″	0.0003″
• 36″ × 30″ elbow	0.0148″	0.0013″
• 30″ × 42″ elbow	0.0544″	0.0066″
• RA damper (full open)	0.0584″	0.0042″
	0.1741″	0.0149″
• OA intake duct loss = ΔP (1) − (2) − (3)	0.25″	0.25″
• Additional RA duct pressure loss req'd =	0.0759″	0.2351″
• Corresponding angle of rotation for 24″ × 28″ parallel blade damper	3°	60°

Pressure Variation Analysis

• Multiple packaged VAV units connected to a common OA shaft.
• Constant minimum OA supply to each VAV unit regardless of flucuation in supply and return air flows.
• In this system, the RA pressure fluctuates continuously at point (3), therefore affecting the duct pressure inside the outdoor air duct (1) − (2) − (3). This in turn affects the amount of OA introduced to the air-handling unit. The amount of fluctuation at point (3) is 0.0759″ w.g. at maximum flow to 0.2351″ w.g. at minimum flow.
• To compensate this variation, the RA damper needs to be rotated from 3° (maximum flow) to 60° (minimum flow), and control may become somewhat unstable between 50 and 60°. See Fig. 11.30.

Figure 11.29   Outdoor air duct pressure variation for a VAV system without return fan.

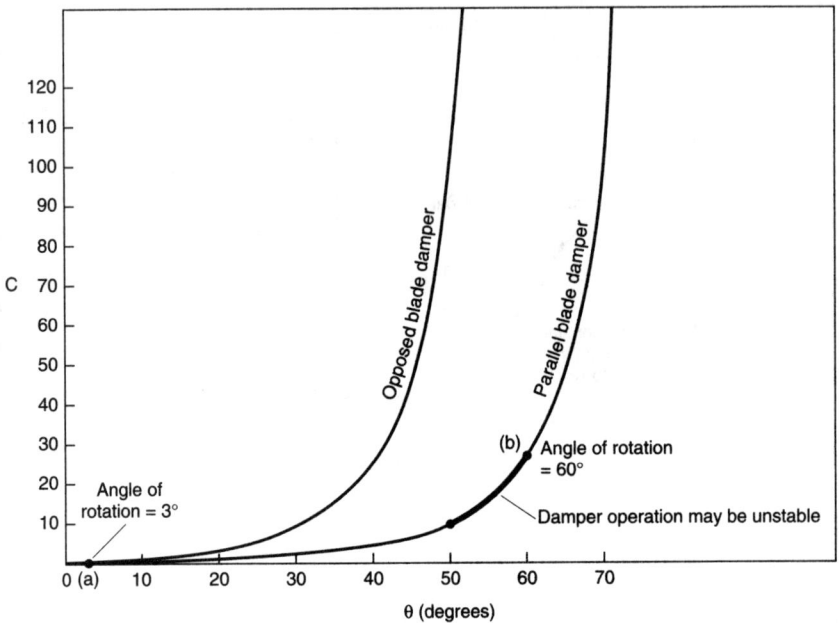

Figure 11.30 Dynamic loss coefficient (c) versus damper angle of rotation in degrees (θ).

and simpler VAV systems. It is especially effective when combined with looped trunk ductwork.

The T-method simulation (Ref. 14, p. 32.16) determines the flow in each duct section of a known design and fan performance. Strictly speaking, it is not a duct design method or technique. However, it can predict airflow and pressure variation at each duct section and can be very effective in analyzing duct design and its component selection.

11.8.3 Method of analysis: a step-by-step procedure

The general procedure for duct pressure variation analysis follows.

1. Before conducting a pressure variation analysis, the duct design and component selection must be completed. Information required for analysis includes:

- Duct layout, size, and fitting types and dimensions
- Damper type, size, and flow characteristics
- System's supply air variation, maximum and minimum design flows
- Outdoor air variation, minimum constant flow, demand-controlled variable flow or economizer cycle for free cooling
- VAV terminal airflow, maximum and minimum for each terminal

Some of the information, especially the last item, may not be easily identified or estimated. However, a detailed analysis is not always required and a simplified worst-case scenario analysis is usually sufficient. This is demonstrated by the example in Fig. 11.28.

2. Based on the design layout and component selection, calculate air pressures at key points in the duct system under specified operating conditions using a manual or computer-aided system model. This model should be capable of describing the flow/pressure relationship of the system, mathematically based on the user input.

3. Analyze the results for unstable damper and fan operation, noise, and comfort-level changes. Again, detailed analyses are generally not necessary. A simple trend analysis based on minimum and maximum flows can identify problem areas, such as oversized dampers and terminals, undersized ductwork, and excessive turndown ratios (maximum/minimum flow ratio). Once problems are identified the original design can be modified to correct the problems.

This design and analysis interaction is a self-learning process that trains the system designer to avoid system problems in advance, without actually encountering such problems in the field. In this way, the designer gains valuable design experience in a very short time. Generally, the system designer can learn useful design tips quickly by going through a few cycles of design/analysis interactions.

11.9   System Reliability and Redundancy Analysis

11.9.1   VAV system reliability analysis

The VAV system is an all-air system with fans as major system components. VAV terminals and outlets, temperature and flow-sensing and control devices as well as air transporting and distributing ductwork are other system components (Fig. 11.31). Of all these system components, fans are the only component with its moving part running continuously during the entire operating period, and therefore should be considered as the major object of reliability analyses.

The reliability of VAV systems can be improved considerably by installing two fans in parallel operation. In Fig. 11.31, two identical fans with a fan curve (A–A) are operating in a system environment, or system curve (C–C), producing a combined fan curve (B–B). In this case, when one fan fails, the system still delivers approximately 75 percent of the design air supply ($Q_1 = 75\%$ of Q).

Figure 11.32 shows normalized percentage load curves plotted against percent of cumulative annual hours of operation for three different building applications. Curve A indicates that for this particular hospital, the cooling load exceeds 75 percent of the design load for 23 percent of annual operating hours. This level of reliability is generally not satisfactory for hospital applications, especially in certain critical areas.

Curve B represents the annual load versus operating hour curve for a hotel in relatively a mild climate. In this case, the 75 percent load level is exceeded by this specific hotel for 20 percent of annual operating hours. Considering a relatively low rate of fan failure, the system designer may decide this level of reliability is acceptable for most hotel applications. However, the level of reliability can be easily upgraded by slightly oversizing the fan capacity required.

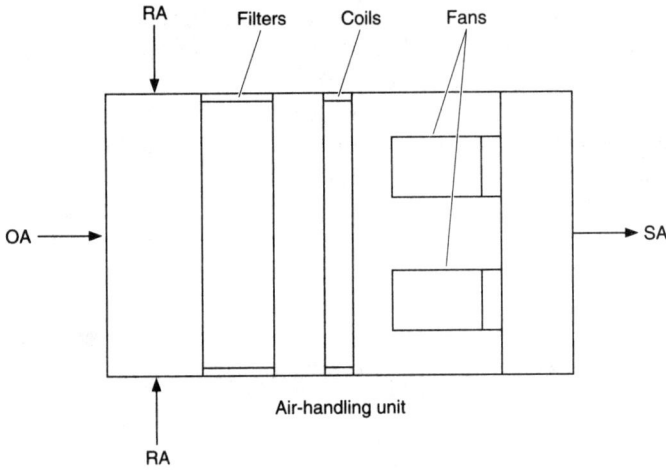

Fan and system curves
Curve A: One-fan operation
Curve B: Two-fan operation
Curve C: System curve

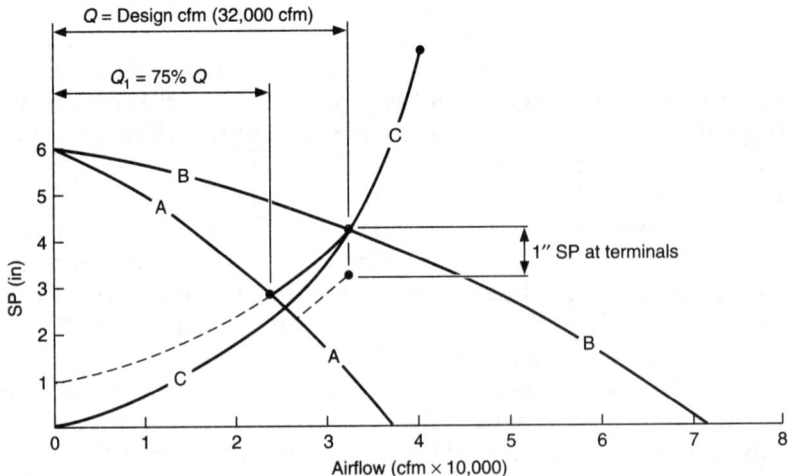

Figure 11.31 Two identical fans in parallel operation.

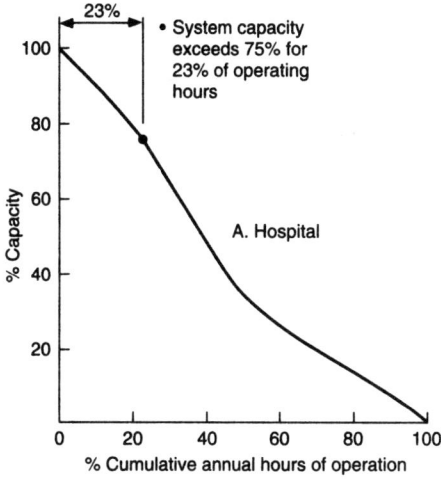

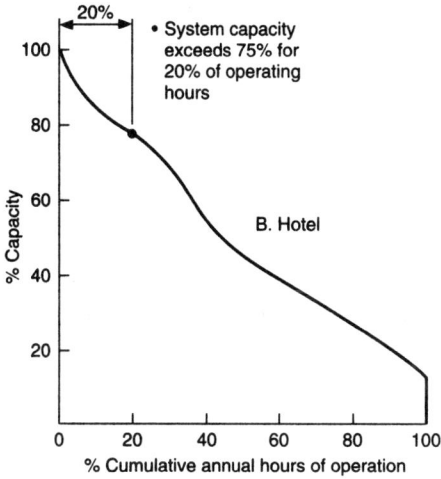

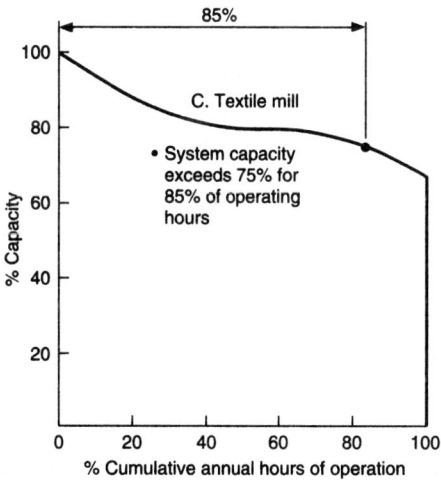

Figure 11.32 Capacity versus annual operating hours.

Curve C, the load profile for a textile mill in a hot, humid climate indicates that two fans in parallel operation are not sufficient to provide the desired reliability level for textile mills, which usually require very high reliability. Possible design options in this case would be:

1. Oversize each fan slightly.

2. Increase number of fans in parallel operation.

3. Use CAV fans and a separate humidification system to maintain the required humidity level when one fan fails to operate.

One more common solution to improve the reliability of two fans in parallel operation is to reduce the supply air temperature a few degrees if the refrigerating system has a sufficient surplus capacity to do so. Also, fan failure may not always occur during the peak load period. Still, system reliability analysis provides necessary quantitative insights into the art of reliability improvements for VAV fan operation, and should be considered for all VAV system designs.

11.9.2 VAV system redundancy analysis

If very high reliability is required for certain airflow-sensitive applications, the system designer may consider a redundancy capacity to be added to each fan in the previous example. With a 34 percent capacity added to each of the two fans, one fan still delivers 100 percent design supply air ($1.34 \times 75\% = 100\%$). The other fan then provides 100 percent redundancy, or standby capacity in this case.

Thus, the modified two-fan VAV system with an additional 34 percent capacity will provide the full design flow throughout the entire annual operating period even if only one fan is operating. However, this is true only when the original fan curve and system resistance remain the same. If system redundancy is a critical design issue, the system designer must conduct reliability as well as redundancy analyses. Typical applications in this category are certain types of laboratories where highly toxic or hazardous chemicals are handled, or isolation wards in hospitals. In these design situations, the system designer must design ductwork and select fans carefully to ensure the system will deliver the sufficient conditioned air when one fan fails to operate.

11.10 Integrated Design Analysis

11.10.1 Interactive building, system, and control design

A VAV system serves three major functions. It controls the space temperature by circulating and modulating the conditioned supply air to each zone. It also introduces outdoor air for ventilation. It pressurizes the building by pumping outdoor air into the building. These three functions, temperature control, building ventilation, and pressurization are accomplished through complex building, system, and control interaction. Figure 11.33 depicts diagrammatically the processes involved and the interaction taking place to accomplish these functions with a VAV system.

Process 2-3-4-5-6 provides cooling and temperature control in each VAV zone (VAV terminal control not shown). This is the base function of any VAV system. However, VAV systems must also serve other functions.

Process 1-3-4-5-9 constitutes a basic ventilation cycle, beginning with outdoor air introduction (1), circulation through spaces by supply and return fans (3 and 5), then exhausted to outdoor through the exhaust air opening (9).

Process 1-3-4-8 pressurizes the building by forcing air through building cracks and openings (elevators, doors, and other openings), thus building up the pressure inside the building.

Process 1-3-4-7 does not contribute directly to any of the three functions described earlier. However, it affects the air balance inside the building and must be carefully considered as a part of the interactive design process.

Process 10 is the infiltration of outdoor air caused by wind pressure and stack effect and can be prevented by the proper pressurization of the building.

Figure 11.33 clearly indicates that the building, system, and control are closely interrelated and highly interactive. For example, a reduction in outdoor air supply caused by the reduced fan output or initiated by demand-controlled ventilation may lead to a loss of building

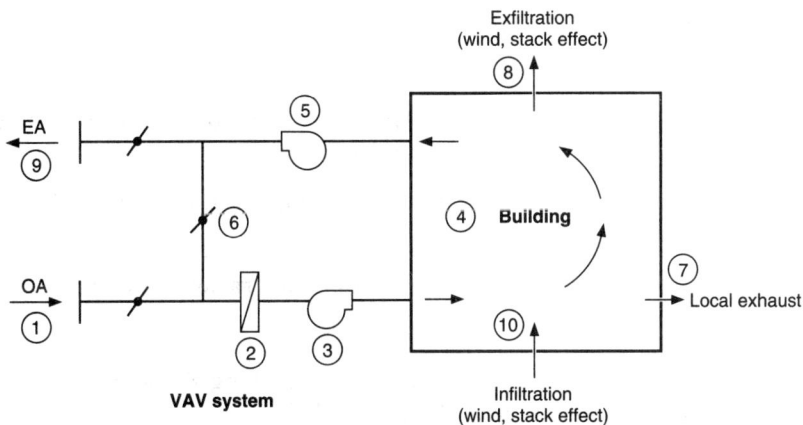

Figure 11.33 Airflow balance in a building air-conditioned by a VAV system.

pressure and an increase in outdoor air infiltration (process 10). The local exhaust system (process 7) may lose its proper source of makeup air, further increasing the infiltration of outdoor air. On the other hand, even with a sufficient outdoor air supply, a porous and leaky building construction combined with high wind pressure and winter-time stack effect may make building pressurization even more difficult to achieve by a VAV system—especially at a reduced load, when the outdoor air supply can and should be reduced. The solutions to all these problems must come from the team effort of all the design professionals involved (architects, engineers, and control professionals) and should begin at an early stage of design and continue to the end of the design process.

The conventional approach of completing building and system design first, then adding controls later rarely produces a well-coordinated, trouble-free design automatically. Ventilation deficiency, unstable operation, insufficient (or lack of) building pressurization, and noisy and drafty space conditions are some of the commonly encountered problems resulting from this design approach.

In an integrated design approach, building, system, and control are considered simultaneously and suitable mathematical models and analytical tools are used to measure the interaction among these design elements. Currently, the measurement of interaction is most effective in the following areas:

- *Damper analysis.* Damper sizing, operation, and interaction with ducted system and control.

- *Stability analysis:*

    Minimum to maximum outdoor air supply

    Damper and fan in series

    Damper and damper in series

- *VAV economizer cycle analysis.* Damper operation and interaction with building, ducted system, and control.

- *Pressurization analysis.* Wind pressure and stack effect on building pressurization.

- *Minimum outdoor air analysis.* Damper operation and minimum outdoor air control.

The first three topics will be covered in the following sections. The last two topics were covered in Secs. 11.6 and 11.7.

11.10.2 Damper analysis

Figure 11.34 shows control dampers most commonly used for airflow control in VAV systems. The discussion in this section is limited to the damper interactions with related building and system components in a VAV operating environment (Fig. 11.35). For other aspects of damper applications, the reader is referred to Refs. 17 and 18.

Venturi-type air valves are also used for airflow regulation in VAV systems. However, their applications are not included in this section (Fig. 11.36).

Oversized dampers. It is not unusual to see oversized control dampers in VAV systems. Oversizing occurs when dampers are sized for the duct dimensions where it is installed. It affects the sensitivity of damper control which in turn impacts the performance of VAV systems. Figure 11.37 shows the effects of oversizing on control sensitivity. Control sensitivity is much higher at the first 10 percent of damper opening (from 0 to 10 percent), while the same damper shows a much lower sensitivity at the last 10 percent of damper opening (from 90 to 100 percent). The control characteristic curve of this damper shows a considerable deviation from the ideal linear curve which gives a constant sensitivity throughout the damper's entire opening range from 0 to 100 percent opening. This change in control sensitivity is a major source of unsta-

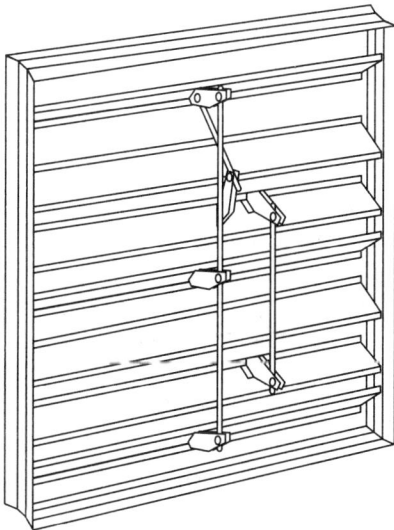

Multiblade damper   Round damper

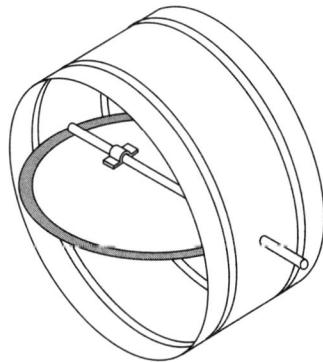

Figure 11.34 Typical dampers used for VAV systems. *(Source: Honeywell Inc. Reprinted with permission.)*

System components

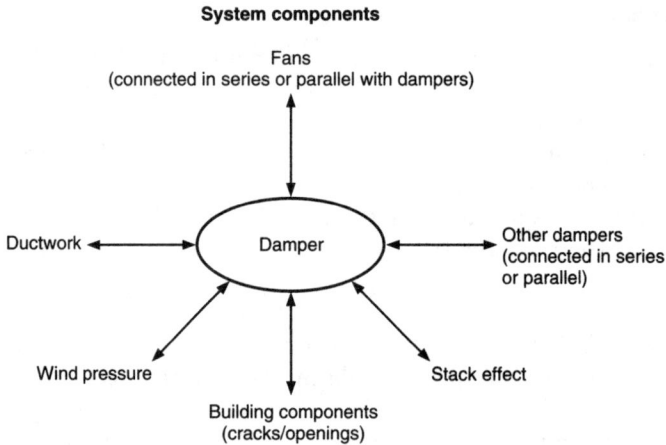

Figure 11.35 Damper interaction with building and system components.

ble system operation. Different methods and techniques for damper sizing are suggested by various damper manufacturers (see Refs. 17 and 18) and produce different results for the same application. Changing sensitivity or nonlinearity of a control damper affects the stability of VAV system operation. Under certain system arrangement and operating conditions, the nonlinearity of a damper interacts with other system components with nonlinear characteristics and further magnifies the problem of unstable control and operation. Section 11.10.3 (following) discusses the subject of stability analysis for three design situations which fall into this category.

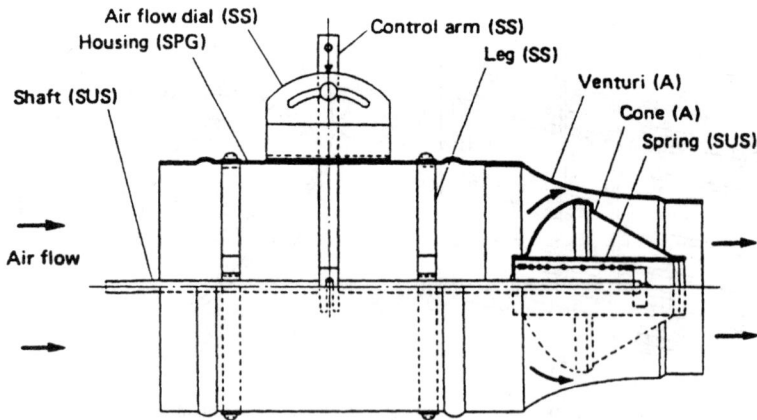

Figure 11.36 Venturi valve for airflow control. *(Courtesy of Japan Mitco Co., Ltd.)*

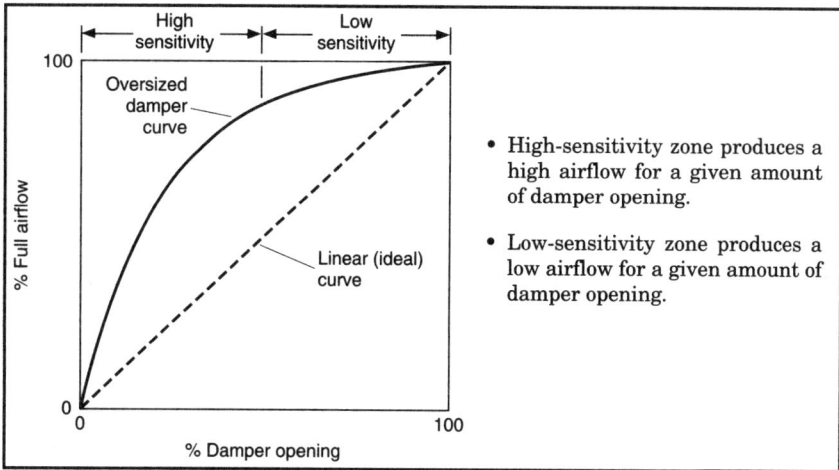

High-sensitivity zone produces a high airflow for a given amount of damper opening.

Low-sensitivity zone produces a low airflow for a given amount of damper opening.

Figure 11.37   Effects of oversized dampers on control sensitivity.

11.10.3 Stability analysis

Minimum to maximum outdoor air supply.   In the VAV system shown in Fig. 11.38, the amount of outdoor air entering the OA intake opening is a function of the pressure difference between points O and D. As the pressure at point O is an atmospheric pressure modified by the prevailing wind pressure, it is essentially an uncontrolled quantity. The outdoor air quantity entering the VAV system is controlled by adjusting the mixing plenum pressure at point D. This adjustment is provided by damper D_{ra}. The control stability of D_{ra} depends on the linearity of the damper control characteristics and also the stability of the duct pressure at point E, which in turn is affected by other system components such as return fan and exhaust air damper. The interactions among these components are complex and often nonlinear, and result in considerable uncertainty about the stability and predictability of outdoor air supply control. Various control schemes have been suggested in an attempt to stabilize outdoor air supply (Fig. 11.39). Before applying any one of these control schemes, the projected stability of outdoor air supply should be carefully analyzed using specific design parameters, system layout, and components, as well as key quantitative data such as minimum and maximum air quantities, desirable pressure ranges, and setpoints. Section 11.7 discussed this subject in detail.

Damper and fan in series.   In this damper and fan arrangement for VAV systems, there are essentially three variations, shown in Fig. 11.40*a, b,* and *c.* Each damper application poses unique design and control prob-

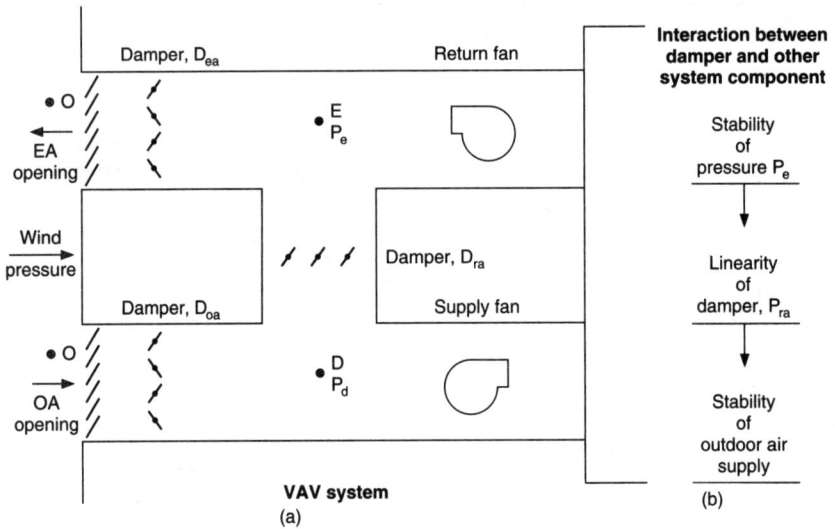

Figure 11.38 Plenum pressure and outdoor air supply. (a) VAV system layout; (b) Control stability.

lems and must be analyzed for its soundness in duct layout, damper, and control selection for the intended application.

Damper and damper in series. In Fig. 11.41, Dampers a and b as well as a and c, a and d, and a and e are in series operation. When these dampers are operating in the unstable or high-gain zone, each damper starts to interfere with other dampers' operation and the whole system becomes unstable. The seriousness of this domino effect, or mutual interference, depends on system layout and component selection, degree of oversizing, intended mode of operation, as well as actual operating conditions and cannot be easily generalized. However, if a VAV system is physically described, and its operating conditions are specified, then its operating stability can be analyzed. For example, the system described in Fig. 11.41, damper a with its parallel-blade configuration, will probably operate in the high-gain, unstable region when the airflow is reduced to 4000 cfm. The VAV terminals are also oversized and will be forced to operate in the unstable region at a low flow condition, especially for those terminals close to the fan, which in this design run at a constant speed.

11.10.4   VAV economizer cycle analysis

Dampers play a critical role in the VAV economizer cycle, though fans are also an important factor. The dampers in this case need to perform the following functions:

Scheme 1:   Maintain a suitable mixing plenum pressure to ensure a proper pressure drop across points O and D for the introduction of the desired amount of outdoor air.

Scheme 2:   Install an outdoor air injection fan to ensure the introduction of the minimum amount of outdoor air.

Scheme 3:   Provide a demand-controlled ventilation system to introduce only the amount of outdoor air required for indoor air quality control.

Scheme 4:   Design and install an economizer cycle to introduce the maximum amount of outdoor air whenever indoor and outdoor conditions are feasible.

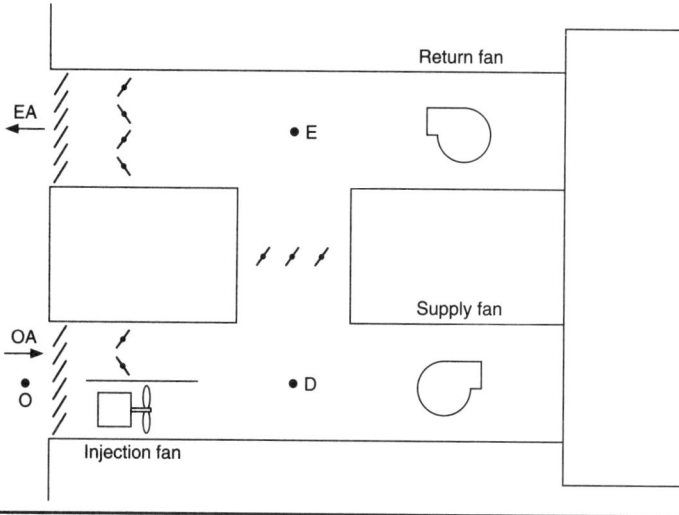

Figure 11.39   Various schemes for outdoor air supply.

1. Introduction of outdoor air from a prespecified minimum to the maximum of 100% outdoor air.

2. Stable control of the mixed air temperature entering the air handling unit.

3. Proper building pressurization.

Again, many factors affect the functioning of an economizer cycle, such as selection and control of dampers, return fans or relief fans, methods of fan capacity, and building pressure controls. A wide assortment of dampers with distinctive flow versus opening characteristics are used in a VAV economizer operation. These dampers are subject to the same selection and operational problems described in Secs. 11.10.2 and 11.10.3. Especially critical is the selection and operation of the mixing damper.

The damper must operate steadily over a wide range of airflow—from a large quantity to no flow—without adversely affecting the other functions of the economizer cycle, namely, proper ventilation and build-

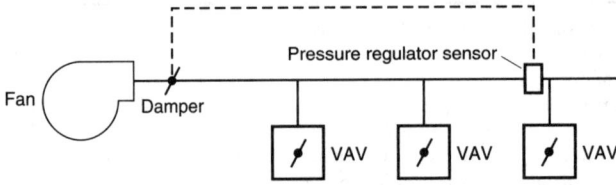

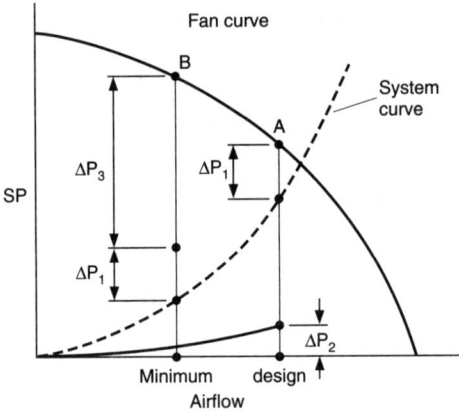

- ΔP_1: Design inlet pressure to VAV terminals

- ΔP_2: Main duct pressure drop

- ΔP_3: Additional ΔP to be absorbed by pressure
    regulator at minimum flow

(a)

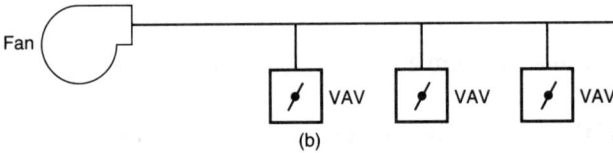

(b)

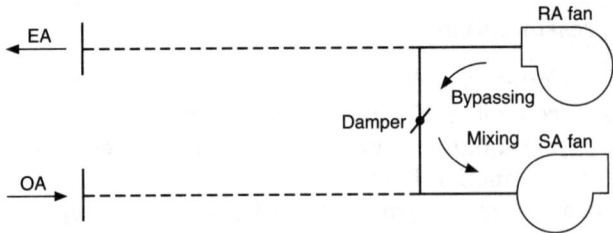

(c)

Figure 11.40 Damper and fan in series operation. (*a*) Fan at a constant speed. Damper used as a pressure-reducing valve. (*b*) Fan under VAV control. Damper used as an airflow valve. (*c*) Fan under VAV control. Damper used as a bypass or mixing valve.

Simple VAV system

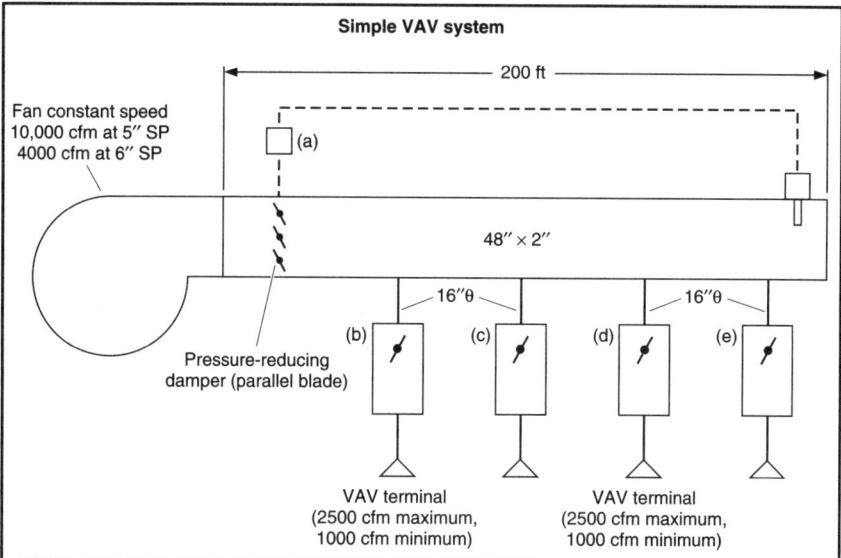

Table A – Damper pressure drops for various damper angle and velocity combinations.

Damper rotation angle (G)	Damper pressure drop corresponding to velocity shown.			
	3000 fpm	2500 fpm	1790 fpm	1000 fpm
10	0.56	0.39	0.20	0.063
20	0.90	0.624	0.32	0.1
30	1.52	1.05	0.54	0.17
40	3.04	2.11	1.08	0.34
50	5.63	3.9	2.0	0.63

Table B – Minimum and maximum velocities for the system shown above.

Duct Type	Max. Flow (cfm)	Duct Vel. (fpm)	Min. Flow (cfm)	Duct Vel. (fpm)
Main duct	10,000	2500	4000	1000
Branch duct	2500	1790	1000	720

Note: Damper pressure drop calculations are based on ASHRAE parallel blade damper Co values (L/R = 1.5).

Figure 11.41 Damper and damper in series.

ing pressurization. Consequently, any problem associated with this damper must be treated and solved as part of "system design problems" involving other key components in the system.

The pressure gradient diagram can be utilized to analyze pressure variations in ducted air systems. The technique is especially effective for economizer cycle analysis.

11.10.5 Pressurization and minimum outdoor air analyses

Building pressure and outdoor air supply are closely interrelated. An increase in outdoor air supply will increase the building pressure unless the building is too porous to build up a pressure, or the pressure is relieved by local exhaust or by system exhaust. Yet conventional HVAC systems are often designed to return 90 percent of the supply air minus local exhaust to the air-handling system, leaving 10 percent for building pressurization. This design approach ignores all other parameters affecting the building pressure and will not ensure the desired building pressure for VAV systems. Using the figure 10 percent of the supply air is arbitrary—it is in reality a continuously changing quantity, and will not be adequate to maintain the desired building pressure.

In the new integrated approach, all design parameters are considered simultaneously, and their interactions with building and system components are measured quantitatively as well as qualitatively by suitable analytical techniques.

11.10.6 Integrated building and system design

The same design principle described in the preceding section applies to other building environmental parameters, such as thermal comfort, IAQ, and acoustics. Ideally, building and system design should be totally integrated, analyzing all building and system interactions concurrently, thus fully automating building and system design. Today building and system design are only partially integrated, though several automated building design packages are currently being developed (Ref. 19). Still, as illustrated by numerous examples in Part 2 (especially in Chap. 11), sufficient information and techniques are available to perform integrated building and system design either manually or by computer and should be used for all VAV designs.

References

1. *ASHRAE Cooling and Heating Load Calculation Manual,* ASHRAE GRP 158; and *ASHRAE Fundamentals,* chaps. 25 and 26.
2. ASHRAE 62-1989 Standard, p. 11.
3. International Standard ISO 7730-1984 (E), Annex D.
4. Carrier Corp., "Total Environmental Quality," Technical Bulletin 92-19 by Dan Int-Hout.
5. Kennedy, C. R., and A. Distefeano, "Improving Building IAQ Reduces HVAC Energy Costs," *Heating/Piping/Air Conditioning,* November 1991.
6. *ASHRAE Handbook 1984,* Systems.
7. Industrial Acoustics Company, SNAP (System Noise Analysis Procedure).
8. Carrier Corporation, Applied Acoustics Program, 1992.

9. Trane Company, Acoustics Program, 1986.
10. Holness, G. V. R., "Pressurization Control: Facts and Fallacies," *Heating / Piping / Air Conditioning,* February 1989.
11. Meckler, M., "Demand Controlled Ventilation Strategies for Acceptable IAQ," *Heating / Piping / Air Conditioning,* May 1994.
12. Avery, G., "The Myth of Pressure-Independent VAV Terminals," *ASHRAE Journal,* August 1989.
13. T-DUCT Duct Analysis Computer Program, based on the ASHRAE T-method, by NETSAL and Associates, 16182 Mount Lowe Circle, Fountain Valley, CA 92708.
14. ASHRAE Fundamentals, Duct Design, 1993.
15. *Loop Duct Design,* a publication of Acutherm Division, Ebco Enterprises.
16. Hull, D. D., "VAV Duct Design with Static Regain and a Microcomputer," *Specifying Engineer,* March 1985.
17. Honeywell Inc., *Engineering Manual of Automatic Control for Commercial Buildings,* 1988.
18. Johnson Controls, Inc., *Damper Manual.*
19. Integrated building and system design packages:

CABDS
APEC, Inc.
40 W. Fourth St., Suite 2100
Dayton, OH 45402

LORAN-T
Taisei Corp.
Tokyo, Japan

HVAC/HWD
Chaina Air Conditioning Research Institute
Beijing, China

Construction of VAV Systems

12

Conformance to Design

A major source of problems with VAV systems is the failure of the construction process to produce a completed system that conforms to the design requirements. There are many deviations in construction that ultimately can lead to a failure of the system to perform properly (Table 12.1). No system changes should be made during construction that are not coordinated with the design engineer and properly described in as-built system documentation.

The conformance to design issue includes other systems that the VAV system supports. For example, the use of a space may change or equipment can be changed. If the VAV design was based on a load that now has been changed, then it is likely that the VAV system design and installation must be changed. Common changes include the substitution of window glass that has a different heat transmission factor, or the increase in electrical consumption in a space, creating more heat.

12.1 Value Engineering

After the construction documents have been completed, the project developer, owner, construction manager, or contractor may elect to modify the specified systems through a process generally known as *value engineering*. This terminology is misleading, as the motive of the process is typically a reduction in the initial cost of the system, and thus, it is *price engineering*. A concern that should be raised when this activity is undertaken is the potential for reducing the quality of the system, or its ability to function properly as designed and intended (Table 12.2). Even the substitution of equipment that appears to be equal to the original design intent may yield a less than satisfactory system. For example, the original design may call for DDC controls, but

TABLE 12.1 Typical 500-Square-Foot Exterior Zone Design Values Compared to Actual As-Built Conditions for a Problem Project

Design item	Design value	Actual value
Transmission load (Btu)	15,000	25,000
Equipment heat (Btu)	10,000	20,000
Ventilation required (cfm)	100	200
Supply system (cfm)	550	350

the value engineering phase proposes to substitute pneumatic or analog electric controls. The contractor may claim that the older technology is just as good and can provide similar control functions. However, the flexibility, reliability, and precision of control may be lost in order to gain only very minor first cost savings.

To properly pursue value engineering, the owner or contractor should proceed only with the assistance and support of the original designer of the systems. Second, all proposed changes should be evaluated on life cycle and initial cost, and any performance or functional limitations created by the change should be documented in detail. Specific features of VAV technology that are subject to harm in the value engineering phase include:

■ Use of smaller ducts, higher pressures, and greater velocities to lower the initial cost of ductwork. This change results in higher fan energy consumption, higher operating noise, and more duct pressure fluctuation.

■ Substitution of lower-cost fan volume/speed control devices that consume more energy and are less effective.

■ Use of smaller or less costly air terminals with less insulation, lighter-gauge metal, higher velocities, and more noise.

■ Use of smaller (or elimination of) sound attenuators, traps, and similar devices.

TABLE 12.2 Typical Value Engineering Conditions for a Problem Project

Design item	Design requirement	As-built condition
Branch duct size	2.25 sq. ft.	1.80 sq. ft.
Fan pressure	5.0 inches w.g.	4.0 inches w.g.
Controls	DDC	Pneumatic
Control zones	10 zones	3 zones
Room sound level	35 dBA	40 dBA
Sound control	Attenuators	No attenuators
Humidity control	Included	Deleted
Fan tracking	Included	Deleted
Economizer	Included	Deleted

- Reduction/elimination in automatic branch control dampers, causing difficulty in balancing or excessive noise at terminals.

- Reduction/elimination of access doors in ductwork that can be used for maintenance and service.

- Substitution of less costly but less capable and less flexible controls, or a reduction in the precision and accuracy of the controls.

- Use of flexible ducts to replace hard ducts in order to save labor, with the general creation of more noise and pressure loss.

- Elimination or reduction in the return air ductwork, using open or plenum returns.

- Reduction in the number of control zones to save initial cost.

- Use of smaller air-handling equipment by considering more system diversity, with the potential undersizing of the system and/or loss of flexibility. This includes changing the coil rating points, raising supply temperatures, and similar minor changes that can affect total capacity.

- Elimination of relative humidity control, and filtration efficiency reduction that reduces initial cost but diminishes IAQ.

12.2 Field Engineering

As the HVAC system is being installed, contractors may by accident, oversight, or necessity make changes to the installed system. These changes can have virtually no effect on the system or they can have a profound effect, depending on the interaction of the change with the design. Common problems are as follows:

- Specified equipment does not fit in the physical space provided and smaller models or versions of the equipment must be substituted.

- Specified equipment is not available.

- The equipment is mixed up or tagged wrong. A smaller model is accidentally placed where a larger unit was specified.

- The contractor likes to install the equipment his or her own way and views the design as incorrect.

- The contractor is making a conscious effort to save labor and materials and is intentionally deviating from the design to maximize profits.

- The installation quality is substandard. Duct leaks are the most common problem of this type.

These problems can yield a VAV system that lacks the design's intended performance. For example, the original design was marginal and several changes are made, resulting in a system that cannot function properly. Therefore, it is important that all field engineering changes be reviewed and approved by the original designer, who has the full appreciation of the required system performance.

12.3 Unauthorized Design Changes

As the construction progresses through its various stages, it is possible that unauthorized *unengineered* changes will occur. Without the engineering participation, proper coordination and system modifications may not be made. One frequent mistake that best exemplifies this problem is moving walls or changing wall materials. A wall may be added to subdivide a space, but the HVAC designer is unaware of this change and the design of the VAV system is not changed. A room is created without a control zone and the design airflow may be incorrect. The other problem is a change in the outside wall material that dramatically changes the transmission or solar load.

Therefore, it is important that design calculations specifically identify all design parameters and assumptions. Documentation of the original design intent allows these later changes in the building to be more easily identified and corrected.

12.4 Faulty Design

One of the most difficult situations is the one where the contractor decides that the design is faulty and must be changed. If a design error or omission is perceived by the contractor, it is imperative that this information be communicated concisely to the designer with supporting documentation. Consider these examples:

- A room lacks a return register but has supply grilles.

- A fan or air-handling system appears too large to be installed in the space provided.

- The control sequence proposed has proven itself unreliable before.

- The ducts or terminals do not appear to be sized correctly for the airflow indicated.

In each instance, the objective of the contractor should be to inform the designer and document the problem. However, the contractor should be willing to follow the design if the designer insists. Avoiding conflict may ultimately prove the best alternative.

12.5   Shop Drawings

The prompt issuance of detailed shop drawings is one means of making a construction project go smoother with fewer problems. The shop drawing submittal should always highlight any deviations from the proposed design, regardless of their size. Minor substitutions or changes could have a major influence on the design. Any changes in duct size, materials, or routing should be identified. Common problems with shop drawing submittals include:

- Failure to list all information required and to verify submittal is in conformance with contract documents.

- Failure to provide coordination information for other trades on duct layout drawings. This includes location of devices such as pressure sensors, dampers, access doors, and duct smoke detectors.

- Failure to identify items that differ from contract documents. All deviations from requirements should be shown whether they are of an equipment or installation method nature.

Table 12.3 summarizes commonly encountered shop drawing errors and omissions.

TABLE 12.3   Common Shop Drawing Errors and Omissions Checklist

Design item	Common problem	Issue
Fan performance	No fan curve or incorrect fan curve	Fan capacity and stability
Fan performance	No sound data	Can not determine if there is a noise problem
Fan performance	Data is not accurate	Vendors do not test all sizes
Fan size	Smaller fan is substituted but not identified	Fan capacity
Ductwork	Location of control items	Needed for correct installation
Ductwork	Conflict with other trades	Shop drawings must include trade coordination
Control device	Not correct range or selection	Must show range of device with sizing criteria
Specified quality	Quality information is omitted	Item of inferior performance may be substituted
Control sequence	Not correct	Control vendor ignores specific requirements

12.6   Periodic Inspection

The construction process covers up many installed items and it is important to inspect these components while they are easily visible and accessible for correction. Duct interiors, linings, joints, and insulation must be inspected as they are installed. An example of this problem is the lunch box that is left in the duct and blocks airflow, only to be discovered during air balance. The insulation may not be vaportight, and the sweating duct produces a stained ceiling from water dripping from the condensation caused by the lack of proper sealing.

12.6.1   Common installation errors

The VAV system must be inspected on a periodic basis as it is being installed. Too often, errors are caught at a time when correction is not feasible or too costly. During construction, these key items should be inspected immediately upon delivery and installation:

- Model number and size of fans, coils, air-handling units, sound attenuators, dampers, VAV terminals, and filters. Each major piece of equipment should have a permanent tag attached that describes the size and model number.

- Size, gauge of metal, and construction of main ducts. Large plenums are often too lightly constructed and bracing must be rigid.

- Quality and thickness of duct insulation. The proper sealing of the ends of internal duct liner is often inadequate.

- Size and location of outside air and exhaust louvers. Too often, changes are made that restrict the size or location of the louvers. Adequate free area of the louvers compared to the physical size is important to proper operation.

- Selection and installation of vibration isolators and expansion joints. Isolators may have to be adjusted to level equipment properly. The attachment of conduits and piping may restrict the correct movement of the isolators.

- Size and position of duct supports. The duct should be rigidly supported in all areas with flexible connections installed at all fans and air-handling devices. If ducts can move, the movement may transmit excessive noise and vibration.

- Installation of fire and volume-control dampers and access doors according to drawings. All dampers must be accompanied by access doors in the duct and in any walls/ceilings that may block access.

- Location and proper installation of thermostats, temperature transmitters, and flow-measuring and pressure-indication devices. The

VAV system typically uses a supply duct static pressure sensor that must be located two-thirds or more of the way down the duct to accurately measure duct pressure.

- Excessively long, bent, or collapsed flexible duct sections (Fig. 12.1). Flexible duct can save time and make terminal installation easier, but it should not be used just because it can fit around obstructions. Using flexible duct improperly causes excessive pressure drop and noise. One common problem is at the inlet of the VAV terminal, where duct bending may cause turbulence at the terminal flow sensor, leading to inaccurate flow measurements.

- VAV terminal control software often involves the setting of a pickup or calibration factor for the airflow sensor. These factors are often incorrect from the factory or set wrong in the field, leading to errors in the airflow delivery.

A common problem with air-handling-unit installation is the proper construction and support of connecting air plenums, mixing chambers, and large ducts. Too often, specifications refer to SMACNA standards for construction, but the sheet metal gauge, reinforcements, and supports are inadequate as supplied by the contractor.

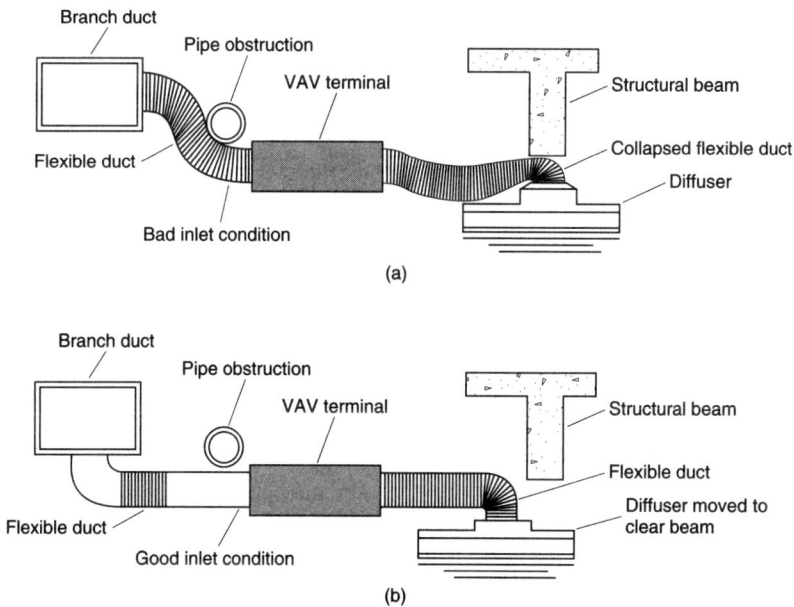

Figure 12.1 VAV terminal installation using flexible duct. (*a*) Improper installation using flexible duct; (*b*) Proper installation using flexible duct.

Accessories such as turning vanes, extractors, and similar devices may be omitted, and insulation can be poorly installed. Adequate fastening of interior liners in plenums is a frequent problem. Large sheet metal sections may lack cross-bracing or supports to prevent vibration during operation. Larger ducts and plenums are typically the biggest problem.

Periodic inspection of these major items at the time of installation prevents delaying the project if these problems have to be corrected at later stages (e.g., during air balance or commissioning).

Coordination of Trades

The typical project offers very limited space for the installation of the mechanical and electrical equipment and associated appurtenances. A common problem that occurs in the field is the use of available space on a first-come-first-served basis. Trades like to be first in an area to avoid working around the installed items of the other trades. Too often, the construction drawings and specifications allow the installing contractor to make minor changes in locating equipment that can cause subsequent trades substantial problems in installation. A common result is the air terminal that has a 6-foot-long flexible duct connecting it to the branch duct that is routed around conduits and light fixtures with four or five big bends—causing it to partially collapse due to limited space.

13.1 Mechanical

The HVAC ducts should be installed before the electrical conduits and immediately after the plumbing. It is important to prevent alteration of the duct size or shape to avoid installed electrical conduits. Small bends in electrical conduit runs may not affect their performance significantly, but additional turns or small bends in VAV supply ducts can significantly change the pressure loss and air delivery.

Ceiling-mounted air supply and return grilles and diffusers should be installed at approximately the same time the ceiling light fixtures are installed. Thermostats and control devices should be installed as late in the construction process as possible to prevent damage from other trade activity.

Filters should not be installed until the building construction area is cleaned and final finishes are in place. VAV terminal controls should not be installed until the ducts are complete, cleaned of debris, and operated for some time to remove any dust that could clog the filters in the VAV terminal controls.

Any lined, rigid ducts that show exterior damage or partial collapse should be carefully examined for separation of the interior liner. Duct leakage should be measured and all leaks corrected before the system is operated.

13.2   Electrical

Power wiring serving mechanical equipment should be installed only after the equipment is set, to ensure that adequate clearance and access is provided. The electrical power need not be available until it is time to test the system. If the air-handling equipment is operated on temporary power, the voltage should be checked to make sure that the equipment is not operated at too high or low a voltage. Excessive voltage drop with temporary services is a common problem that can damage electric motors.

Care should be exercised to keep electrical conduits out of the way of ducts and any associated piping. A typical problem is the conduit that is routed in front of access doors, preventing the door from opening fully or blocking access to the opening after the door is removed/opened.

When the VAV system uses large electronic variable-speed drives, there is a potential EMI/RFI problem that can interfere with the operation of the drives, the controls, and other equipment in the building. Drive wiring should include proper isolation in the form of transformers, filters, and dedicated feeders to protect power panels from noise. Good grounding practice and bonding of enclosures are also important.

The proper location and installation of disconnects and lockouts for all motors are items that often cause problems. Disconnects must be readily accessible and visible. Problems with space for VAV terminal electric heater disconnects above ceilings often occur.

13.3   Control

Wiring for HVAC control should be installed by technicians who are familiar with proper termination techniques. Generally, control wiring should be installed after power wiring. If necessary, raceways should be installed for the later addition of the control conductors and termination. Most control failures result from improper wire termination, inadequate grounding, or noise caused by nearby power devices. Use of a metal raceway or shielded cables reduces RFI problems. Control devices are sensitive to the characteristics of the cable used. Though a cable may have the right number and size of conductors, it may have the wrong impedance for the control devices. Control cable selection should rigidly follow the manufacturer's recommendations.

The control devices are very easily damaged and should be installed as late in the construction process as the schedule permits. It is advisable to rough in control wiring, tubing, panels, and piping wells, with installation of the transducers and panels coming later. Pneumatic damper motors are very rugged, but their connecting linkages are easily damaged. The major concern with the terminal controls is the flow-measuring-device transducer that can become fouled with construction dust if the air-handling system is operated during construction without proper cleaning and filter installation. The typical DDC controls for a VAV terminal include a filter in the airflow-measuring circuit that can become clogged, causing an erroneous reading and operation.

Pneumatic control problems are generally caused by physical damage to the control lines, and all exposed flexible tubing must be protected from damage by raceways. Rigid copper lines seldom have damage problems. If oil or water contaminates the lines, the connected control devices can be damaged beyond repair. This can happen if air compressors are operated during construction without proper care and maintenance.

13.4   Plumbing

The installation of the plumbing can interfere with the air-handling equipment in several areas. A major concern is roof drain piping above ceilings that also must contain the main supply and return/exhaust ducts. Another area is the installation of condensate drains that often must fit into very limited space yet remain serviceable. If possible, the plumbing, air ducts, and air-handling equipment should be installed simultaneously so that the installation can be carefully coordinated.

With cooling coils on the suction side of the fan, the proper trapping of the coil condensate line is essential. To avoid problems, the trap depth and fall of the line should be twice the pressure rating of the fan. Furthermore, this line should be accessible and easily dismantled for cleaning.

14

Protection of Equipment

The scheduling of many projects is purposely as short as possible and contractors may be forced to receive and install equipment at a time when the building and site conditions pose a risk to the equipment. The mechanical equipment is often damaged during construction, causing poor performance or other problems that ultimately require equipment replacement. (See Table 14.1.)

The most frequent problem is physical damage to the insulation on the piping and ducts. Vapor barriers and insulating materials cannot survive impact or crushing, yet during construction it is common for other trades to damage these materials. Water damage to insulation is another problem that occurs when ducts and equipment are installed in a building before it is completely watertight. Lack of heat and a controlled environment can lead to corrosion of control contacts and wiring devices. Sites near the ocean or in industrial areas are especially vulnerable to this corrosion problem.

14.1 Fans and Motors

The typical damage to fans involves the dropping of debris or tools into the fan itself, which leads to bent blades or an unbalanced condition. If possible, the air-handling unit and all fans should be protected from any debris falling through the ducts or outside the ducts until they are ready to operate. Dampers and coils are also damaged in a similar manner.

Electric motors can be damaged if they are stored in a way that allows water to enter and freeze. Another problem with motors is that during construction they may be connected to a temporary power supply that may not have adequate overload protection or proper voltage.

TABLE 14.1   Common Construction Damage Items

Item	Condition	Problem
Motors	Incorrect storage	Water in bearings and early failure/noise
Air-handling units	Bent and damaged casing	Leaks and inadequate performance or excessive noise/vibration
Finish on equipment	Chipped, peeled, or burned	Excessive corrosion and early failure
AHU/duct coil fins	Coil face damaged/fins bent	Excessive pressure loss and loss of capacity
Insulation on duct or piping	Vapor barrier damaged and water is condensing	Loss of capacity; corrosion; internal duct insulation collapses, blocking airflow
Flexible duct	Excessive bend or collapse	Inadequate airflow
Automatic dampers	Bent or sticking	Inadequate control and loss of energy/capacity
Valves	Dirt in system	Damaged seats that won't shut off
Filters	Clogged or missing	Dirty system may not work properly and can cause IAQ problems later
Strainers	Clogged or missing	Loss of capacity/damage to controls
Ducts	Collapsed by physical contact	Loss of capacity and noise or leakage
Vibration isolators	Collapsed	Excessive noise and vibration
Manual dampers	Damaged	Inadequate capacity
Controls	Paint contamination	Do not work
Controls	Dirty sensors	Inaccurate or failed
Controls	Broken by physical contact	Controls do not work

Fans that are operated without the proper ductwork can become overloaded and burn out the motors. VAV fans that are operated in a manner that bypasses the pressure controls can lead to duct damage from excessive pressure.

14.2 Controls

The controls should be installed as late as possible, and the contractor may want to operate the mechanical equipment before the controls are complete. Typical problems encountered include blowing out dampers from excess pressure, damage to control linkages caused by manual overriding of the controls to make the system operate, and operation of a dirty system that contaminates the control devices with construction dirt.

One concern is the life safety function the controls may provide, and operation of the systems without proper controls can pose a life safety risk. For example, fire stats or smoke detectors may not be installed or working, yet the contractor operates the equipment. Without these controls in place there is a risk that the air-handling system could quickly spread smoke and fire throughout the building.

14.3 VAV Terminals

The damage to VAV terminals during construction takes several forms. One problem is the damage to the flow-measuring devices caused by operation in a dirty system. Another is the damage to the inlet supply duct if it is flexible duct. Too often, the flexible inlet/outlet duct is bent, crimped, or partially collapsed during construction. In areas where workers can walk or stand on the duct and terminals, damage takes the form of bent casings, sticking dampers, and torn/crushed insulation.

14.4 Filters

During construction the most overlooked and abused component in the air-handling system is the air filter. Units are often operated without filters, with the wrong filters, or with less efficient filters. Filters also can become clogged very quickly under construction conditions, and they may collapse before they are changed. For example, the contractor may use only the low-cost pre-filters and omit the expensive bag filters. These conditions lead to the introduction of dirt and debris in the coils, fans, casings, ducts, and controls. Later, this contamination can cause poor performance, failure, or bad IAQ.

Debris and dirt in ducts are easier to keep out than to remove after accumulation. The daily construction routine should include the sealing of all duct openings as the duct is installed. Ducts should not be left open. Animals, birds, lunch boxes, tools, and trash can accumulate in ducts that do not have the openings closed immediately. Temporary closures can consist of sheet metal, plastic, or other materials that should be removed only when the duct opening is permanently terminated.

15

VAV Controls

The control contractor or subcontractor typically employs technicians who are well trained on the type of controls being installed. However, each project is unique, and special (as well as basic) requirements of the project may be overlooked. Table 15.1 lists some of the problems commonly encountered. Habits of the technicians can be both good and bad, and it is essential that the design be explained to the technicians carefully. Understanding the controls themselves is very different from understanding the complete HVAC system.

Problem areas include the location and installation of duct static pressure sensors. Some companies and technicians do not use pitot tubes but merely a short section of rigid tubing. This tube if installed incorrectly can measure total pressure instead of duct static pressure, leading to poor performance. Another problem is installing sensors without calibration. Many projects require little precision and the out-of-the-box accuracy of sensors is adequate. However, control technicians may follow the same procedures in a laboratory where the accuracy is critical.

Software-driven control systems represent a tremendous flexibility that can be both beneficial and very harmful. On large projects, thousands of lines of code may be input into the control system, and many control errors can be unknowingly introduced through bad code, input errors, and similar mistakes. The debugging and thorough checkout of this code can require literally years. Therefore it is essential that all software be developed and installed in a manner that controls and manages the error process. Control simulation techniques using simulation at the factory, downloading of code via modems rather than field input, or the installation of chips (that cannot be mixed up) all provide extra margins of reliability. If possible, all code should be proven before

TABLE 15.1   Common As-Built Control Problems

Pneumatic controls	Issue	Problem
Air compressor	Capacity	Runs too often/too small
Air supply	Dry and clean	Dryer too small or system contaminated with oil/water
Dampers	Modulation and tight shutoff	Bad linkage or motors too small
Dampers	Control range	Poor modulation/sticking/binding
Valves	Modulation and tight shutoff	Wrong size/leak
Zone terminal	Temperature control	Thermostat in wrong location
Airflow monitor	Fan tracking	Bad location/poor accuracy
Static pressure controller	Stable fan pressure down duct	Incorrect range and location
Temperature control at fan discharge	Averaging element	Averaging element not used

DDC controls	Issue	Problem
Panels	Power supply	Unreliable or EMI noise
Panels	Erratic problems	Grounding, shielding and wiring practice
Sensors	Accuracy/reliability	Not calibrated and/or wrong range
Control loops	Stability	Not tuned correctly
Valves	Modulation/shutoff	Incorrect size
Dampers	Modulation/shutoff	Incorrect size and linkage/ mounting of motor
Airflow monitor	Fan tracking	Bad location/field conditions for use
Supply temperature control	Averaging element	Averaging element not used
VAV terminals	Correct cfm	Pickup factor wrong

installation; canned pre-engineered and tested control sequences help reduce the number of errors.

15.1   Limitations

Two common faults found in many systems are (1) the use of the control system to correct a basic fault in the primary or secondary air-handling system and (2) control design. (See Table 15.2.) Controls are sometimes expected to perform tasks that are not possible given the

TABLE 15.2   Control System Limitations

Control item	Function	Condition
Fan static pressure and capacity	Stable pressure	Fan curve does not permit stable control, bad fan selection
Parallel fan control	Use one or more fans depending on load	Wrong type of fan for parallel operation, fan may stall on start-up
Fan tracking/airflow monitors	Space pressure and correct introduction of outside air	Incorrect control design and need for control accuracy that is not feasible
Space temperature	Stable temperature	Inadequate supply airflow or temperature beyond design
Space temperature	Correct and accurate control	Transmitter incorrectly located
Space comfort	Occupant comfort	System is designed for dry bulb only and not for total comfort/IAQ
Indoor air quality	Good IAQ	Ventilation and IAQ were not part of the system design and controls will not provide function alone
Energy conservation	Reduced energy consumption	System is not capable of supporting energy conservation sequences due to capacity problems
Relative humidity	Proper humidity for comfort, too high/too low	Chilled water temperature is too high or there is no humidification system furnished

nature of the controlled equipment. For example, control sequences cannot correct for an oversized valve or damper. If the damper/valve is incorrectly selected, the precision and speed of response of the DDC control system will not provide any better control than an old pneumatic system under the same conditions. If a fan or pump is operated in the unstable region of its curve, then controls will not correct the problem. If a fan is too small or the system is designed with too low a diversity factor, then the controls cannot maintain duct static pressure. VAV terminal units cannot supply adequate air if the total static pressure available is too low.

The most common control failure is the room thermostat/temperature transmitter that is incorrectly located. The occupants will remain unhappy no matter which setting is used. The problem is not the control itself but the design/construction that improperly located it.

15.2 When to Install

The controls are often installed based on a schedule that is payment- and progress-driven. To earn money, the control contractor installs the controls. This fact, though, may lead to the installation of the controls at a time that is too early for adequate protection. Duct airflow sensors and VAV terminal controls should be installed only after the ducts are clean and the proper filters are installed to prevent the entry of dirt and debris. This implies that the duct system and all fans have to be completely installed.

Relative humidity sensors, room thermostats, and temperature transmitters should be installed after the room finishes are installed. Painting activities should be coordinated with the control vendor so that paint or paint fumes do not damage delicate devices. Pneumatic devices should be installed after the air compressors and filter dryers are operating continuously on permanent power. DDC panels should be installed after the space is watertight, properly conditioned, and provided with permanent power.

Damper linkages and duct-mounted devices should be installed after all other trades have finished work in the area. Pipe-mounted sensors and flow switches should be installed immediately before pipe insulation installation and after all other trades have essentially completed work in the area. Central computers and monitors should be installed when the rooms containing them are complete and have been supplied with permanent power and air-conditioning.

Though the industry habitually ignores field calibration and setting of the sensors, this important step must be performed. Even slight errors in calibration can cause a system to function improperly. The extra time spent testing and verifying the accuracy of the sensors is small compared to the time that can be spent trying to troubleshoot problems caused by miscalibration.

Chapter

16

Air Balance: Testing and Adjusting

The test and balance contractor (T&B) or technician faces a dilemma on each project. If the testing finds problems, it can be expensive to correct them. Therefore, pressure is often put on the T&B contractor to issue a report that shows no problems. The T&B contractor must resist this pressure and provide accurate reports.

16.1 Importance of Air Balancing

The VAV systems intrinsically change airflows during operation. The system can operate incorrectly yet maintain space conditions except for periods of peak load. If a system is not carefully balanced, the system problems may not be uncovered until the contractor has been paid and left the job. For example, the outside airflows could be grossly incorrect and yet the building occupants may not initially notice a comfort problem. The building could be under severe negative pressure and yet in mild weather the problem would not represent a major problem to the occupants.

Air balancing has the responsibility to verify that the system is functioning as designed, and to properly distribute the airflow under all operating conditions. Components that need calibration and adjustment to perform properly depend on the testing performed during balancing. T&B verifies the control sequences and settings of the controls.

16.2 Critical Items

Many VAV systems have several unique properties that make air balance testing more difficult than it is for constant volume systems. One of these unique properties is the calibration and use of the pressure-independent VAV terminals. These devices measure airflow, and they generally must be calibrated and proven accurate. This testing generally requires the cooperation and assistance of the control vendor, the terminal manufacturer, and the air balance contractor. Another issue is the interaction with constant volume systems such as toilet exhausts, exhaust hoods, and similar systems. Generally, all constant volume systems should be balanced and set before balancing the VAV system.

If the VAV system was designed with a supply fan that has a smaller airflow capacity than the arithmetic sum of the terminal devices, then the terminals must be balanced in groups, with some set at full flow and others at partial flow, to prevent exceeding the system capacity. A common error that occurs with system balancing on systems that use return fans is balancing the systems with the return damper open. Generally, all balancing of the return system should be accomplished with the return fan isolated by closing the return damper. The supply fan should be isolated from the return fan while it is initially being balanced. This recommendation applies to systems that have 100 percent outside air capability. If the supply fan and return fan systems are designed for less than 100 percent outside air, then the return damper must be opened during balancing. To prevent the supply fan from pulling too much air, the return damper should be adjusted to maintain a positive static pressure at the discharge plenum of the return fan. Failure to properly adjust the return damper can lead to incorrect balancing of the return fan.

The air balance procedures should include a complete verification of the operation of the control system and control sequences. Thus, the air balance contractor is working with the control contractor to verify correct operation. The data obtained from the air balance instrumentation should be used to calibrate the control settings. During air balance, there are three key parameters that must be addressed: airflow, noise, and control stability.

16.2.1   Airflow

When most people think of air balancing, they envision the technicians setting control dampers, taking readings at diffusers, and measuring the pressures in the ducts. With the VAV system, the airflows are typically set for maximum design day conditions and the control system automatically adjusts the system flows downward to meet conditions. Therefore, the critical airflow parameter is the maximum available flow. If the air balance data indicates lower-than-design maximum is available, then corrective action is imperative.

Another flow issue is the difference in exhaust and outside air quantities. The outside air available from the VAV system must never be less than the sum of the constant volume exhaust systems, and an adjustment should be made to ensure adequate outside airflow. This adjustment is essential to maintaining good IAQ.

Some air balance tests have used the arithmetic sum of the terminal and/or diffuser deliveries as the airflow for the air-handling unit. This procedure should never be permitted because it conceals the potential duct leakage. Traverses of main supply and return ducts are mandatory to compare with the sum of the outlets in order to determine if leakage is excessive or terminal pickup factors are correct. Furthermore, final readings of all flows on an air-handling system must be taken at one time. This means that the readings of outside air, return air, and supply air must be concurrent with the terminal readings. Even on relatively large systems, this requirement is not an excessive burden.

16.2.2 Noise

The air balancing contractor often takes sound-level readings while the system is being balanced. If excessive noise is encountered, the noise source should be identified. The noise problem can be isolated if the return and supply systems are individually turned off while the readings are taken. Initially, VAV terminals should be cycled through their airflow range to detect whistles and vibrating components. Final noise readings should be repeated after the controls are fully functional and

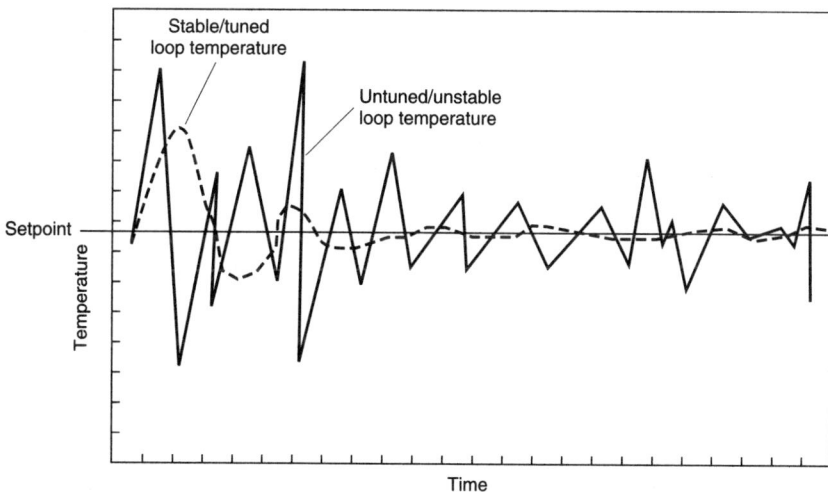

Figure 16.1 Example of a trend of a stable and unstable loop.

the airflows are correctly set. All noise readings should be annotated with the operational condition of the supply and return systems.

Initial sound readings can be taken using a sound meter that reads sound pressure on the A-weighted scale, but if problems are suggested by this reading, an octave band analyzer should be used to identify the sound level in each band. Knowing the band of the sound allows an engineer to better determine what measures may be needed to correct the problem.

16.2.3 Control stability

The control loops initially require tuning. Tuning is the process of adjusting the time response of the control system so that it can maintain setpoints without excessive error, delay, or cycling of the control devices. A major source of problems on VAV projects is the lack of proper tuning of the control loops that affect space temperature, humidity, fan tracking, down-duct static pressure, and discharge temperatures. Tuning and development of stability is difficult because many loops interact in a VAV system. For example, lowering the supply air temperature should cause the VAV terminals to require less supply air, and they will close their dampers and raise the duct static pressure, thus causing an adjustment of the supply fan. If one of the control loops is unstable, the other loops can be affected.

The trending function of DDC systems helps identify and correct unstable operation, and all major control points should be trended to verify stability (Fig. 16.1). This trending includes both inputs and outputs, and it should be done over several seconds, minutes, and hours to identify if the instability is of a short-term "hunting" nature or long-term drift. Where control system performance permits, the input and corresponding output points should be concurrently trended and displayed graphically for easy comparison.

17

Commissioning

Commissioning is the process of achieving, verifying, and documenting the performance of buildings, HVAC systems, and other systems. This process detects any latent system defects, provides an opportunity to make any required adjustments to operation, and trains the operators.

17.1 Purpose

Any new system may not operate as well as it is capable of operating. The purpose of the commissioning process is not the determination of how well the contractor installed a system to meet the design document requirements. It is not meant to replace the final system inspection and the corrective punch-list process. Instead, commissioning is a process of investigation and learning directed at fine-tuning and adjusting the as-built system. The performance of the system is investigated to determine if it is adequate for meeting the operational requirements. The learning process is directed toward developing user/operator confidence in operating the system correctly.

Commissioning should be used constructively to uncover potential problems before they lead to major operational deficiencies. Design limitations, errors, omissions, constructions faults, and new requirements may be discovered during the process. It is important that the design engineer, contractors, and operator personnel participate in this process as a team.

One element of all commissioning exercises is the physical validation of the system documentation, including:

- As-built system drawings, including the location and access to all operable components (valves, sensors, motors, etc.)
- Control sequences and control documentation

- T&B reports, with spot checking of flow, pressures, temperatures, and sound readings
- Maintenance manuals to verify the equipment descriptions are complete and accurate

17.2   Common Mistakes

The most frequent mistake made with commissioning is treating it like another final construction inspection. The commissioning team should expect the system to be complete with all punch-list items resolved before the start of the commissioning. Another error is not having a detailed plan that describes each step of the process with anticipated results. As in any good meeting, the agenda is the tool that guides the progress.

Commissioning should occur at a time that allows all systems to be exercised without concern for interruption of services. No tenants should occupy the building if their presence prevents exercising the systems.

If possible, the commissioning should not be conducted by a party that has a self-interest in the outcome. Though the team can include the design engineer, the contractors, and the operator of the system, the team should be led by a third party who is paid to direct and certify the process. This independence of the process is critical to prevent any covering up of potential deficiencies or limitations that may be found. Objectivity is a critical element in selecting a commissioning agent. The project contractor and the design engineer should be avoided if possible.

17.3   Cost

The cost of commissioning has varied widely due to the variations in the level of detail required on various projects. As the complexity of the systems varies widely between a simple office building and a medical center or laboratory, so too will the commissioning cost. To determine if the proposed cost is fair, the level of expected detail and number of labor hours should be submitted in a breakdown. Blanket quotations on a square-foot basis with general statements such as "commission the HVAC system" should be avoided. Generally, for mechanical systems the commissioning cost should be no more than one-fourth to one-half the T&B cost on the project.

Operation of VAV Systems

18

Understanding Design Intent

The VAV system designer seldom meets the operator of the HVAC system. The original design is based on a large number of assumptions and operating requirements that may bear little resemblance to the final facility the operator encounters. The storeroom may have become an executive office, and the heavy mainframe computer load may become only a few PCs on desks. With so many people involved in the construction process and so many differing objectives, the only common links between the designed facility and the actual facility are the construction documents and design calculations.

The design documents define what the designer thought should be provided to meet the operational requirements. Understanding these documents provides a clear snapshot of how the systems are intended to function. If the facility was properly commissioned, these documents are annotated with as-built information, and operational data has been added to support the correct operation of the systems. Given a thorough commissioning, there should be no obvious design flaws or construction deficiencies remaining.

Design documents should be easy to understand and read by the operator, but if the operator is unclear on what the system should do, the original designer or another with the required technical expertise should be consulted to provide an explanation. All system operators should have access to the design documents and be familiar with their intent.

If the operator or owner of the facility plans to modify or expand the facility, the original calculations of the HVAC system designer are beneficial in determining what loads were assumed, what diversity was provided, if any, and other key issues that would impact the revised system. Some HVAC designers may be reluctant to provide calculations, and this issue should have been resolved when the designer was retained. Design calculations reveal many mysteries that may appear in the design documents. For example, the original owner may have specified a minimum load per square foot, or may have mandated a particular system configuration, and this information should be revealed in the design calculations.

The design calculations should be reviewed only by competent professionals who understand the technology. Preferably, the original designer should be used, but the design information should be in a format that allows others to easily understand the design basis.

In summary, the HVAC design documents and calculations along with updates provided by the commissioning process form the foundation for operating, modifying, and expanding the HVAC system throughout its life. The system operator should have access to these documents and have a good understanding of their content.

18.1   Zoning

As time passes, the zoning of an HVAC system may not be appropriate for the operational requirements. A typical problem is a load imbalance in a zone that causes one part of the zone to be much cooler or hotter than the other. The zone thermostat cannot properly sense the zone temperature, and complaints are generated.

Another frequent problem is the relocation of room partitions or the installation of new partitions that prevent the circulation of air within the zone. Sometimes a diffuser that was meant to serve 600 square feet ends up in a 150-square-foot room, or one large room has now become three offices. This disturbance in the original zoning plan is one of the most common sources of comfort complaints.

To reduce or eliminate complaints, the VAV system must be zoned properly with respect to the space use and load. There are no exceptions to this rule, and it does not matter whether the original design or later changes cause the zoning problems. To function properly, the VAV system must be correctly zoned.

This zoning process involves alignment of the thermostats, the diffusers, and the adjustment of the airflow to match the load requirements. Zones may have to be added or rearranged on a dynamic basis to match the occupancy requirements. For example, if remodeling involves moving or adding partitions, then the VAV system should be adjusted by the revision of its zoning to match the remodeling. Seldom are too many control zones a problem, and adding zones is usually the best way to address complaints.

18.2   Actual versus Design

Over time, the HVAC design ages just as the facility ages. New office computers, changes in staffing, or changes in space use make the HVAC design obsolete. The concept of an obsolete design does not involve the equipment itself or the technology used as much as it does the configuration of the equipment. Older buildings may generate complaints not because the equipment is failing but because the design is obsolete. The VAV system is very flexible, and it is relatively easy to periodically update the VAV design by evaluating the changes in the zoning and air delivery. Keeping the design of the system up to date also helps in determining when the equipment is actually failing. Design updates reduce and eliminate complaints.

Before any modifications or changes are made to a VAV system, the actual versus design conditions must be evaluated and the design updated. Repairing or replacing equipment without updating the design most often fails to resolve the perceived problems.

Saving energy has become a national objective that has high visibility and high priority. Energy savings translate into dollar savings, and every Btu saved is rewarded by lower operating cost. However, the saving of energy can have a price in IAQ and occupant comfort that may not be intuitive or readily identified.

Because VAV systems dynamically attempt to adjust their operation to save energy, many energy-savings schemes can prove counterproductive. For example, refrigeration energy can be reduced by raising the chilled water supply temperature. This refrigeration savings, though, may be offset by more VAV system fan energy to deliver more of the warmer air to satisfy space conditions. Refrigeration savings are lost on the airside portion of the system. Likewise, the higher chilled water temperatures may produce high space humidity conditions because the VAV systems are no longer able to dehumidify the air with the higher chilled water and coil temperatures. The result could be an IAQ problem generating complaints, or damage to the facility such as mold on the wallpaper and in the carpets.

The most common energy-savings measure is the reduction in outdoor air. Too many dampers are closed or control sequences modified to generate this savings. Loss of proper ventilation is seldom instantaneously noticed, and the final discovery of IAQ problems may take a long time. Operators should not reduce outside air quantities for the purpose of saving energy unless a design professional reviews the change and determines that the resulting building ventilation will be adequate.

Many occupant complaints have been generated through the use of several energy-savings techniques. Duty cycling, where the whole air-handling unit is turned off for a brief period, has been one method. Duty cycling with VAV units is seldom productive because the system will automatically try to recover immediately after the air-handling unit is turned on. The load is stored in the space and the energy saved is immediately lost during the recovery cycle. However, while the unit is off and during the recovery period, the occupants may be uncomfortable.

Another technique is the automatic setup and setback of the terminal temperature setpoints based on time of day. For example, at 4:00 P.M. the VAV terminals might reset the control point to 80°F with the assumption that the office workers will be going home before they notice the rise in temperature. Another version is the weekend setback followed by the Monday morning warm-up. With these schemes, zones with limited reserve capacity may not recover correctly. The temperature in the space or ventilation may remain incorrect for extended periods even though the majority of the system sees no problem. Any use of these methods creates a thermal storage and recovery problem that needs to be examined space by space.

In developing energy-savings schemes and cycles, the energy savings should be classified according to the impact on IAQ, comfort, and equipment operation. If the energy-savings method is not potentially harmful to IAQ, comfort, or the equipment, it should be implemented. However, if there are doubts, the measure should be carefully studied, and modeled if necessary, before implementation.

19.1 Techniques

Energy savings fall into three categories, based on technique:

1. Primary (energy plant) and secondary (HVAC system) equipment efficiency

2. Process changes

3. Control or operating changes

Each of these categories has a different effect on the VAV system.

19.1.1 Primary and secondary equipment efficiency

This category involves replacing a piece of the HVAC system in a way that does not change the operational parameters of the design. These changes seldom affect comfort, IAQ, or the effective operation of the VAV system. Typical changes include:

1. Installation of a more efficient refrigeration or heating system without changing the chilled or heating water setpoints

2. Replacing fan vortex dampers with VSDs

3. Installing high-efficiency motors on fans and pumps

4. Installing more efficient fans or pumps

5. Installing new VAV terminals that operate at lower duct static pressure

6. Replacing leaking dampers or valves

7. Upgrading duct and pipe insulation

8. Installation of higher-mechanical-efficiency air filters with corresponding less pressure drop

19.1.2 Process changes

This category involves altering the design of the HVAC system in some manner to gain efficiency and lower energy consumption. These changes can be combined with the changes in primary and secondary equipment. These changes can affect comfort, IAQ, and the proper operation of the VAV system. These changes include:

1. Installation of heat recovery equipment at the air-handling unit. The typical impact involves the quality and quantity of exhaust and outdoor air. For example, the installation of a heat wheel may reduce the amount of outdoor air, its temperature, and relative humidity, resulting in less energy consumption but possibly poor IAQ unless the air quantity is increased.

2. Installation of thermal storage equipment with a change in the chilled or heating water setpoints. Typical problems with chilled water involve cooler water that causes coils to become oversized. Control and IAQ problems can result. Many heating water changes involve using cooler water that may not meet the needs in all zones.

3. Installation of economizer operation, either airside or waterside. A common concern with adding an economizer is the protection of the chilled water coil from freezing and the interaction with other control sequences.

4. Replacement of a chilled water air-handling unit with a DX unit, or replacement of a draw-through air-handling unit with a blow through-unit. Though the Btu capacity and air volume may be similar, the part load performance, or discharge conditions could change enough to cause problems.

5. Installation of fan coils, radiation or other auxiliary/local heating/cooling system within the space. Typical problems are control coordination and simultaneous heating and cooling.

6. Conversion of a zone from constant volume to VAV, with IAQ or pressure control a potential problem. Constant volume zones are easier to balance and typically will have more air moving within them. If the diffusers and grilles do not mix the air well, the lower VAV air volumes can result in poor circulation and mixing, leading to IAQ problems.

7. Conversion of ventilation control to pollution-sensor operation. The fixed ventilation cycle is now made dynamic, with potential ventilation problems occurring. Too often, a ventilation sensor measures a bulk or average condition, and it can lower the amount of outdoor air below safe levels for specific areas.

19.1.3 Control or operating changes

The controls on HVAC systems receive the most attention due to many factors, and the industry has been prone to emphasize how effective small, low-cost control changes can generate large energy savings. However, small and insignificant control changes can generate big IAQ and comfort problems along with large energy savings. Almost all of the control changes need to be carefully examined before implementation. Typical changes include:

1. System shutdown during unoccupied periods. Typical problems include loss of proper ventilation with a build-up of pollutants unless purge cycles are correctly included.

2. Adjustments to AHU discharge temperatures and chilled water temperatures, with the typical problems being loss of humidity control and bad IAQ.

3. Adjustments to AHU static pressure control to reduce fan energy, with the potential for VAV terminal airflow starvation in critical zones, causing poor IAQ and loss of comfort control.

4. Changes to outdoor air control, with the typical reduction effects including poor IAQ and loss of proper ventilation. Energy-driven proportional or adjustable control schemes often fail to include IAQ safeguards.

5. Setback or setup of zone control points with the potential for occupant comfort complaints.

6. Changes in VAV terminal minimum airflows to reduce need for reheat, with loss of proper air changes and ventilation in the space.

20

Interdependence with Other Systems

The VAV system is a subsystem within the overall HVAC system, and it is dependent on the other facility systems, such as the electrical system, for its proper performance. Many problems identified as VAV problems may actually originate in other systems. The VAV system can also cause other systems to malfunction if it is not operated correctly. Problems that have been encountered are:

1. Failure of the VAV system to maintain space conditions due to problems with the refrigeration system. Chilled water supply volume and temperature are common problems that lead to space temperature and humidity control problems.

2. Drafts and comfort problems in exterior zones caused by a building envelope that has too many leaks. Sweating windows or walls are problems that result from inadequate insulation value or too high an interior humidity level.

3. Comfort problems caused by miscalibrated or malfunctioning controls. The control system could be incorrectly controlling temperature at the air terminal setpoint, causing the room air to rise 10°F beyond the true setpoint.

4. Low voltage or noise in the electrical system that interferes with the air-handling system VSDs. The VSDs may cause the air-handling system to provide too little or too much pressure because of this noise.

5. Noise from the VAV terminals or ducts caused by a lack of sound control from the partitions or ceilings. The air-handling system

may be operating correctly, but the acoustic design anticipated better sound control from the walls and ceilings.

6. Lack of ventilation or odors in areas caused by the inadequacy or failure of exhaust systems that are not part of the VAV system. All air-moving systems have to work together and properly.

7. Time-clock or control computer time set wrong, causing the controls to operate as though the building is unoccupied when it is occupied. These time settings often affect the air-handling-unit operation, ventilation, and the terminal setup/setback.

8. Doors blocked open that are intended to provide zone control. The VAV system may be trying to condition space that should be unconditioned.

9. Shading systems such as awnings removed from windows, energy control films removed from windows, or blinds that are not in place. These systems typically limit solar gain, and when they are changed or removed the VAV system may appear to be malfunctioning.

10. Failure or shutdown of supplemental air-conditioning systems such as fan coils or local cooling units that are intended to share the cooling load with the VAV system.

Making Modifications and Correcting Problems

There are three possible outcomes to modifying a VAV system. The first is successful attainment of the anticipated change or improvement; the second is no change; and the third is a deterioration in performance and failure to achieve the goals of the modification. These outcomes are generally controlled by how well the modification is planned and implemented.

Too often a system change is made with insufficient information available and/or inadequate analysis of the problem. An example is the classic too-hot-or-too-cold complaint. The technician's traditional first step is the reset of the zone thermostat or space temperature. However, if possible, the first step should be a trending of the reported temperature to determine what the response of the system is according to time. Trending often reveals information that allows the symptoms of the problem to be separated from the real causes.

Using the too-hot-or-too-cold complaint as an example, the following steps should be taken:

1. The source of the complaint should be documented. It is probably not an HVAC problem if, for example, the person is wearing clothing that is inappropriate, or the person is sitting in a room with six others who are comfortable. The individual human subjective nature of the complaint must be recognized and accounted for throughout the process.

2. If the control system can trend the zone temperature and setpoint, trending should be established to document the system operation. It is helpful to include parameters such as terminal damper posi-

tion, chilled water temperature, and other factors that may affect the system.

3. The zone and air-handling equipment along with controls should be examined for broken components or miscalibration. Incorrectly adjusted VAV terminal dampers, or tape on diffusers, or cardboard stuffed into a diffuser are common user modifications that lead to complaints.

4. After all available data is gathered, the source of the problem should be easily determined by analysis of the information. Once the source of the problem is established, the remedial alternatives should be planned.

5. The easiest and least costly remedial action should be attempted first. It should be recognized that the final solution may not be timely, but interim modifications may limit the problem to an acceptable level.

The modification of a VAV system can result in by-products that lead to new problems. Thus, a problem can be fixed, and the fix can cause another problem. Examples of this cycle include:

1. Changing the zone setpoint, or adjusting the terminal to deliver more air, and the air-handling unit runs out of capacity. The previously satisfied zones now begin to generate complaints.

2. The increased airflow from the terminal resolves the space overheating problems, but now the occupants complain about the terminal noise or the drafts.

3. The decreased airflow from the terminal resolves the space overcooling, but now IAQ problems and complaints develop from a lack of air changes and ventilation.

4. The modification was successful but the occupants continue to complain. They have become overly sensitive to the space conditions, and they expect problems to continue due to the previous problems.

The correction of problems and the making of modifications is an iterative process that may not have a satisfactory or immediate conclusion. Simple problems can be difficult to identify if the data needed to solve the problem is not available. Therefore, if the problems become chronic or the expected success is not obtained after expending the anticipated resources, it may be advantageous to have a third party collect the data and recommend solutions. Additional test equipment or people more skilled in troubleshooting systems may be able to find the cause of problems that internally could not be found.

Indoor Air Quality

Operating a VAV system while maintaining low operating cost, comfort, and energy goals can be challenging since many operational factors are in conflict. Good IAQ requires attention to specific system components, and IAQ can be lost long before occupant complaints provide an opportunity to investigate and correct the problem. Therefore, IAQ demands a proactive approach to the VAV system.

The exhaust fans should be periodically tested for correct volume delivery, and any exhaust ducts that are under positive pressure should be checked for leaks. The use or presence of potentially harmful chemicals and materials in the interior should be under constant monitoring. If exhaust fans are increased in flow or added, the outdoor air ventilation system should be adjusted to provide additional makeup air. Adjustments in the air balance should be made to keep zone and building pressures in proper balance.

The ductwork should be inspected periodically and cleaned. Insulation tears, loose liners, broken/defective dampers, and leaking ducts should be corrected. The control system calibration and proper execution of the control sequences should be verified at least once at the start of each cooling or heating season.

Installed air filters should be of the highest efficiency established by the design and operating conditions. Filters should be changed based on the manufacturers' recommendations, including either time or pressure drop. Coils and drain pans should be inspected quarterly for dirt and contamination. If products to inhibit bacteria, microbes, or fungi are used, seasonal testing should be done to verify the effectiveness of these materials. Any moisture/water accumulation in drip pans and ducts should be eliminated. The system should be inspected for unauthorized modifications such as taped-up louvers, diffusers, or grilles that would cause interruption of proper operation. Even the practice of users opening windows or leaving doors open could provide an opportunity for contaminated/untreated air to enter the facility.

The minimum flow of all VAV terminals or the flow of a separate ventilation system should be verified to ensure the space is receiving the proper amount of ventilation and air changes. Exterior louvers and ducts should be inspected quarterly for clogging/damage to insect screens, birds nests, and other sources of contamination.

If the facility is large, previous IAQ problems have occurred, or the process within the space is critical, IAQ testing should be performed that includes measurement of carbon dioxide in the space, sampling for noxious gases, collection of air samples for inspection of dust/spores, and other tests needed to ensure quality.

Resolving Problems

A problem with HVAC can be quickly and easily solved or it can be a chronic and erratic problem that seems to have no cause and no solution. The successful solution to many problems depends upon the availability of data and records that too often are ignored until the problems develop. Good operating practice demands the maintenance of the design documents, as-built information on all changes to the system, and operating parameters such as chilled water temperature. When problems occur, the time they occur is often just as important as what happens. It is generally not possible to go back and reconstruct data that was not collected at the time the problem occurred. Real-time data gathering must be available for the analysis of operating problems.

Record keeping can be labor-intensive or it can be inexpensively obtained through the HVAC control system. A well-designed DDC control system includes a data-collection and archiving utility that collects and stores the needed data. Too often, design shortcuts and first-cost savings delete or limit the HVAC control system features that support data collection and documentation of system problems. Because these features are so important, the use of the previous pneumatic and electric controls that lack this feature are discouraged and not recommended. A modern facility needs this data, and any facility without this capability should be considered for upgrading. All steps in the problem-resolution process require good records. Record keeping is the key to resolving problems.

23.1 Trending

The trending capability of DDC HVAC control systems is limited by the availability of input points and system memory. The trending tools generally have the ability to provide sampling according to any desired time frame, from once a minute to once an hour or day. Typical DDC disk storage systems and graphical tools provide a powerful means of presenting the trended data for later analysis.

Variables such as space temperature is often trended along with other parameters that could affect the space temperature. Trending related points such as valve or damper positions also reveals how the control system is responding.

23.2 Troubleshooting

The recommendation to use DDC controls exclusively is a product of experience developed by troubleshooting many HVAC systems. Troubleshooting requires information and analysis of the data that in many instances can be effectively obtained only from a full-featured DDC control system.

23.2.1 Comfort complaints

Investigation into HVAC problems involves careful research and analysis to separate the symptoms from the true problems. Interaction of system components and systems themselves provides an opportunity for confusion unless a methodical plan is followed.

1. The first step involves carefully defining the complaint according to *who, what, when,* and *where.* Information such as weather often plays a critical role. Therefore, it helps to have a written complaint form that describes the complaint in these terms. There is no need to provide technical information with the complaint.

2. The second step is determining if the complaint is an isolated event or part of a pattern. History and records of other complaints/problems can play a role here. For example, the time of the complaint might indicate that it occurred when the chiller was down for maintenance.

3. The third step involves screening out the obvious problems from the not-so-obvious causes. For example, the complaint may be caused by and noted as the space thermostat being broken. An obvious solution is to replace the thermostat, but a hidden cause may be too many people adjusting the thermostat because of temperature problems.

4. Once the written complaint is studied for possible remedial action, the next event should be an inspection of the physical conditions by a technician to evaluate the system. Careful inspection of the physical equipment should be combined with discussion with the occupants. Further insight into the problem is usually obtained, and the complaints may grow once the occupants provide more input.

5. Most complaints can be resolved by the technician at the time of inspection. The common VAV system faults are easy to detect and repair:

   a. Broken, miscalibrated, or incorrectly set space sensor/thermostat.

   b. Diffuser dumping, drafts, or noises from ductwork, with a correction to the manual damper setting required.

   c. Slipping or stuck damper on terminal, or damper motor defective.

   d. Wiring or pneumatic tubing problem, with bad wiring connections a leading problem with DDC.

   e. Controls defective or software corrupted. (Software corruption occurs when the working instructions the DDC system uses suddenly for no reason change due to hardware problems or unknown causes.)

   f. Problem with air-handling unit, such as high chilled water temperature, valve closed, fan off, belts slipping, filters clogged, dampers stuck, and so on.

6. If the physical inspection reveals no obvious problems, the next step should be a trending of the conditions to determine if a pattern is occurring. Typical problems uncovered by trending include:

   a. The space has insufficient airflow for the load, and the terminal cannot deliver an adequate amount of air to satisfy the zone demands.

   b. The space is incorrectly zoned and the space thermostat/sensor is not detecting a problem. The people are complaining while the space temperature readings look good. An auxiliary temperature recorder placed in the space near the complaining occupants usually can document this problem.

   c. The system lacks capacity to satisfy the VAV terminal. This includes the air-handling fan being out of capacity and starving the terminal for air, and the chilled water temperature being too high or the flow too low.

23.2.2 Noise complaints

The noise generated by an HVAC system is usually uniform throughout its life. However, there are problems that develop as part of normal operation, aging, and repairs. Initial complaints result from:

1. Fans that are improperly selected, silenced, or located, resulting in noise that is beyond acceptable levels

2. Ductwork that has too high a velocity or VAV terminals that are selected incorrectly

3. Lack of proper equipment isolators and sound transmission control in the structure

Complaints from noise that result later in the life of the system include:

1. Noise caused by slipping belts
2. Variable-speed electronic drive motor noise caused by adding drives to existing fans
3. Noise caused by fans speeding up to meet increased loads
4. Noise caused by components in the duct system that have become loose and are rattling
5. Sound caused by duct leakage due to joints, gaskets, or other seals failing
6. Sound generated by VAV terminals moving more air to meet increased loads
7. Incorrectly adjusted dampers that may be inadvertently closed, causing higher pressure and velocities
8. Physical damage to fans, ducts, or terminals

23.2.3   Indoor air quality (IAQ) and sick building syndrome (SBS)

Indoor air quality and the sick building syndrome complaints generally are related to ventilation rather than comfort. As described previously, HVAC systems have many functions, but comfort is the primary consideration. Most IAQ complaints are the result of some ventilation deficiency in the system or a result of some harmful contamination that overwhelms the ventilation capabilities. Common IAQ problems are easily resolved, but in all cases the health aspects of IAQ complaints should cause the operator to be more responsive and thorough.

Troubleshooting IAQ complaints involves a study of not only the HVAC system but the building and surrounding environment for potentially harmful conditions. The people involved in the complaint should be carefully interviewed, and their sensitivity to specific environmental agents should be determined. The obvious ventilation problems should be investigated first:

1. Faulty or closed outdoor air dampers
2. Failed/damaged ventilation and exhaust fans
3. Dirty ducts and/or filters
4. VAV terminals that close off completely
5. Lack of purge cycle after HVAC system shutdown
6. Faulty/incorrect fan-tracking control on air-handling system

7. Incorrect air balance, building under negative pressure

8. Air-handling systems functioning correctly, but outdoor air introduction is too low

9. Bird droppings/nests in air intake ducts

If the HVAC system is determined to be working correctly, the next step should be aimed at eliminating the following environmental causes:

1. Air contamination caused by remodeling, painting, new carpet, and other construction-related activities

2. Air contamination caused by external sources such as construction, vehicle exhaust, and vents from other facilities

3. Internal process activities that are not properly exhausted by hoods (e.g., equipment cleaning, solder stations, photographic or other chemical processes, blueprint machines, printing machines, copiers, or other similar devices that can release fumes)

4. Plants that release pollen or odors that people are allergic to

5. Cigarette, cigar, or pipe smoking

If all steps reveal there are no obvious system malfunctions or readily detectable environmental sources of the complaints, testing of the air must be done:

1. The carbon dioxide and carbon monoxide content should be sampled over a period of days in the affected areas.

2. Air samples should be obtained for inspection of biological contamination from dust, mold, fungi, bacteria, and other agents.

If biological testing reveals no problem, and all the other inspection reveals no problem, the IAQ complaint may not have an origin that can be detected by traditional methods. This IAQ mystery is caused by the variability of the causes and the sensitivity of the persons affected. The IAQ problem may come and go based on season, weather, or schedule of work. Thus, repeating the testing at periodic intervals may uncover the cause of the problem that one-time testing will not reveal. IAQ problems may also go away as mysteriously as they developed if the cause is a one-time event.

24

Control Setpoints

The control setpoints are initially determined by the system designer and they may be correct or incorrect for the existing operating conditions. The guidelines that follow apply to the common system setpoints.

24.1 Outdoor Air and Ventilation

ASHRAE ventilation guidelines are a minimum standard that should be met for all occupancies. In addition to following the recommendations (e.g., 15 or 20 cfm per person), the operator also needs to consider the location of the personnel, distribution of the personnel, sources of internal air contamination, the need for exhaust makeup air, and the effectiveness of the terminal devices in distributing the air. It should be anticipated that good air quality may demand substantially more air than the minimum guidelines. For example, there may be only 30 people in an office building, but the building contains a print machine that needs 600 cfm makeup for the exhaust. The building may have toilet exhaust of 400 cfm. Furthermore, the air distribution could be heavy near the heat-producing equipment and limited in the personnel areas. Thus, the fresh air need is not 30 times 15 cfm or 30 times 20 cfm according to ASHRAE minimum levels, but about 1500 cfm to provide makeup and building pressurization.

Outdoor air setpoints should be held constant during occupancy and reduced during periods of vacancy. The suggested or code-derived minimums may fall far short of providing good IAQ, and testing of carbon dioxide may assist in providing better values for the setpoint. In no case should outdoor air setpoints be reduced below any ASHRAE or code minimums or the minimums shown on record drawings.

24.2   Economizer: Dry Bulb and Enthalpy

Airside economizer operation and savings vary by climate and location of the building. Generally, the enthalpy economizer has a theoretical advantage in saving energy, but the control complexity often outweighs its potential advantages. The maintenance of accurate humidity and dew point sensors has been a traditional problem. Generally, all enthalpy changeover setpoints should be based on the return air enthalpy as (opposed to the outdoor air enthalpy). Thus, enthalpy economizer setpoints are dynamic and vary with the weather conditions and indoor conditions.

Dry bulb economizers should be set at a point where the humidity level of the outdoor air will have little effect on the space conditions. Generally, most systems will switch over at 55 to 58°F.

24.3   Chilled Water Temperature

The air-handling system coils are designed around a specific chilled water entering temperature which should not be changed without analysis of the effects. However, if the building is experiencing low relative humidity or lack of ventilation from the VAV terminals, it may be advisable to raise the chilled water supply temperature. Trending of VAV terminal performance, space relative humidity, and VAV fan volumes can lead to an ideal setpoint.

Control sequences can be established that change the chilled water supply temperature based on outdoor temperature and relative humidity to save refrigeration energy on days when the colder water is not needed. Care should be exercised, though, not to lose the refrigeration energy savings by having to move more air through the VAV system. Thus, chilled water setpoints can be raised as long as there is no substantial effect on the VAV air-handling-system discharge conditions.

If the building is experiencing overcooling or lack of ventilation, it may be beneficial to raise the air-handling-unit discharge setpoint and the chilled water temperature to cause an increase in the air changes provided by the VAV system. Chilled water setpoint changes should be made slowly to permit observing the effect on space relative humidity, since too high a setpoint will cause loss of humidity control.

24.4   Heating Water Temperature

Heating water temperature affects the VAV system at the air-handling unit and at the terminal reheat coils. Generally, the heating water setpoint can vary with outdoor air temperature to save energy during mild periods, and the lowest possible setpoints provide energy savings and control stability. If too low a water temperature is suspected, terminal valve positions can be trended to discover whether the heating is inadequate.

24.5   Space Comfort

The energy crisis of the 1970s brought about space setpoints that were beyond the ASHRAE comfort range. Saving energy has been associated with setting space temperatures higher in the summer and lower in the winter. Though these policies are a conservation measure, they may also be a source of occupant complaints. All seasonal setback and setup policies should be directed toward reducing complaints and saving energy. It is also possible that a single setpoint is unsatisfactory.

Modern DDC controls can vary the space setpoint based on time of day, occupancy, and outdoor air temperature and humidity. Nights may require a higher setpoint to accommodate radiation losses through glass areas and the naturally lower metabolic rate of the human body at night. Furthermore, the absolute number set on the sensor or thermostat may not reflect the actual conditions. Sun shining in an office may warm up the occupant directly, causing a need for more cooling. Ideally, the control system can be programmed to adjust itself for these effects.

The flexibility of DDC controls in maintaining space setpoints that are flexible and easily changed is one major reason these systems are recommended over the other systems. Trending and logging setpoints and actual conditions also is another feature available only through DDC systems.

24.6   AHU Discharge Temperature

The AHU discharge temperature is usually established by the system designer and should not be varied unless problems occur. If the system overcools the space or the terminal minimums are inadequate for ventilation needs, the discharge temperature can be raised. However, with elevated discharge temperatures, the following effects should be considered:

1. Loss of humidity control caused by elevated coil temperatures and more moisture in the supply air.

2. Increased fan energy and increased noise from the system due to moving more air.

3. Loss of space temperature control in critical zones since the warmer air has less cooling effect. Terminal volume limits may have to be raised to compensate for the higher discharge temperature.

4. Lack of air-handling-system capacity caused by the reduced cooling effect of the warmer air.

24.7   Relative Humidity

Humidity control is rarely provided except in critical applications. If there are relative humidity controls, it is generally not advisable to change the setpoints because these setpoints are process-driven. For example, a museum may demand high humidity to protect the artifacts, and changing the humidity could immediately cause damage to the art.

24.8   Terminal Minimum Airflow

In the past, terminal minimum airflows were seldom a concern. However, with IAQ now as important as comfort was in the past, more attention should be paid to this setpoint. Low terminal minimum airflow is a primary cause of IAQ complaints. Minimum air changes and the introduction of outdoor air depend on this value.

If the design minimum terminal airflow produces overcooling in the space, the space loads and ventilation requirements should be reviewed. If possible, the discharge temperature can be raised to prevent overcooling. In no case should the terminals in occupied areas be allowed to shut off completely, or all ventilation will be lost. (This does not apply if a separate ventilation system is installed.)

24.9   Duct Static Pressure

Duct static pressure requirements are commonly a function of the VAV terminal minimum requirements and the duct design (each duct system has a design pressure that was determined to be acceptable by the designer). Energy savings can be obtained by lowering the duct static setpoint. Innovative DDC control sequences have been developed that dynamically change the duct static setpoint based on real-time terminal requirements. Though these sequences can work well, they should not be used alone to control duct static pressure. For example, one bad terminal could cause the entire system to operate at too high a static pressure. Therefore the control system should measure the duct static pressure and compare it to the effect on the worst-case terminal. If raising the static pressure does not raise the terminal airflow, then the static pressure should not be raised.

Duct static pressure should be controllable separately from the terminals for troubleshooting purposes and for loop tuning. Dynamic routines that are dependent on the terminals have been known to cause problems if the terminal pickup values are in error or the ductwork has restrictions that cause one or more terminals to call for excessive pressure. These dynamic routines vary in sophistication and each should be examined carefully before application to a VAV system.

24.10   Zone Pressure

Zone pressure is best controlled by regulating or tracking the flow of supply air with the exhaust and return air. DDC systems can dynamically support positive or negative control, but sensing zone pressure is an art that should be carefully implemented.

Zone pressures may fluctuate with doors opening, and the control sequence may need to disable the control loop when doors are opened. Door alarms may also help prevent incorrect pressure. Trying to measure typical pressures that are at or below 0.01 inches w.g. may prove impractical. Critical applications such as those in laboratories may need industrial-grade sensors and redundant systems to ensure safe operation. Specialized mass-flow room/space sensors must be used to detect very low pressures/flow and the direction of the flow compared to adjacent areas.

24.11   Building Pressure

Building pressure control is a function of the HVAC system, building size, shape, weather, envelope quality/tightness, and the opening and closing of exterior doors. Buildings can experience a wide range of pressures on a simultaneous basis. The windward and higher portions of the building might show positive relative pressure, while the leeward and low portions show negative pressure. Building pressure measurement thus is an art, requiring properly located instrumentation that is very sensitive and accurate.

For most applications, building pressure setpoints should be established by the direction of flow and relative flow through the main entrance doors. Ideally, the building will be in positive relative pressure to the outdoors of about 0.01 inch w.g. or less on a calm day. ANSI Standard A117.1-1985 recommends a maximum opening force of about 3.5 pounds for a 3×7 foot door, which equates to an air pressure difference across the door of 0.08 inch w.g. Generally, 0.06 inch w.g. is considered the upper limit. Interior non-fire-rated doors should be limited to 5 pounds opening force or about 0.04 inch w.g. Thus, maintaining a 0.01-inch-w.g. pressure is safe and limits the building exfiltration through the door, and a setting of no more than 0.05 inch w.g. should be used.

24.12   Shutdown/Start-Up

Energy-saving routines establish setpoints for VAV system start-up and shutdown. However, these energy-saving control systems many times have a comfort or temperature function rather than an IAQ function. Shutdown and start-up times need to be adjusted not only to provide comfort but good IAQ as well.

If IAQ complaints are being experienced, shutdown and start-up times should be revised to provide more ventilation, and these routines may have to be eliminated.

25

Performance Tests

The performance of VAV systems can change over time, and therefore periodic testing is needed to ensure the system can perform adequately under all conditions.

25.1 Air-Handling Units

The air-handling unit should be tested annually for airflow and cooling/heating capacity. Static pressure readings, outdoor ventilation airflow, and discharge temperature should be measured.

25.2 Exhaust Systems

All building exhaust systems should have their flow and pressure measured and compared to design and historical values on an annual basis. The bulk flow out of the building should be added up and compared to the makeup air quantity.

25.3 Ducts

Duct layout should be compared to as-built drawings to detect any changes. Leakage testing should be performed if leakage is suspected. Annual checking of the pressure profile in the system can detect blockages caused by loose liners or dampers that may be mispositioned.

25.4 Terminals

Minimum and maximum terminal airflow should be measured annually and compared to the control system and design values. Pickup factors (calibration constants that are used to convert the airflow sensor readings to cfm) should be adjusted if necessary. Space room sensor calibration should be tested at the same time.

25.5   Diffusers/Grilles

These devices are often subject to tampering by the occupants, and flows should be measured annually. Flow hood measurements typically take very little time, and comparison to historical or previous air balance can easily identify problems.

25.6   Controls

Control sensors should be sampled on a routine basis for accuracy and repeatability. The performance of these sensors should be supported by regular calibration testing.

25.7   Refrigeration

The refrigeration source should be measured for performance compared to the air-handling-system needs. Maximum available tonnage should be recorded and compared to design and historical requirements. Flow through valves, strainers, and control valves should be verified at least annually.

25.8   Heating

Available heating system capacity should be tested annually before each heating season. Flow through strainers, coils, and control valves should be verified by either temperature or flow measurements. Control valve leakage should be tested along with maximum flow.

Chapter

26

VAV Routine Maintenance

There are many books available on HVAC system maintenance, and the following description is provided as a starting point for a comprehensive preventative maintenance program. These items are identified as the minimum standard of care necessary to prevent comfort and IAQ complaints. The manufacturers' instructions and recommendations for all equipment should be rigidly followed.

26.1 Air-Handling Units

Annual inspection and cleaning include:

1. Casings, for physical damage, leaks, and corrosion
2. Motors, especially any cooling fan blades or passages
3. Fan blades, dirty, bent, or missing
4. Insect screens on louvers, dirty or damaged
5. Torquing/tightning electrical connections
6. Calibration of internal control devices
7. Repair of defective corrosion protection such as paint, galvanizing, or plating

Seasonal or semiannual inspection and cleaning include:

1. Intake and exhaust louvers
2. Coils, for contamination and corrosion
3. Lubrication of bearings and all moving parts such as linkages
4. Vibration testing on large systems to detect early bearing failure or vibration causes

5. Steam traps, water/steam strainers

Quarterly inspection and cleaning include:

1. Dampers, including motor full stroke, slippage, binding, and correct control response
2. Filters (even though they are generally changed on a time basis or by pressure drop, the correct type and condition should be verified)
3. Filter racks for bypass air and damaged seals
4. Controls for correct sequences and operation
5. Cooling coil drip pans and drains, inspection for biological contamination with remedial action
6. Slipping belts with retensioning as needed
7. Fan guards and motor overloads
8. Freeze stats, fire stats, and smoke detectors for correct operation
9. Control valve motors, linkages, seats, and packing
10. Leakage of shutoff/isolation valves

26.2 Ducts

At least annually, the ductwork should be inspected for:

1. Leakage, including broken expansion joints, loose gaskets, and access doors removed/open
2. Physical collapse or damage to duct exterior
3. Separation, erosion, or other damage to internal liner
4. Damage to external vapor barrier on insulation, or crushing of insulation
5. Internal dampers, including smoke control and fire dampers, in the wrong position, sticking, or inoperable
6. Unauthorized modifications or changes not identified on as-built drawings

26.3 Terminals

VAV terminals should have their controls checked, damper positions/response trended, and space setpoints validated at least twice a year. Preferably, this would occur at the start of the heating or cooling season. If trending shows the damper to be full open and/or the space beyond setpoint, the airflow should be physically measured. At least annually, damper motor shafts and linkages should be inspected and tightened if loose. Some terminals have the damper blades attached by screws and bolts that can loosen. Liners should be inspected, and the airflow pickup tubes checked for contamination. If the airflow-sensing system uses filters, these filters should be checked for clogging. At least quarterly, the terminals should be cycled through their full range of air delivery to determine if the terminal can respond to the controls properly.

Terminals that have water reheat coils should have an annual inspection to check the control valve packing and proper stroking. Inspection of piping and coils for corrosion should be performed annually. Electric reheat coils should be examined for physical damage and the airflow switches tested at the start of the heating season. All electrical connections should be inspected and properly tightened annually. If the terminal is the fan-powered type, proper lubrication and cleaning of the fan should be performed. If the terminal is connected by flexible duct, the duct connections should be checked for leakage, and any duct damage (e.g., loose duct tape) should be repaired.

26.4 Diffusers and Grilles

For good IAQ, diffusers and grilles should be cleaned on a quarterly basis. Any physical damage should be repaired or the grille/diffuser replaced. Rusting diffusers should be cleaned and repainted or replaced. Fasteners should be checked annually to prevent the device from failing. If complaints have been noted, airflow tests should be performed to determine the correct setting of dampers.

26.5 Controls

Modern DDC controls contain self-diagnostic routines and can provide so much information that the routine use of the system is the best maintenance tool. However, the sensors can fall out of calibration or be damaged. Also, output devices such as valves and dampers experience wear and damage. Therefore, the input/output devices in a DDC system require routine attention. Setpoints and control sequences should be examined for the correct values. If errors are found they should be corrected immediately. With DDC, the wiring and power supplies may need routine inspection. Some systems contain backup batteries for power-failure protection, and these batteries have to be replaced.

The trending of control inputs and outputs may reveal that the control loops need to be retuned. Tuning of proportional (P-only) loops can easily be accomplished by the operator, but tuning of proportional-integral (PI) and proportional-integral-differential (PID) loops may require a specially trained technician or loop-tuning software/hardware routines to determine the correct values. (Tuning involves the input of constants that control the ability of the control loop to develop the correct output value from the input values without excessive hunting, overshooting, or offset.)

Older pneumatic and analog electronic systems require much more intense maintenance. The most critical item in a pneumatic system is the cleanliness and dryness of the air supply. Thus compressors, air dryers, and air filters need very close attention or the entire system can be damaged. Few pneumatic systems receive a level of maintenance adequate to retain their accuracy or reliability. Older, non-DDC hybrid pneumatic and supervisory electronic systems are especially vulnerable to poor maintenance. If possible these older systems should be scheduled for replacement.

27

Space Function

The use of space changes with time, and it is not unusual for the initial VAV system to become a problem if ventilation needs increase. Many space uses demand constant volume technology. For example, if a constant number of air changes are required, or space pressure must be precisely maintained, the VAV system may need to be replaced by CAV. This change can be accomplished with DDC systems by reprogramming the VAV terminals to CAV, though other system modifications may be required (e.g., adding a reheat capability if the CAV use overcools the space).

Mixing CAV and VAV zones on a single system can be very successfully implemented if the system was originally designed for the VAV application. CAV conversion generally will require a reheat source by either dual duct or terminal reheat using hot water or electric sources. One application that is becoming popular is the time-of-day system. For example, a zone may be CAV during periods of occupancy to maintain ventilation rates, and then switch to VAV at times when the space is unoccupied. This CAV/VAV usage is totally feasible with today's DDC controls, and it illustrates how a VAV design can be adapted to meet a wide range of IAQ, space comfort, and ventilation requirements.

Laboratories that use exhaust hoods for hazardous materials were at one time all CAV, but VAV systems have been designed to support the ventilation of the laboratory with the hood system using a variable flow. The CAV laboratory systems can be converted to VAV by the addition of special controls that provide tracking of the exhaust and supply systems, along with sensors in/on the hoods to measure the hood flow to keep it within safe limits. This conversion process can save substantial amounts of energy while preserving adequate ventilation and safety.

27.1   Space Function Changes and IAQ

Many times the space conditions are established for comfort or to meet requirements that did not consider IAQ. If, for example, a large space is storage or general office space and then it becomes a conference room, the change in space function may not affect the load on the air-conditioning system as much as the IAQ. The existing HVAC system could have thermal cooling capacity, but it may lack the ventilation capacity to provide good IAQ in a conference room. In the same manner, items such as machines that make blueprints may be added to a work space with an exhaust fan to remove harmful fumes. The addition of this small exhaust fan may not alter the HVAC space loads considerably, but it could unbalance the space pressure and cause ventilation and IAQ problems. Simple, small changes in space function that appear relatively minor can lead to major IAQ problems if not fully evaluated.

The VAV system is almost always a dry bulb comfort system that will automatically meet thermal conditions if it has the capacity. However, the VAV system does not have the same flexibility in changing outdoor air ventilation rates or indoor air changes to preserve IAQ. There are DDC control sensors available that can measure indoor pollution and carbon dioxide levels on a dynamic basis, and they offer a partial solution to this problem. These devices, though, can seldom detect a circulation or local problem in a zone. However, even when these devices are installed on a system, the final and best protection of space IAQ is a periodic test to confirm the ventilation rates are adequate for all the occupied spaces under existing conditions. With test data, it is then possible to fine-tune the HVAC system to meet comfort, ventilation, and IAQ requirements. Therefore, to be an IAQ-sensitive system, VAV has to be continually updated by the operator to meet changing conditions.

Glossary

Construction

EMI Electromagnetic interference is generally a high-frequency noise that travels within the electric power system and can disrupt the proper operation of control equipment or other devices that depend on power quality for proper operation.

pickup/calibration factor A dimensionless constant that is placed within the control parameters of the air measurement control loop to permit accurate conversion of the pressure signal received from the probe and transducer into a correct representation of the airflow.

RFI Radio frequency interference is a higher-frequency noise that travels through the air and can disrupt the proper operation of electrical equipment.

Control

control stability Controls are stable when the output variables and controlled variables are due to change at a rate that is too fast compared to the desired controlled state. Control stability is also demonstrated by controls that provide accurate readings and no drift of the controlled variables.

DDC system A control system consists of electronic sensors, direct digital controllers, electric or pneumatic actuators, and analog-to-digital or digital-to-analog converters if required, all linked together in a network by communication lines. A DDC is a system that employs digital logic and digital processing of input/output signals, with the computer connected directly to the sensors and actuators. This is in contrast to a supervisory control system where the computer is indirectly linked to the sensors and actuators through controllers (and therefore, control in indirect), and to an analog system that processes data that is not digitized (a proportional thermostat connected to an analog valve is such an analog control device).

fan tracking Whenever a variable volume system uses two or more fans, a fan-tracking problem occurs. The operating of one fan in the system affects the operation of the other fan, and some means of coordinating the action of the two fans is required. Changing flows and/or pressure in the system caused by one fan will affect the performance of the other fan, and a fan-tracking system is used to control this interaction to maintain the desired flow and pressure produced by both fans. Fan tracking is especially important when return fans or exhaust fans are used with supply fans.

traverse of supply and return ducts The operation of the fan systems and controls are best determined by a traverse of the supply and return ducts that is performed in accordance with accepted ASHRAE and air balance association practices. The traverses should provide the duct airflow in cfm and static and velocity pressures. This information is useful in checking the calibration of the controls, verifying fan tracking, verifying fan performance, checking duct leakage, and even estimating system effect. Traversing the main return and supply ducts with the system operating at varying loads provides critical information that can be used to solve any potential problems that may develop. This process is also relatively simple, as it requires a minimum amount of equipment, such as a pitot tube and manometer, but it clearly demonstrates the operational capability of the system.

trending Trending is the process of control inputs and outputs as selected over a period of time. Trending for short periods such as a few seconds or minutes can be used to verify control stability. Longer periods of hours, days, or months can be used to identify long-term problems such as control drift or external factors. Trending is important because it allows the operator to compare control information over time and to see how input values vary in relation to output values and setpoints. Trending also provides permanent records that can be used as historical information to document problems for later analysis or resolution.

tuning Tuning is the process of selecting coefficients for proportional integral and derivative (PID) equations used in control loops. The tuning process involves finding coefficients that raise the loop to operate with a minimum number of overshoot cycles while maintaining control of the controlled variables. The tuning process can use a computer to measure and calculate the coefficients required by the process, or the operator can input values based on experience or observation of the loop behavior. After tuning is completed properly, the control loop will be stable and will provide smooth control that is free of excessive resetting of the output device (hunting), free of excessive swings in the controlled variable, and capable of maintaining the desired setpoints throughout the operating range.

Design

design analysis Analysis is an indispensable element of the design process. Quality design requires both synthesis and analysis. A design is synthesized first in a given situation, but always analyzed to verify the design goals originally set. Analysis is especially important in design when environmental quality is emphasized, because environmental factors are mutually interactive and must be analyzed simultaneously. (Chapter 11 discusses design analysis in detail; the 10 sections are each dedicated to a specific subject ranging from load to integrated design analyses.)

design considerations The architect and the system designer share a common goal of designing quality buildings with functioning systems, yet their concerns and interests are not exactly compatible with each other and often result in incompatible design considerations. This problem must be resolved by merging them into an integrated design in which building and system problems are solved jointly and interactively.

design documents Good design documents include design intent, system description, construction documents, as-built drawings, operation and maintenance instructions and manuals, and control documentation. The first three items are prepared during the various stages of design; the rest are prepared during the construction phase and should be finished by the time the construction is completed and the building is handed over to the building owner.

design procedure The explicit description of design procedures assists the system designer in understanding the process of design, the flow of information, specific activities, and resulting design documentation. Design is a science as well as an art. The ability to design cannot be learned through understanding the design procedure alone, but it is a key element in improving design quality. When it is combined with experience and problem-solving techniques, the system designer will be able to design high-quality VAV systems.

integrated design The idea of integrated design is to consider building systems and their interactions with the building and among themselves at a very early stage of building design. The interactions are analyzed and measured constantly until the design is completed. The integrated design approach recognizes the fact that the building is an integral part of system design and therefore cannot be isolated from system design. The traditional design approach treats each building system as an independent entity, assuming that while a system is being designed it will not have any impact on other systems and the building itself. Integrated design corrects this misconception and enables the system designer to design high-performance and building-integrated systems.

VAV design for environmental quality HVAC design is a creative process for arranging various system components in a new way. Thus, VAV systems can be designed with a special emphasis on environmental quality. In this case, the design's ability to provide a quality environment is measured and analyzed through experience and knowledge based on design considerations, component selection, and system layout. The resulting system is simulated to prove its performance for environmental quality.

Environmental Quality

acoustics Noise is an undesirable sound. The function of HVAC acoustics is to minimize the noise generated and transmitted by HVAC systems and equipment. It is particularly critical for VAV systems because changing airflow may produce noticeable sound pressure changes in the occupied space. (Chapter 6, Sec. 6.4 discusses this subject in detail.)

ADPI Air Diffusion Performance Index is a comfort index for space air diffusion based on air velocity and effective draft temperature, which is a function of local air temperature and velocity as well as average room dry bulb temperature. ADPI is the percentage of traverse points satisfying an effective draft temperature between -3 and $+2$ at a velocity less than 70 fpm. Since VAV systems continuously change supply airflow, which in turn reduces the space air velocity, it is important to consider both PMV/PPD as well as ADPI values in defining thermal comfort for a VAV design.

building pressurization Pumping more air into the building than exhausting from it is the major factor influencing the pressure inside the building, though there are also other factors affecting the building pressure. As the outdoor air for ventilation is the only controllable source for building pressurization, and it is sensitive to VAV operation as well, the minimum amount of outdoor air must be carefully selected and controlled by VAV systems. (Chapter 11 analyzes this subject in detail.)

comfort model Various comfort models have been proposed to describe the occupant comfort by a single number. Predicted Mean Value (PMV) and Predicted Percent Dissatisfaction (PPD) are commonly used to predict the comfort level provided by HVAC systems. The International Standards Organization (ISO) developed the FORTRAN program, and its tabulated output is available for the prediction of PMV and PPD values, or the prediction of occupant's comfort level.

HVAC Acronym for heating, ventilating, and air-conditioning.

IAQ Acceptable indoor air quality is the condition of air in which there are no contaminants at harmful concentrations, and a majority (80 percent or more) of the people exposed do not express dissatisfaction. Inadequate outdoor ventilation is the major source of IAQ complaint. As VAV systems continuously change air supply to each zone, which could result in reduced ventilation, it is important to conduct ventilation analyses for VAV system design. (See Chap. 6, Sec. 6.2 and Chap. 11, Sec. 11.4.)

thermal comfort Condition of mind which expresses satisfaction with thermal environment. Thermal comfort is affected by activity level, thermal resistance of the clothing, air temperature, humidity, and movement, as well as nonuniformity of the thermal environment. VAV systems provide thermal comfort by careful adjustment and distribution of supply air at both maximum and minimum airflows.

total environment quality A concept of designing and operating HVAC systems that pays close and simultaneous attention to thermal comfort, indoor air quality (IAQ), air distribution, acoustics, and building pressurization for the best quality of indoor environment.

VAV system An HVAC system that has the capacity to vary the volume of air supplied to the zone to meet the cooling requirements in each zone.

Index

ABOUT THE AUTHORS

STEVE CHEN is vice president of HC Yu & Associates, a major international HVAC consulting firm. Previously, he developed HVAC software packages and prepared the *Air System Manual* for Carrier International Corp. Mr. Chen has more than 40 years of experience with air conditioning, ranging from system design, installation, operation, and troubleshooting to equipment sales, marketing, and teaching.

STANLEY DEMSTER is president of KJD Services, a firm specializing in resolving HVAC-related air quality and energy problems for commercial, institutional, and medical facilities. Mr. Demster has over 25 years of HVAC experience. As resident engineer for the Federal Aviation Administration, he provided hands-on technical assistance during construction of critical aviation support facilities.